EMERGING NANOTECHNOLOGIES IN RECHARGEABLE ENERGY STORAGE SYSTEMS

EMERGING NANOTECHNOLOGIES IN RECHARGEABLE ENERGY STORAGE SYSTEMS

Edited by

LIDE M. RODRIGUEZ-MARTINEZ

CIC Energigune, Miñano, Spain

NOSHIN OMAR

Mobility, Logistic and Automotive Technology Research Center (MOBI), Department of Electrical Engineering and Energy Technology (ETEC), Vrije Universiteit Brussel, Brussels, Belgium

ELSEVIER

elsevier.com

Elsevier
Radarweg 29, PO Box 211, 1000 AE Amsterdam, Netherlands
The Boulevard, Langford Lane, Kidlington, Oxford OX5 1GB, United Kingdom
50 Hampshire Street, 5th Floor, Cambridge, MA 02139, United States

Library of Congress Cataloging-in-Publication Data
A catalog record for this book is available from the Library of Congress

British Library Cataloguing-in-Publication Data
A catalogue record for this book is available from the British Library

ISBN: 978-0-323-42977-1

Publisher: Matthew Deans
Acquisition Editor: Simon Holt
Editorial Project Manager: Sabrina Webber
Production Project Manager: Nicky Carter
Designer: Greg Harris
Typeset by Thomson Digital

CONTENTS

CONTRIBUTORS

P. Alaboina
North Carolina A&T State University, Greensboro, NC, United States

E. Bekaert
CIC Energigune, Miñano, Spain

A.M. Bittner
The Foundation for Science, Bilbao; CIC nanoGUNE, Donostia-San Sebastián, Spain

P.N. Borza
Transilvania University of Brasov, Brasov, Romania

L. Buannic
CIC Energigune, Miñano, Spain

S.-J. Cho
North Carolina A&T State University, Greensboro, NC, United States

F. di Persio
European Commission, Joint Research Centre (JRC), Directorate for Energy, Transport and Climate, Energy Storage Unit, Petten, The Netherlands

E. Goikolea
CIC energiGUNE, Miñano, Spain

S. Goutam
Mobility, Logistic and Automotive Technology Research Center (MOBI), Vrije Universiteit Brussel, Brussels, Belgium

M. Hernandez
Mobility, Logistic and Automotive Technology Research Center (MOBI),
Vrije Universiteit Brussel, Brussels, Belgium

V. Koroteev
Novosibirsk State University, Novosibirsk, Russia

A. Kriston
European Commission, Joint Research Centre (JRC), Directorate for Energy, Transport and Climate, Energy Storage Unit, Petten, The Netherlands

U. Lassi
University of Oulu, Research Unit of Sustainable Chemistry, Kokkola, Finland

N. Lebedeva
European Commission, Joint Research Centre (JRC), Directorate for Energy, Transport and Climate, Energy Storage Unit, Petten, The Netherlands

B. Lei
Karlsruhe Institute of Technology, Institute of Applied Materials–Applied Materials Physics, Eggenstein-Leopoldshafen, Germany

A. Llordés
CIC Energigune, Miñano; The Basque Foundation for Science, Bilbao, Spain

A. Melcher
Karlsruhe Institute of Technology, Institute of Applied Materials–Applied Materials Physics, Eggenstein-Leopoldshafen, Germany

M. Messagie
Mobility, Logistic and Automotive Technology Research Center (MOBI)
Vrije Universiteit Brussel, Brussels, Belgium

R. Mysyk
CIC energiGUNE, Miñano, Spain

L. Oliveira
Mobility, Logistic and Automotive Technology Research Center (MOBI),
Vrije Universiteit Brussel, Brussels, Belgium

N. Omar
Mobility, Logistic and Automotive Technology Research Center (MOBI), Vrije Universiteit Brussel, Brussels, Belgium

A. Pfrang
European Commission, Joint Research Centre (JRC), Directorate for Energy, Transport and Climate, Energy Storage Unit, Petten, The Netherlands

S. Rangaraju
Mobility, Logistic and Automotive Technology Research Center (MOBI),
Vrije Universiteit Brussel, Brussels, Belgium

M. Rohde
Karlsruhe Institute of Technology, Institute of Applied Materials–Applied Materials Physics, Eggenstein-Leopoldshafen, Germany

V. Ruiz
European Commission, Joint Research Centre (JRC), Directorate for Energy, Transport and Climate, Energy Storage Unit, Petten, The Netherlands

J. Salminen
Boliden Kokkola, Kokkola, Finland

J. Sanfelix
Mobility, Logistic and Automotive Technology Research Center (MOBI),
Vrije Universiteit Brussel, Brussels; Vrije Universiteit Brussel, Anderlecht, Belgium

H.J. Seifert
Karlsruhe Institute of Technology, Institute of Applied Materials–Applied Materials Physics, Eggenstein-Leopoldshafen, Germany

M.-J. Uddin
North Carolina A&T State University, Greensboro, NC, United States

P. Van Den Bossche
Mobility, Logistic and Automotive Technology Research Center (MOBI), Vrije Universiteit Brussel, Brussels, Belgium; Vrije Universiteit Brussel, Anderlecht, Belgium

J. Van Mierlo
Mobility, Logistic and Automotive Technology Research Center (MOBI), Vrije Universiteit Brussel, Brussels, Belgium

W. Zhao
Karlsruhe Institute of Technology, Institute of Applied Materials–Applied Materials Physics, Eggenstein-Leopoldshafen, Germany

C. Ziebert
Karlsruhe Institute of Technology, Institute of Applied Materials–Applied Materials Physics, Eggenstein-Leopoldshafen, Germany

PREFACE

The internal combustion engine (ICE), used for vehicular propulsion since the beginning of the automobile era, has come under pressure as a significant contributor to air pollution and greenhouse gas emissions linked to global climate change. These growing environmental concerns, as well as rising petroleum prices, have strained the global economy and spurred research into the development of various types of clean energy transportation systems using electrically propelled vehicles, such as battery electric vehicles (BEV), plug-in hybrid electric vehicles (PHEV), and hybrid electric vehicles (HEV).

Development, sizing, optimizing, and modeling of the energy storage system are among the main challenges for the development of the BEV, PHEV, and HEV.

In particular there is a strong interest in the emerging energy storage technologies, such as lithium–sulfur, solid state, zinc–air, lithium–air, and hybrid capacitors. Research is focusing on higher performances, durability, safety, lifetime, and cost, with specific accents according to the application, traction or stationary.

A lot of scientific research work has been performed to provide solutions for these challenges.

Emerging Nanotechnologies in Rechargeable Energy Storage Systems provides an extensive overview of the technology development process for traction and stationary applications from material to system level.

The initial chapters describe in detail the present technology progress from materials points of view and in particular for nanotechnologies.

The second half of the book covers the system-related topics, such as safety, thermal management, life cycle assessment and applications.

The book deals with key topics in the field of rechargeable energy storage technologies that are interesting for researchers, PhD students, master students, R&D centers, and battery system and application designers.

Lide M. Rodriguez-Martinez
Noshin Omar

CHAPTER ONE

Electrolytes for Li- and Na-Ion Batteries: Concepts, Candidates, and the Role of Nanotechnology

E. Bekaert*, L. Buannic*, U. Lassi, A. Llordés*,†, J. Salminen‡**
*CIC Energigune, Miñano, Spain
**University of Oulu, Research Unit of Sustainable Chemistry, Kokkola, Finland
†The Basque Foundation for Science, Bilbao, Spain
‡Boliden Kokkola, Kokkola, Finland

Contents

1 INTRODUCTION AND ELECTROLYTE CONCEPT

All electrochemical devices, such as batteries, capacitors, electrolytic cells, or fuel cells contain electrolytes, which is the ion transport media; its role is identical irrespective of the selected chemistry or device. The electrolyte is commonly "sandwiched" between two electrodes and is responsible for ionic charge transfer between them.

Good interfacial contact combined with chemical stability is needed between electrodes and electrolyte to ensure an effective ionic transfer and should be continuously evaluated for emerging materials. In fact, these electrified interfaces have been the center of interest since the rise of modern electrochemistry and continue to be in the lithium-based and beyond-lithium rechargeable battery technologies [1].

Emerging Nanotechnologies in Rechargeable Energy Storage Systems

In batteries, the energy output depends on the chemical nature of the positive and negative electrodes (also called cathode and anode, respectively; this convention is only correct during the discharge process) [1]. Ideally, the electrolyte should not undergo chemical changes during battery operation, that is, when the faradaic processes take place within the electrodes. Therefore, regarding chemical reactivity, the electrolyte can be considered as an inert component of the battery requiring stability against both positive and negative surfaces. In the actual device, the electrochemical stability of the electrolyte is made in passivation (kinetic) rather than thermodynamic [2]. This is particularly important for rechargeable battery systems but is often a challenge due to the strong oxidizing and reducing properties of the cathode and anode, respectively. Indeed, the voltage window of operation for Li- and Na-ion batteries is given by the nature of the electrolyte and electrode materials and its chemical potential. Typically, liquid electrolytes have less stability than solids at high potentials, as their organic components can easily become oxidized (Fig. 1.1A).

Conventional Li- and Na-ion batteries (commonly abbreviated LIB and SIB, respectively) are those containing a liquid electrolyte, which provides the ion transport medium between the cathode and anode through porous separator. The lithium or sodium ions exist as dissolved species Li^+ or Na^+ in the electrolyte and carry out charge transfer reactions at the electrode surfaces, namely oxidation or reduction in the cell reaction.

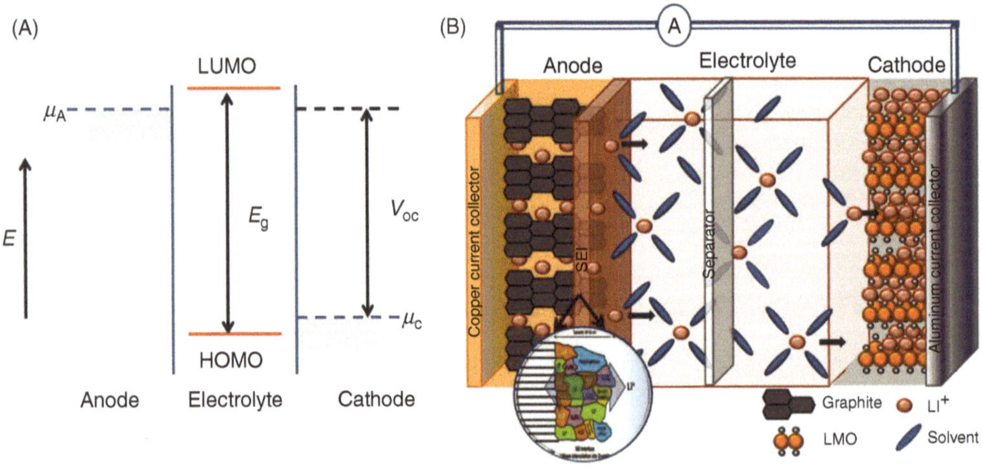

Figure 1.1 (A) Relative energy diagram of electrode potentials and electrolyte energy gap in a typical Li-ion battery where μ_A, electrochemical potential values of anode; μ_c, electrochemical potential values; E_g, energy gap; V_{oc}, open circuit voltage; HOMO, highest occupied molecular orbital; and LOMO, lowest unoccupied molecular orbital. (B) Schematic diagram of lithium intercalation/deintercalation. *(Adapted from K. Xu, Nonaqueous liquid electrolytes for lithium-based rechargeable batteries, Chem. Rev. 104 (10) (2004) 4303–4418 [2], C. Daniel, Materials and processing for lithium-ion batteries, JOM 60 (9) (2008) 43–48 [3], and E. Peled, D. Golodnitsky, G. Ardel, Advanced model for solid electrolyte interphase electrodes in liquid and polymer electrolytes, J. Electrochem. Soc. 144 (1997) L208– L210 [4]).*

The electrolyte between cathode and anode provides a sufficient source of lithium or sodium ions for the electrode reactions, while maintaining electrical insulation. Ion transport within the electrolyte is influenced by the solvation degree and by the presence of the counter ion species (anions for charge balance). When the battery is discharging the negative electrode release some of its lithium ions, which flow through the electrolyte to the positive electrode and vice versa during charging (Fig. 1.1B).

The constant demand for batteries with higher energy density leads to the development of more oxidizing/reducing electrode materials, continuously increasing the stability requirement of the electrolyte [2]. A specific surface chemistry is often crucial for kinetic stability of the electrode/electrolyte interfaces. The redox potential against some reference potential, are commonly used to quantify the redox potential of electrode materials. The range in volts between the oxidative and reductive decomposition limits of an electrolyte is employed to quantify the stability of an electrolyte, which is known as the "electrochemical window." The redox processes occurring at each electrode should occur within this electrochemical window to allow for rechargeable battery operation. Electrochemical stability is not the only property that an electrolyte should meet. An ideal electrolyte should provide the following characteristics:

- *High ionic conductivity* to efficiently transfer ionic charges between electrodes
- *Low electronic conductivity* to prevent self-discharge and short circuiting
- *Large electrochemical stability window* to prevent its decomposition in the range of cell working potentials
- *Chemically inert to the other cell components* including separator, current collector, electrode substrates, and cell packaging
- *High wettability of the electrodes surface*
- *Robustness in harsh conditions*, such as extreme temperatures, mechanical and electrochemical abuses
- *Environmentally friendly*

To date, the most widespread electrolytes and the ones found in commercial batteries are electrolytes based on liquids [2]. However, solid electrolytes are now rapidly emerging as promising alternatives given their wider electrochemical window of stability [5,6]. In addition, unlike liquid electrolytes, solids are nonflammable and do not suffer from leakage, providing a safer option for large-scale application (e.g., electric vehicle) [7].

However, solid electrolytes have not yet been widely used in commercial batteries, owing to materials' limitations, such as low ionic conductivity and poor wetting properties, as well as processing-related characteristics, that limit their integration and assembly into solid-state devices (Fig. 1.2). The high interfacial resistance at the solid–solid interface between the electrodes and electrolyte arises as a key scientific challenge that still needs to be solved for the successful implementation of these materials in commercial batteries (Table 1.1).

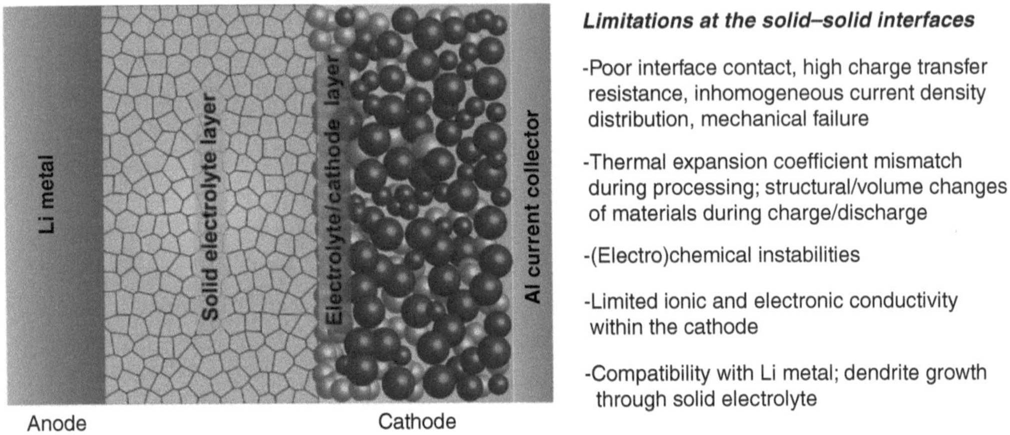

Figure 1.2 Schematic of an all-solid-state battery device and main challenges arising at the solid–solid interfaces. *(Image credit: A. Llordés).*

Table 1.1 Advantages and disadvantages of solid versus liquid electrolytes [8]

Solid electrolyte

Advantages	Disadvantages
1. Excellent chemical and physical stability 2. Perform well as thin film (≈1 μm) 3. Ionic conduction only (excludes electron). Transference number = 1	1. Reduced contact area with electrodes 2. Interface stress due to charging discharging 3. Lower ionic conductivity than liquids

Liquid electrolyte

Advantages	Disadvantages
1. Excellent contact area with electrode 2. Can accommodate volume expansion at the electrodes during cycling 3. High ionic conductivity	1. Poor physical and chemical stability 2. May need of the formation of "solid electrolyte interface" 3. Both ionic and electronic conduction. Transference number typically 0.5

The replacement of traditional liquid electrolytes by solid electrolytes would expand the spread of rechargeable batteries. Not only will it provide a safer system—the flammable and toxic organic solvents will no longer be present—but it will also lead to substantial increase in energy density (2–3 greater than existing battery technologies) [9]. Indeed, the use of metallic anodes (Li metal or Na metal for LiB and NaB, respectively), providing the ultimate energy storage capacities, becomes possible as the solid electrolyte acts both as a separator and medium for ionic charge transfer. The solid electrolyte becomes a physical barrier, possibly preventing the growth of metallic dendrites from the anode to the cathode, a common limitation of current battery systems. The combination of ionic conductor medium and separator in a single component will also facilitate cell processing. Additionally, the absence of low vapor pressure solvents will allow for higher vacuum level during cell sealing, which will have a positive impact on the more resistive interfaces between electrodes and solid electrolyte. However, unlike its liquid counterpart, a solid electrolyte will not "wet" the electrode materials. It is therefore necessary to rethink the composition of the electrodes and incorporate a fraction of solid-state electrolyte in order to ensure good ionic transport within the electrodes and minimize interfacial resistance at the junction between electrodes and electrolytes.

2 LIQUID ELECTROLYTES

Liquid electrolytes are commonly used in batteries and supercapacitors due to their low surface tension and viscosity, and high wetting property [10]. These characteristics provide optimum contact with the materials composing the electrodes, leading to low interfacial resistance [11]. Some storage technologies as Li-S or metal batteries need special requirements in terms of electrolyte due to the high chemical reactivity of battery components as polysulfide or metallic lithium [12,13]. Moreover the electrolyte design also requires considering other factors as the poor electrochemical kinetics for some components and the selection has to consider and raise the electronic and ionic path.

A typical liquid electrolyte is usually a solution of a Li or Na salt dissolved in an organic solvent mixture. In rechargeable devices, water or organic compounds are used as solvents and are referred to as aqueous and nonaqueous electrolytes, respectively. They are straightforward to prepare and homogenize due to the fast ionic diffusion in liquids, which makes them easily scalable for large-scale energy storage applications. The organic solvents include propylene carbonate (PC), ethylene carbonate (EC), diethylene carbonate (DEC), dimethylene carbonate (DMC), and their mixtures. In Li-ion batteries, the most typical salt is lithium hexafluorophosphate, $LiPF_6$, but $LiBF_4$, $LiAsF_6$, LIBOB, LiTFSI, LiFSI have also been commonly used [14]. The latter two, LiTFSI and LiFSI, belong to newer generation of salts providing better chemical stability. For Na-based batteries, equivalent salts have been investigated including $NaClO_4$, $NaBF_4$, $NaPF_6$, NaTf, NaTFSI, and NaFSI [15]. Organic solvent-based electrolytes are flammable and susceptible to thermal runaway phenomenon. A big problem is that EC, one of the most volatile solvents, is also used as an additive to form a stable SEI layer [16]. Nonvolatile room temperature molten salts, known as ionic liquids, have also been investigated intensively within recent years as candidates for lithium-ion battery electrolyte components.

Many studied battery systems have more or less problems with electrolyte interactions. The organic solvents of the liquid electrolyte are required to efficiently dissolve the selected Li salts. Their dissolution power is such that it can also lead to dissolution of the cathode or anode materials. There are many examples in different battery chemistries. Powerful organic solvents can dissolve sulfur and polysulfides, for example, in well-studied *lithium sulfur* (Li-S) battery systems [17]. The dissolved sulfur species can migrate through the cell causing power fade and capacity losses. That problem is severe hindrance in development of this promising technology [17,18]. Ionic liquids can also dissolve lithium metal oxide cathode materials. At the electrode interface, the electrolyte should not react continuously but only provide a small protective layer called solid electrolyte interface (SEI).

Sodium batteries have been tested with similar mixtures of organic solvents and Na-based salts. However, this technology has not yet reached commercialization because the use of common organic solvents, such as PC, EC, DEC, or DMC has not been adequate to enable long lifetime for sodium batteries [19].

Chemical stability is one of the most important properties for the electrolyte especially in terms of safety and performance. The consumption of the electrolyte due to irreversible reactions on electrode/electrolyte interface and chemical instability can lead to continuous capacity fading, loss in power rate and increase of cell internal pressure and generation of gaseous products that can induce explosion of the device [20–23]. Moreover, the electrolyte decomposition might be highly exothermic in some aggressive cases and may lead to a thermal runaway of the battery. Liquid electrolytes require a complex thermal management system [24]. Moreover, the electrolyte decomposition might be highly exothermic in some aggressive cases and may lead to a thermal runaway of the battery. Liquid electrolyte requires a more complex thermal management system [24].

Extensive effort is ongoing to enhance the reliability and safety of battery electrolytes. Different approaches are being studied:

1. Additives to build up stable SEI and/or increase the thermal stability [25].
2. Redox shuttles to protect from overcharge in particular for Li-S systems [26].
3. Shut-down separators to prevent thermal runaway [27].
4. Other type of salts (as imidazolium, $LiNO_3$, $LiBH_4$, LiB_3H_8, $LiCB_{11}H_{12}$) to reduce toxicity [28,29].
5. Use of solid electrolytes (ceramic, polymer, or composites).

2.1 Importance of the SEI layer

Electron movement through the cell can be prevented if the potential is higher than the energy level of the highest occupied molecular orbital (HOMO) of the electrolyte to prevent the oxidation of electrolyte on the anode (Fig. 1.1). Similarly, the potential of the cathode has to be lower than the energy level of the lowest unoccupied molecular orbital (LUMO) of the electrolyte to prevent reduction on the cathode. As a result, the electrode potentials of a battery are limited by the HOMO–LUMO gap of the electrolyte. This cell potential limitation can be partly prevented by kinetic control, that is, keeping the anode potential too low to be stable [30]. During the first charge of the battery, the electrolyte is reduced on the anode, but the decomposition products formed during the reaction form a passivating layer, called solid-electrolyte interphase (SEI) [31,32]. SEI layer is formed around the anode and it prevents any further reduction occurring during the following cycles. If the SEI layer is damaged due to side reactions, many detrimental effects occur for the battery performance. These include electrolyte decomposition, rapid heating and pressure rise in the cell, larger deposition of insulating solids, decrease of energy efficiency, and increased resistivity [20–22,33].

A stable SEI layer is therefore a necessity for a safe battery performance, especially because the most commonly used anode material, graphite, has a potential that is too low (1.1 V vs. Li) for any organic solvent-based electrolyte to operate without a SEI layer. Most cathode materials are within the electrochemical window (~4 V vs. Li) of the organic solvent-based electrolytes. Thus cathode operation does not require an SEI layer in the same way the anode does. SEI layer composition depends on the identity of the lithium salt and the solvent in the electrolyte.

Improving safety of a conventional organic solvent-based electrolyte system can be done by improving the stability of the SEI layer that is formed on the surface of the anode as a result of decomposition products of electrolyte reduction. This interface prevents the fresh electrolyte contacting with the anode, slows down the reaction rate, and thus enables the use of lithium at the anode. The same phenomenon occurs for lithiated carbon anode. Stable SEI layer is therefore a prerequisite for a good battery performance and it also determines the battery voltage limit. Chemical modification of the SEI layer and additives, such as catechol carbonate, alkyl sulfones, alkyl phosphates or phosphazenes [34,35], which might also act as flame retardants, are used to improve stability and also increase the battery safety for lithium-ion batteries with organic solvent electrolytes [36].

Despite these efforts, liquid electrolytes still present severe drawbacks. In this frame, ionic liquids (ILs) are predicted to play a major role as the result of their hydrophobicity and intrinsic safety [37]. ILs could be used as a protective layer of lithium anode against moisture, for example. On the other hand, ILs exhibit a very high viscosity compared to standard organic solvents. Consequently, in order to obtain acceptable conductivity level, proper mixtures with ILs and salts has to be made [38]. Further, low viscosity cations and, especially, anions need to be researched [39]. Last but not least and despite the fact that ILs are good solvents for ionic materials, they present a high reactivity versus metallic lithium. They are mostly used as additives in classical electrolytes due to their high cost [40].

2.2 Additives: general

Additives have been extensively studied to address some of the limitations of liquid electrolytes. The use of additives can enhance cell safety and chemical stability resulting in improved lifetime, better performance, and durability. Additives can prevent or hinder the flammability of organic electrolytes, shut down battery operation under abuse conditions, and protect from overcharge or increase the cell overcharge tolerance. Similar to electrolytes, additives should not be toxic and need to be inexpensive. Overall, additives have an effect on improving Li-ion battery safety performance but can also have a negative impact on the electrochemical properties, and a stable SEI layer requires additives, such as EC, in the electrolyte [36]. Batteries in general must be designed to be as safe as possible even under difficult conditions.

2.2.1 Electrolyte additives used in Li-ion batteries

Use of electrolyte additives is one of the most economical and effective methods for the improvement of battery performances. According to their functions, the electrolyte additives can be divided into five categories: (1) SEI forming improver, (2) cathode protection agent, (3) salt stabilizer, (4) safety protection agent, and (5) other agents, such as solvation enhancer molecules and wetting agents. Usually, the amount of additive in the liquid electrolyte is less than 5 wt.% and is sufficient to significantly improve the battery cycle life [34].

2.2.1.1 Additives for SEI forming improver

SEI layer formation around the anode prevents any further reduction occurring during the following cycles. The main components of SEI are the decomposition products of electrolyte solvents and salts. SEI layer formed before the intercalation of lithium ions is unstable and abundant with inorganic compounds. The SEI formation can be facilitated by chemical modification, that is, by chemical coating an organic film layer onto the graphite surface via an electrochemical reduction of additives. Additives for SEI forming improver can be divided into reduction-type and reaction-type. *Reduction-type additives* (including either polymerizable monomer or reducing agent) assist and facilitate the formation of a stable SEI layer. *Reaction-type additives* act as scavengers of radical anions, or complexes with the decomposition products to form more stable SEI components. These additives are not reduced during cell operation [34].

2.2.1.2 Additives for SEI morphology modifier

The presence of inorganic components, such as Li_2CO_3 and LiF, can lead to SEI instability [21]. Therefore, many boron-based anion receptors have been developed to dissolve LiF [41]. The most well-known receptor is tris(pentafluorophenyl)borane (TPFPB), which is sufficient to improve cycleability and capacity retention of cells containing $LiPF_6$- or $LiBF_4$-based electrolytes. Alkali metal salts have also been considered for improvement of SEI formation. In Ref. [42], it was observed that the SEI formed in the presence of sodium ions ($NaClO_4$) is more uniform and has less resistance than without sodium ions. Pretreatment of graphite materials using insoluble sodium salts in organic electrolytes (such as Na_2CO_3 or LiCl) has also been investigated to improve SEI formation [43]. On the other hand, opposite effects of potassium ions have been observed with different electrolytes [44].

2.2.1.3 Additives for cathode protection

From the electrolyte viewpoint, the performance of cathode materials is dependent on two factors, the presence of water or acidic impurities and the irreversible oxidation of electrolyte solvent [45]. Additives for cathode protection are used for this purpose. In the first proposed mechanism, the electrolyte solvents are chemically oxidized (oxygen released from the cathode) to generate H_2O and CO_2. The produced H_2O can hydrolyze $LiPF_6$, forming acidic products (such as HF), which dissolve cathode materials. Additives for cathode protection, such as amine molecules can form complexes with PF_5 thus reducing the reactivity and acidity of PF_5, and preventing the dissolution of cathode. The other mechanism of these additives is to form a protective film on the cathode surface. This occurs when additive molecules combine with the dissolved metal ions to form insoluble products, which effectively cover the surface to prevent dissolution. Basically, the combination of these two mechanisms seems to be the best solution to protect the cathode [34].

2.2.1.4 Salt stabilizer additives

Salt stabilizer additives are used to hinder the anion reaction effects of lithium salts with dissociating anions like PF_6^- and BF_4^- that show Lewis acid behavior. Thermal and chemical instability of $LiPF_6$ electrolyte is affected by the following factors: (1) high equilibrium constant of $LiPF_6$ decomposition and (2) high reactivity of the resulting PF_5 gas formation. As a result of these reactions, the deposition of solid LiF in the SEI layer is increased leading to higher resistivity (or interfacial impedance) and gas generation inside the battery. Therefore, additives for $LiPF_6$ salt stabilizer are used to tackle these problems [34]. Some of the problems related to Lewis acid type anions and resulting corrosive effects can be avoided by using lithium salts with chemically more stable anions $TFSI^-$ and FSI^-, for example.

2.2.1.5 Additives for safety protection

Safety protection additives are used for overcharge protection and as fire-retardant additives. The overcharge protection additives are classified as redox shuttle and shutdown additives. The former protects the cell overcharge reversibly, while the latter terminates the cell operation permanently. During the overcharge, shuttle molecules are oxidized reversibly at the positive electrode. The oxidized species then diffuse to the negative electrode where they are reduced back to neutral molecules. The maximum current for shuttle additive depends on the concentration of the shuttle molecules in the electrolyte, the diffusion constant of the shuttle molecules, and the number of charges carried by the shuttle species [46]. Shuttle additives should meet several requirements as presented in Table 1.2. Several candidates for redox shuttle additives have been tested, but most of them do not meet these requirements. The first shuttle additives included compounds, such as metallocenes and dihydrophenazine derivatives [46] with a small redox potential

Table 1.2 Typical electrolyte additives used in lithium-ion batteries

Additive function	Need for additive	Additive types	References
SEI forming improver	Formation of SEI layer, which is unstable before the intercalation of Li ions, can be facilitated by chemical modifications	Reduction-type additives (polymerizable monomer and/or reducing agent) Reaction-type additives	[34]
SEI morphology modifier	Inorganic compounds, such as LiF, are reasons for SEI instability	Boron-based anion receptors Alkali metal salts	[34–36,49,50]
Cathode protection	Presence of water and acidic impurities or oxidation of electrolyte dissolve cathode	For example, amine molecules, additives either form complex to prevent cathode dissolution or a protective film on cathode surface	[34]
$LiPF_6$ salt stabilizer	$LiPF_6$ electrolyte is thermally and chemically unstable	Lithium salts with chemically more stable anions are used	[34]
Safety protection	Overcharge protection (redox shuttle and shutdown additives) Fire-retardant additives	For redox shuttle additives: • reaction highly reversible • electrochemically stable • potential higher than normal potential of positive electrode but lower than decomposition potential of electrolyte • oxidized and reduced forms highly soluble and mobile For fire-retardant additives: • chemically terminating chain reactions • physically forming isolating layers	[26,36,38,39]
Other types of additives	Low solubility of Li salts or low conductivity Corrosion protection Low wetting of separator	Ionic solvation enhancer molecules Al corrosion inhibitor Wetting agent/viscosity diluters	[34]

range (2.8–3.5 V), and were therefore only suitable for low-voltage lithium batteries. Recently, anisole compounds have been studied due to their promisingly high redox potential and good solubility in the lithium battery electrolytes [47]. Shutdown additives for overcharge protection were recently introduced by Dantsin et al. [48]. Majority of these shutdown additives belong to the aromatic family.

Additional safety concerns are related to the high flammability of the organic liquid electrolyte. Fire-retardant additives have been developed to face this problem. The mechanisms of these additives are physical, by building isolating layers to stop the combustion process, or chemical, by terminating radical chain reactions [30].

2.2.1.6 Other types of additives

Other additives include ionic solvation enhancers and viscosity diluters. Ionic solvators, such as crown compounds, promote the solubility of Li salts and consequently increase ionic conductivity. Wetting agents can be used to improve the wetting of the separator by the electrolyte. Sometimes the liquid electrolyte cannot wet the nonpolar polyolefin separator efficiently, particularly in the presence of propylene carbonate or ethylene carbonate, which are commonly added to improve thermal tolerance and performances at elevated temperatures. Different types of molecules have been tested including ionic and nonionic surfactants.

As a general rule, additives should not worsen cell performance. Using an additive to address a specific issue can be beneficial but long term cycling should be performed to evaluate the durable benefits of additives over time.

2.3 Electrode–electrolyte compatibility: SEI with ionic liquids

Ionic liquids are molten organic salts of which several cation and anion combinations have shown to be potential electrolytes. Most of these ionic liquids are based on imidazolium cations [51]. Ionic liquids are liquids at room temperature. They have high ionic conductivity, are nonflammable, nontoxic, and have a wider electrochemical window than traditional organic solvent-based electrolytes. Due to the low vapor pressure, ionic liquids can be used in applications, such as Li/air batteries [52]. Most of the studied ionic liquids are not chemically stable below the voltage of 1.1 V, and therefore do not form a stable SEI layer and cannot be used with carbon or lithium anode. To remediate these limitations, they are usually mixed with organic solvents in order to form a protective SEI layer or used in combination with solid inorganic or polymer electrolytes. Recently, ionic liquids containing pyrrolidinium cation and bis(fluorosulfonyl)imide (FSI) anion showed high cyclability with Li/$LiCoO_2$ cell without any additive [53].

2.4 Use of nanotechnology in liquid electrolytes

Battery performance is governed by several properties and complex interactions. Electrolytes used in rechargeable batteries must be chemically stable but reactive enough to form the SEI layer. These side reactions should not proceed beyond providing the protective layer. Any additional solid deposits on the electrodes contribute to increase interfacial impedance and overall resistance of the cell. All additives including nanomaterials must not contribute into harmful and detrimental processes during battery operations. New cell chemistries should be carefully tested and put into long cycling and storage tests with variable conditions to find out if any real improvements were made. It often turns out that promising cell chemistry works well for a short period of time and fails catastrophically before meeting expectations required for commercialization [54].

The search for improved properties in battery chemistries include, for example:

1. large electrochemical stability
2. high thermal stability
3. wide operating voltage range
4. wide operating temperature range
5. low vapor pressure
6. high conductivity
7. high capacity
8. long storage life
9. long cycle life
10. overall safety and abuse tolerance
11. low cost

Many of the previously mentioned points can be addressed by the choice of electrolyte and chemistries used. Ionic liquids, for example, could handle most of the points mentioned previously, except for cost, and in many cases, low temperature performance. The addition of solid powders of Al_2O_3, TiO_2, and ZrO_2, especially in nanoparticulate form, into liquid electrolytes can improve conductivity [45] in some cases. This is due to changes in balance between free ions and ion pairs due to altered physicochemical interactions. Within volume fraction range of 0.2–0.5% the conductivity was doubled. This approach is also investigated in the case of solid polymer electrolytes (Section 3.1). Nanomaterial additives can enhance the charge transfer properties in the electrolyte–electrode interface and in the electrode as well as tune properties in solid polymer or other combination electrolytes [45]. The favorable effects directly in liquid electrolytes are very limited but in the whole cell system, especially in the electrodes, there are more possibilities.

3 SOLID ELECTROLYTES

Battery safety can be improved by replacing the organic solvent-based liquid electrolyte with a nonflammable alternative, such as solid electrolytes. In addition to improved safety operation of the battery, those alternative electrolytes may provide additional beneficial properties, such as higher stable operational voltage (>4.5 V), too high for traditional organic solvents. They also offer compatibility with alternative battery chemistries (Li/O_2 and Li/S), that have very high theoretical capacity and in which conversion chemistry is applied instead of intercalation chemistry [55]. Solid electrolytes also allow the use of some cathode materials, which have poor cycleability with liquid electrolyte (such as sulfur, which dissolves in organic solvents) [56].

Solid electrolytes regroup two main categories of candidates: polymer-based membranes and inorganic-based materials. While polymer-based electrolytes provide interesting mechanical properties due to the flexible nature of the polymer matrix, they often offer ionic conductivities 2 or 3 orders of magnitude below the ones of the more rigid inorganic materials [57]. In order to combine the advantages of both families, new concepts are currently being designed including the preparation of polymer–ceramic composites and polymer–ceramic hybrids (chemically bonded). Each category will now be discussed.

3.1 Polymer-based electrolytes

Polymer-based electrolytes consist of a chemical salt dispersed in a polymer matrix. The salt, once dissociated, can allow transport of charges between electrodes. However, as the salt is composed of one cation (Li^+ or Na^+) and one anion, the charge transfer will be assumed by both positive and negative entities, with possible partial preference for one over the other based on chemical interactions with the polymer matrix. It is therefore necessary to quantify the ratio of charge transferred by the electrochemically involved ion, that is, the cation. This is commonly reported as transference number or ion transport number and is expressed as:

$$t_{\text{cation}} = \frac{\sigma_{\text{cation}}}{\sigma_{\text{cation}} + \sigma_{\text{anion}}} = \frac{D_{\text{cation}}}{D_{\text{cation}} + D_{\text{anion}}} \tag{1.1}$$

with t_{cation}, the cation transference number, σ_{cation} and σ_{anion}, the conductivities of cations and anions, respectively, and D_{cation} and D_{anion} the diffusion coefficient of cations and anions, respectively.

Li^+ mobility in a polymer matrix is provided by the amorphous region of the network [58] and relies on the segmental motion of the entangled polymer chains to diffuse the ions by a solvation–desolvation process [59,60] (Fig. 1.3). The ionic conductivity of polymeric membranes dramatically increases above the glass transition (T_g) of the polymer host as its segmental motion becomes more pronounced. This has led to the quest for polymers with lower T_g [59]. To facilitate the dissociation of the salt, the polymer matrix should have a high dielectric constant (ε) and electron-donor characteristics, which are provided by polar groups, such as ether (—O—), carbonyl (C=O), and cyano (C≡N) [60]. In addition, better salt–solvent complexation can be achieved if the salt contains large anions, which provide a higher degree of charge delocalization [60].

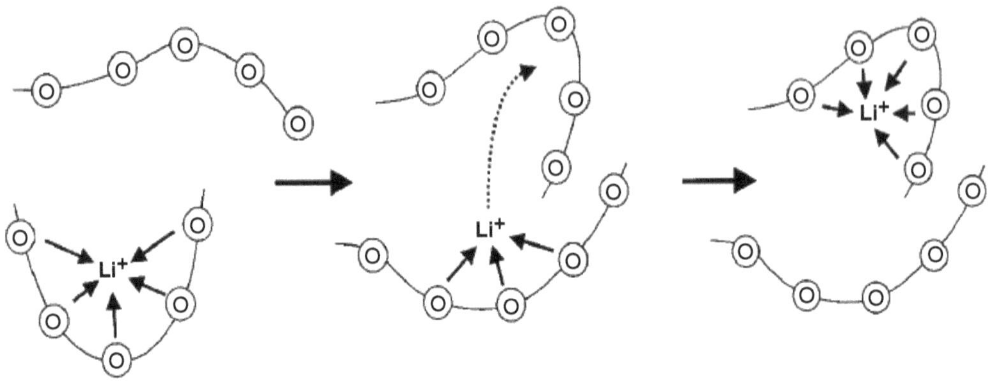

Figure 1.3 ***Schematic of the segmental motion assisted diffusion of Li^+ in the PEO matrix.*** The *circle* represents the ether oxygens of PEO. *(Reproduced from W. H. Meyer, Polymer electrolytes for lithium-ion batteries, Adv. Mater. 10 (6) (1998) 439–448 [59] with permission from Wiley, Copyright © 1999).*

Polymer-based electrolytes are commonly divided into two subgroups: solid polymer electrolytes and gel polymer electrolytes, introduced in the following sections. The chemical structure of the most commonly used polymers for membrane preparation are gathered in Table 1.3.

3.1.1 Solid polymer electrolytes

Solid polymer electrolytes (SPEs) are composed of Li or Na salts dispersed into a polymer matrix, usually based on polyethylene oxide (PEO). Polypropylene oxide (PPO) has also been considered for such applications but leads to lower ionic conductivities due to the methyl groups, which prevent the complexation of Li^+ [59]. The chemical structure of PEO and PPO can be found in Table 1.3. The preparation of SPE membranes involves dispersion of the salt–polymer couple in a solvent, such as acetonitrile or methanol followed by casting of the slurry and evaporation of the solvent. The salt dissociation is dependent on the salt to polymer concentration: optimum dispersions—and therefore ionic conductivities—are commonly found for a [EO or PO units]: [Li^+] ratio of 25:1 [59].

High molecular weight (MW) PEO-based SPEs are polycrystalline membranes with randomly oriented micron-size domains [62]. They provide excellent mechanical properties but poor ionic conductivity, in the order of 10^{-8} S/cm at room temperature (MW = 0.9×10^6 g/mol) [63]. Following reports showed increased room temperature ionic conductivities due to more homogenous SPE preparations, with σ_i in the range of $\approx 10^{-6}$ to 10^{-5} S/cm [64–66]. Lowering of the PEO molecular weight and modification of its termination group can dramatically increase its ionic conductivity: a room temperature Li^+ mobility of $\approx 5 \times 10^{-4}$ S/cm was reported for polyethylene glycol (PEG) with MW = 200 g/mol [67]. However PEO with such low MW

Table 1.3 Polymer hosts commonly used for the preparation of polymer-based electrolytes

Polymer	Acronym	Repeating unit	Glass transition temperature T_g (°C)	Melting point T_m (°C)
Poly(ethylene oxide)	PEO	$—(CH_2CH_2O)_n—$	−64	65
Poly(propylene oxide)	PPO	$—(CH_2CH(—CH_3)O)_n—$	−60	—[a]
Poly(acrylonitrile)	PAN	$—(CH_2CH(—CN))_n—$	125	317
Poly(methyl methacrylate)	PMMA	$—(CH_2C(—CH_3)(—COOCH_3)_n—$	105	—
Poly(vinyl chloride)	PVC	$—(CH_2CHCl)_n—$	85	—
Poly(vinylidene fluoride)	PVDF	$—(CH_2CF_2)_n—$	−40	171
Poly(vinylidene fluoride–hexafluoropropylene)	PVDF–HFP	$—(CH_2CF_2)_n—(CF_2CF(—CF_3))_m—$	−90	135

[a]Amorphous.

Source: Reproduced from A.M. Stephan, Review on gel polymer electrolytes for lithium batteries, Eur. Polym. J. 42 (1) (2006) 21–42 [61] with permission from Elsevier, Copyright © 2006.

does not provide enough mechanical cohesion for further processing. Irrespective of the molecular weight, a drop of ionic conductivity is commonly observed below the melting temperature of the hosting polymer. It results from a decrease in amorphous (i.e., conducting) phase volume fraction, a restriction in chain mobility, and an increased chain tortuosity [60].

Several strategies have been developed to lower the crystallinity of high MW PEO, including the design of comb-like polymers with short polymeric side chains attached to the main PEO chain [68–71]. The presence of short chains leads to an increase in Li^+ mobility ($\approx 10^{-5}$ to 10^{-4} S/cm). Unfortunately it is also associated with a lowering of mechanical properties. Strategic bonding of the PEO chains, also known as cross-linking, [72–74] can prevent crystallization of the main chains. This ensures a high fraction of amorphous polymer matrix with $\sigma_i \approx 10^{-5}$ to 10^{-4} S/cm while keeping the mechanical properties of the original polymer chains. However the nonconducting cross-linking agent should represent only a limited amount of volume fraction or it will result in a decrease of the ionic conductivity [59]. Finally, the careful design of block copolymers can provide rigid reinforcing segments from high MW chains combined with softer sections based on low MW chains, ensuring both excellent mechanical properties and high room temperature ionic conductivity with $\approx 10^{-4}$ S/cm [75–79].

Inorganic fillers, such as nanoparticles of TiO_2, Al_2O_3, or SiO_2, were first introduced by Weston and Steele [80] in 1982 to improve the mechanical properties of SPEs. The membranes change from a gel-like to a solid-like film due to the interaction of the fillers with the polymer matrix. In addition to mechanical property increase, optimum amounts of ceramic nanoparticles (usually around 5–10 wt.%) can enhance the membrane ionic conductivity by 1 order of magnitude [81–84], with up to $\approx 10^{-5}$ S/cm at 25°C [84]. Finally, the presence of nanofillers increases the chemical stability at the electrode/SPE interfaces [61].

Decoupling the ionic transport by limiting the charge transferred by the anionic entity is an interesting approach to increasing the cationic transference number. In a few PEO complexes, such as PEO_6:$LiXF_6$ (X = P, As, Sb), crystallization of the PEO chains forms open tunnels inside which Li^+ complexes with the polymer matrix. The anions are forced to remain outside the tunnel due to size constraints, creating a free pathway for the cationic diffusion. However, the resulting ionic conductivity remains quite poor, in the order of 10^{-7} S/cm [85], and this particular crystallization behavior is only observed for low MW PEO, which limits the practical application of such complexes. PEO_6:$NaPF_6$ and PEO_8:$NaPF_6$ show a similar crystallized structure, with a maximum Na^+ mobility at 7.7×10^{-7} S/cm for the latter [86]. The cationic transference number can also be increased to unity by anchoring the anion to the polymeric chain, preventing its contribution to ionic motion. Such membranes are commonly referred to as single ion conductor SPEs [87–90]. However, their ionic conductivity at room temperature remains in the same range as a traditional SPE ($\sigma_i \approx 10^{-7}$ to 10^{-5} S/cm), once again due to the crystallization of the polymeric chain.

3.1.2 Gel polymer electrolytes

Gel polymer electrolytes (GPEs) consist of a polymeric membrane onto which a minimum amount of classic "salt–solvent" combination is added. The salt and organic solvent can be picked from the list of common Li/Na salts and carbonate solvents used in liquid electrolytes. The solvent acts as a plasticizer and its addition results in the swelling of the polymer matrix, which physical aspect changes from a solid to a gel. The gellification of the membrane facilitates the interfacial contact with the electrodes in comparison to SPEs. The most common polymers used to prepare GPEs are polyacrylonitrile (PAN) [91–93], polymethylmethacrylate (PMMA) [94–96], polyvinylchloride (PVC) [97], and polyvinylidene fluoride (PVDF) [98–102]; their chemical structure are represented in Table 1.3. The room temperature ionic conductivity of GPEs generally falls in the range of 10^{-4} to 10^{-3} S/cm. While polymers with an amorphous character like PMMA provide higher ionic conductivities, their mechanical properties are usually lower due to the lack of rigidity in their matrix. PAN and PVDF provide higher mechanical properties due to a higher degree of crystallinity but it also implies a loss in ionic mobility [103]. Copolymers, such as poly(vinylidene fluoride–hexafluoropropylene) (PVDF–HFP) [61,104–106] can help overcome the loss of mechanical properties while keeping a promising ionic conductivity. The presence of CF_3 groups on the HFP monomer increases the amorphocity of the copolymer in comparison to the more crystallizable PVDF. The higher ratio of amorphous phase enhances the ionic conductivity of the pure

Table 1.4 Composition (wt.%) and conductivity of PVDF–(PC + DEC)–$LiClO_4$ gel polymer electrolyte at 30°C

PVDF–HFP	PC	DEC	$LiClO_4$	σ_i (mS/cm) at 30°C
25	35	35	5	0.17
25	32.5	32.5	10	0.22
25	30	30	15	0.75

Source: Reproduced from D. Saikia, A. Kumar, Ionic conduction in P(VDF-HFP)/PVDF–(PC + DEC)–$LiClO_4$ polymer gel electrolytes, Electrochim. Acta 49 (16) (2004) 2581–2589 [105] with permission from Elsevier, Copyright © 2004.

PDVF-based GPE up to 7.5 × 10^{-3} S/cm (Table 1.4), while the crystallinity provided by PVDF eases the cohesion of the film and its processability [105].

The design of copolymers, blends, and cross-linked polymers has been of latest interest to (1) reduce solvent leakage that can occur in more crystalline GPEs, (2) increase the ionic conductivity getting close to the 10^{-2} S/cm range, and (3) ensure good mechanical properties, relying on a homogenous distribution of crystalline and amorphous phases [105,107–109]. Blending has been additionally investigated for controlling the porosity of the GPE membranes and optimize solvent uptake [110,111]. Electrospining has proven to be a successful method for preparation of blends, copolymers, and core–shell structures [112–115]. As for SPEs, inorganic nanofillers are commonly added to the GPEs' composition in order to enhance their ionic conductivity [116–122]. Finally, the addition of ionic liquids to the GPEs has been considered for ionic conductivity and electrochemical stability enhancement [123–125].

Only a limited number of reports currently address Na-based GPEs [118,126–131], but their number will most likely increase in the upcoming years, following the growing interest in Na-based technology. Good ionic conductivities in the 10^{-3} S/cm range are obtained for Na-based GPEs; however with a low Na^+ transference number ($t_{Na+} \leq 0.3$) [127,129,131]. A considerable increase in transference number to 0.52 was reported following the incorporation of 15 wt.% SiO_2 nanofillers [129].

In comparison to SPEs, GPEs benefit from increased ionic conductivities, close to the values observed for liquid electrolytes. The addition of a small amount of solvent facilitates the dissociation of the salt in the polymer matrix, the larger its polarity the more pronounced the effect. While GPEs provide excellent ionic conductivity, their processing into thin films is more complex than for SPEs but not impossible. In 1997, Bellcore patented the fabrication process of Li-ion cells containing PVDF–HFP-based GPE membranes [132]. The electrochemical stability of the chosen solvent could be a limitation for GPEs application in high voltage batteries. However, if the polymer of the GPE is used in the electrode formulation as a binder, it will further reduce the interfacial resistivity between electrodes and electrolyte, in particular if the gellification step is performed after cell assembly, as this can lead to "fusion" of the different layers.

3.2 Inorganic electrolytes

Inorganic glasses as well as crystalline materials share the same ion transport mechanism, that is, they conduct through defective or interstitial sites at the local atomic scale. The disorder in an amorphous network presents numerous sites available for mobile cations to migrate, leading to ionic conductivity typically higher than in the crystalline counterpart [133,134]. However, to achieve superionic or fast-ion conducting properties in inorganic materials, that is, high-mobility and long-range diffusion, similar than in liquid electrolytes, the atomic structure must have long-range structural order. Therefore, crystalline materials are more promising candidates than glasses, for application in solid-state batteries. This section will be mainly focus on crystalline inorganic electrolytes, though examples of high-conducting glass-ceramics will be also discussed.

3.2.1 *Metal oxides*

The high ionic mobility in Na β-Al_2O_3, with chemical formula $Na_2O{\cdot}11Al_2O_3$, was first reported in 1967 by Yao and Kummer [135]. Na β-Al_2O_3 has a hexagonal layered structure with consecutive planes of cations and anions along the *c* direction, Na^+ diffusion occurring through the planes perpendicular to the *c*-axis. Exchange of Na^+ by Li^+ leads to slower diffusion rates [135] due to a decrease in ionic radius, which confined Li^+ in a potential well with a dramatic increase in activation energy [136]. Na β-Al_2O_3 provides very high ionic conductivities ranging from $\approx 10^{-2}$ to 10^{-1} mS/cm at 25°C, depending on the synthesis method, combined with very low activation energy of ≈0.2 eV [137,138]. Na β-Al_2O_3 has been extensively studied for applications in molten salt electrochemical cells, such as high temperature Na/S and Na/$NiCl_2$ (also known as ZEBRA) batteries characterized by high energy and high power densities [139]. Both systems operate at elevated temperature (>200°C) with melted electrodes ensuring a good electrode/electrolyte contact with liquid/solid interfaces. When decreasing the operating temperature, wetting of Na β-Al_2O_3 by Na becomes an issue leading to an increased interfacial resistivity. This explains the limited application of Na β-Al_2O_3 electrolyte at room temperature which implies overcoming highly resistive solid/solid interfaces. Due to the particularly high ionic conductivity provided by Na β-Al_2O_3 and its early discovery, only one additional inorganic material has been considered for Na^+ solid electrolyte application, and has given its name to an entire family of isostructural ionic conductors, NASICON [140].

The term NASICON stands for "Na Super Ion Conductor." It was given to $Na_{1+x}Zr_2Si_xP_{3-x}O_{12}$, a solid solution of $NaZr_2P_3O_{12}$ and $Na_4Zr_2Si_3O_{12}$, after the discovery of its high ionic conductivity by Goodenough and Hong in 1976 [141] who reported an ionic conductivity of 0.2 S/cm at 300°C, similar to the one of Na β-Al_2O_3. However this value decreases rapidly with lowering of the temperature due to a higher activation energy (E_a = 0.29 eV), reaching $\approx 10^{-3}$ S/cm at room temperature [141], 1–2 orders of magnitude below the one of Na β-Al_2O_3. Additional materials belonging to the NASICON family have been considered as Li solid electrolytes. They have the general formula $LiM_2(PO_4)_3$ with M = Zr, Ti, Ge, Sn [136,142]. An increase in charge carrier concentration, hence in ionic conductivity, can be achieved via the substitution of M^{4+} by M'^{3+} leading to the extensively studied $Li_{1+x}Al_xGe_{2-x}(PO_4)_3$ (LAGP) and $Li_{1+x}Al_xTi_{2-x}(PO_4)_3$ (LATP) [143,144]. Bulk conductivities can reach $\approx 10^{-3}$ S/cm for $Li_{1.3}Al_{0.3}Ti_{1.7}(PO_4)_3$ at room temperature; however the grain boundary resistivity is high which results in a lower total ionic conductivity ($\approx 10^{-4}$ S/cm) [143–145]. NASICON conductors are usually stable in air, in water and at high potentials [136]. However, they do contain Ti^{4+} or Ge^{4+} which can both be reduced at lower potentials [146].

The perovskite $La_{2/3-x}Li_{3x}\square_{1/3-2x}TiO_3$ (LLT) has been extensively studied as a solid-state electrolyte, where □ represents cation vacancies in the crystal structure. From the parent structure $La_{2/3}\square_{1/3}TiO_3$, the aliovalent substitution of La^{3+} by Li^+ consumes parts of the available cationic vacancies. It is the presence of remaining vacancies that renders the diffusion of Li^+ through the lattice possible. The resulting cell is composed of alternating La poor and La rich layers along the *c*-axis, Li^+ diffusing through the La poor layers, perpendicular to the *c* axis [147]. The perovskite is well known for providing high bulk conductivity ($\approx 10^{-3}$ S/cm), unfortunately coupled with much lower grain boundary conductivity ($\approx 10^{-5}$ S/cm) [148,149]. Some studies have been dedicated to lowering of the grain boundary resistivity [149–152]. However, similar to LATP, the material is unstable at lower potentials due to the presence Ti^{4+}, electrochemically active at about 1.5 V [153].

Garnets crystallize in a cubic unit cell (space group Ia3d) with general formula $A_3B_2(CO_4)_3$, the A^{2+}, B^{3+}, and C^{4+} cations being in dodecahedral, octahedral, and tetrahedral coordination, respectively. Stoichiometric Li-garnets $Ln_3Te_2(LiO_4)_3 \equiv Li_3Ln_3Te_2LiO_{12}$ (Ln = Y, Pr, Nd) show very poor ionic mobility ($\approx 10^{-5}$ S/cm at 600°C) with a high activation energy (>1 eV) suggesting a large energy barrier for the ionic motion between distant tetrahedra [154,155]. The arrangement of the tetrahedral units leads to the existence of interconnecting octahedral sites which can be occupied by Li^+ leading to $Li_xA_3B_2O_{12}$ with $5 \leq x \leq 7$. Those compositions are often referred to as "Li-stuffed garnets." Commonly investigated compositions include $Li_5La_3M_2O_{12}$ (M = Nb, Ta), $Li_6ALa_3M_2O_{12}$ (A = Mg, Ca, Sr, Ba, and M = Nb, Ta), and $Li_7La_3Zr_2O_{12}$ [156]. The gradual increase in Li concentration is accompanied by the redistribution of Li over the tetrahedral and octahedral sites, the ratio of octahedral to tetrahedral units increasing with Li. The resulting contiguous network of charge carriers gives rise to higher ionic conductivities with on average $\approx 10^{-5}$, $\approx 10^{-6}$, and $\approx 10^{-4}$ mS/cm at 25°C for $Li_5La_3M_2O_{12}$, $Li_6ALa_3M_2O_{12}$, and $Li_7La_3Zr_2O_{12}$ (LLZ), respectively [156]. Various substitutional schemes have been examined to evaluate the ionic mobility of intermediate Li concentrations. The best results are reported for $Li_{7-x}La_3Zr_{2-x}M_xO_{12}$ (M = Nb or Ta) and $Li_{7-3x}M_xLa_3Zr_2O_{12}$ (M = Al or Ga) with Li nominal content of ≈6.5 [157]. The fastest Li mobility recorded at 25°C, to date, reached 1.3×10^{-3} S/cm for $Li_{6.55}Ga_{0.15}La_3Zr_2O_{12}$ [158]. Care has to be taken when preparing those materials. Thermal treatments per-

Table 1.5 Summary of oxide-based solid electrolytes

Charge carrier	Structure	Electrolyte material	σ_{total} (mS/cm) at 25°C	Limitations for application in SSBs	References
Na^+	Na β-Al_2O_3	$Na_2O{\cdot}11Al_2O_3$	25	Poor wetting by Na metal	[138]
Na^+	NASICON	$Na_3Zr_2Si_2PO_{12}$	0.5	Poor ionic conductivity at 25°C	[141]
Li^+	NASICON	$Li_{1.3}Al_{0.3}Ti_{1.7}(PO_4)_3$	0.7	Unstable with Li metal	[145]
Li^+	Perovskite	$La_{2/3-x}Li_{3x}\square_{1/3-2x}TiO_3$	0.08	Unstable with Li metal	[149]
Li^+	Garnet	$Li_{6.55}Ga_{0.15}La_3Zr_2O_{12}$	1.3	Moisture sensitive	[158]

formed in Al_2O_3 crucibles often lead to unintentional Al substitution, which impacts the final composition of the garnet and its ionic conductivity. Due to its intrinsic high ionic conductivity, processing in air results in fast Li^+/H^+ exchange, which dramatically reduces the ionic mobility [158–160]. Finally, $Li_7La_3Zr_2O_{12}$-based garnets have been tested in Li-ion batteries showing good electrochemical performances with $LiCoO_2$ [161]. However, metallic Li deposition has been recently reported in the pores of garnets after cycling [162]. It was also observed in a hot pressed sample with 99% density [163]. The chemistry at the grain boundary is most likely involved in this issue and will need to be overcome to ensure future industrial applications of the garnets.

The main candidates for metal oxide electrolytes in Li and Na- based technologies are gathered in Table 1.5 along with their respective ionic conductivities and limitations for application (Table 1.5).

3.2.2 Metal sulfides

Sulfur is a more polarizable chalcogenide than oxygen, meaning that the electronic cloud can be deformed when the mobile cation migrates from site to site in the crystal lattice. That is, sulfide ions attract Li ions more weakly than oxide ions. In addition, the larger size of sulfide ions provides more open spaces in the anion sublattice, facilitating the transport of charged species. As a consequence, sulfide-based electrolytes are typically faster ion conductors than oxide electrolytes. For instance, thiolated versions of a LISICON-oxide type structure show ionic conductivity values up to 3–4 orders of magnitude higher than the nonthiolated [57,164]. From the viewpoint of integration into solid-state battery devices, another advantage of sulfides is their ductility and low grain boundary resistance, which allows intimate contact with electrodes by simple cold pressing [165,166]. Despite the promising characteristics of sulfide electrolytes, their application in solid-state batteries is severely hampered due to their poor chemical and electrochemical stability. Best performing (liquid-like conductivity) sulfide systems readily react with the oxygen and moisture from atmosphere, releasing toxic H_2S gas. In addition, they are not thermodynamically stable against Li metal anodes nor against cathode materials, such as transition metal oxides and phosphates [167,168]. Great scientific efforts are being carried out in the last years aiming to overcome the aforementioned limitations. The approaches include development of new materials with improved stability and ion conductivity via compositional/structural modifications as well as surface protective coatings [57,169–172].

This section will cover the most promising families of sulfide electrolytes (thio-LISICON and glass-ceramics) focusing on ion transport mechanisms, stability enhancement, and their integration into full devices.

The thio-LISICON family was first introduced by Kanno and coworkers at the beginning of the 21st century [173,174]. The structure of these first systems consisted mainly of hexagonal closed packed (*hcp*) sulfide ions arrays with group 13, 14, and 15 elements (M) in tetrahedral coordination, which form the framework structure. The MS_4 tetrahedral units were isolated from each with Li ions occupying both tetrahedral and octahedral sites [175,176]. Parent compositions based on stoichiometric Li_xMS_4 usually showed ionic conductivity below 10^{-6} S/cm. Aliovalent cationic substitutions and multielement composition have been the most used strategies to increase ionic conductivity in this family of electrolytes, yielding to a wide range of solid solutions. These advancements are summarized in Table 1.6.

One of the most outstanding progresses in this field was achieved in 2011 by Kamaya et al. [57], who reported a new Li superionic conductor $Li_{10}GeP_2S_{12}$ (LGPS), with an impressive conductivity value of 12 mS/cm at room temperature, which is comparable to that of liquid electrolytes. The exceptionally high ionic conductivity of tetragonal LGPS is attributed to a nearly isotropic Li hopping processes in the bulk lattice of LGPS with activation energy, $E_a \approx 0.22$ eV [180]. Replacing the costly Ge, in LGPS, by relatively cheap Sn or Si has been shown to also give high conductivity and similar activation energy [169,170]. First principles calculations have also been used to predict the phase stability, electrochemical stability, and Li^+ conductivity of $Li_{10\pm1}MP_2X_{12}$ family of superionic conductors (M= Ge, Si, Sn, Al, or P, and X = O, S, or Se). This computational study [181] confirmed the experimental observation that substituting Ge by Si or Sn, has limited impact on stability and conductivity, while the oxide derivatives $Li_{10}MP_2O_{12}$ were predicted to be significantly less stable and have lower conductivity.

The record conductivity value of 12 mS/cm, in LGPS, has been recently surpassed by a new composition $Li_{9.54}Si_{1.74}P_{1.44}S_{11.7}Cl_{0.3}$. In this material, double substitution with aliovalent elements Si and Cl, yielded values as high as 25 mS/cm at room temperature [171]. The ion transport mechanism is characterized by three-dimensional (3D) conduction pathways (1D along the *c*-axis + 2D in the *ab* plane), the latter contributing to the enhanced conductivity compared to LGPS. Although this low activation energy path is still not well understood. It is believed that chlorine doping (Cl populating unique sites in the PX_4 tetrahedra) cause a different lithium distribution in LGPS [171].

Table 1.6 Summary of sulfide-based solid electrolytes

Electrolyte material	Structure	Ion conductivity at RT (mS/cm)	Limitations for application in SSBs	References
$Li_{10}GeP_2S_{12}$	Thio-LISICON	12	High cost of Ge Unstable versus Li metal Air and moisture sensitive	[57]
$Li_{10}SiP_2S_{12}$	Thio-LISICON	2.3	Unstable versus Li metal Air and moisture sensitive	[169]
$Li_{10}SnP_2S_{12}$	Thio-LISICON	7	*Same as above*	[170]
$Li_{3.25}Ge_{0.25}P_{0.75}S_4$	Thio-LISICON	2.2	High cost of Ge Unstable versus Li metal Air and moisture sensitive	[174]
$Li_{9.54}Si_{1.74}P_{1.44}S_{11.7}Cl_{0.3}$	Thio-LISICON	25	Unstable versus Li metal Air and moisture sensitive	[171]
$Li_{9.6}P_3S_{12}$	Thio-LISICON	1.2	Air and moisture sensitive	[171]
$Li_{3.83}Sn_{0.83}As_{0.16}S_4$	Thio-LISICON	1.4	Unstable versusLi metal	[172]
$70Li_2S$-$30P_2S_5$ ($Li_7P_3S_{11}$)	Glass-ceramic	3.2	Unstable versus Li metal Air and moisture sensitive	[177]
$70Li_2S$-$30P_2S_5$ ($Li_7P_3S_{11}$)	Densified glass-ceramic	17	Unstable versus Li metal Air and moisture sensitive	[176,178]
β-Li_3PS_4	Thio-LISICON	0.16	Unstable versus Li metal Air and moisture sensitive	[179]
Na_3PS_4	Thio-LISICON	0.18	Unstable versus Li metal Air and moisture sensitive	[166]

Glass-ceramics can also yield superionic behavior with values as high as 17 mS/cm and $E_a = 0.18$ eV at room temperature for the system Li_2S–P_2S_5, in which the high conducting phase $Li_7P_3S_{11}$ crystallizes within the precursor glass upon thermal treatment [176,178]. The grain-boundary resistance in this material was reported to be 5 times higher than that of a previously reported Li_2S–P_2S_5 system [177]. This difference in conductivity is related to the fact that a better densification process increases the overall conductivity, not by enhancing ionic conduction in the bulk, but by reducing the grain-boundary resistance. The grain-boundaries around the $Li_7P_3S_{11}$ crystal domains are amorphous regions that have the effect of lowering the grain-boundary (GB) resistance in contrast to polycrystalline systems.

A key factor determining superionic conducting properties in the thio-LISICON family materials is related to the topology of the anion sublattice. It has been recently shown, by first principles calculations, that the body-centered cubic (bcc) sulfide sublattice, present in most of the LGPS-based superionic conductors, allows the presence of energetically equivalent Li sites within the diffusion network and thus, lower activation energy for migration [182].

Despite the high ionic conductivity of thio-LISICON-based electrolytes, they have serious limitations that strongly limit their practical application in batteries. These limitations include: chemical reactivity in ambient air, incompatibility with Li metal, as well as electrochemical instability. The origin of the air/moisture instability is related to the hard acidity of the positively charged elements P^{5+}, Ge^{4+}, and Si^{4+} present in these sulfide compounds. Based on the hard and soft (Lewis) acids and bases (HSAB) theory, such hard acids will react faster and will form more stable compounds with a hard base like oxide ion O^{2-} rather than a softer sulfide ion S^{2-}. Therefore, in contact with oxygen or moisture, thio-LISICON-based electrolytes containing hard bases will undergo displacement reactions, forming oxide phases and releasing sulfur or H_2S. A promising approach to avoid the high air sensitivity of these materials is to replace the hard acid elements with softer ones like Sn or As. Following this approach, Liang et al. reported an air-stable sulfide electrolyte with composition $Li_{3.83}Sn_{0.83}As_{0.16}S_4$ and ion conductivity of 1.39 mS/cm at 25°C [172].

3.3 Composite solid electrolytes

As shown throughout this chapter, great scientific efforts have been devoted to develop ceramic and polymeric electrolytes for application in solid-state batteries. However, none of the existing materials satisfy all the criteria for application in large-scale solid-state batteries (Table 1.7). For instance, realizing a nontoxic and stable solid electrolyte with high ionic conductivity requires the use of ceramic oxide-based conductors, which are hard materials. However, achieving low charge transfer resistance (R_{ct}) at the solid electrode–electrolyte interface requires at least one solid of that interface, to be a soft material, for example, a polymer or sulfide-based electrolyte. From the point of view of manufacturing and industrial scalability, the solid electrolyte is preferred to be compatible with solution/slurry processing or lamination techniques and the thicknesses obtained should be easily tunable. These characteristics, are, in turn, incompatible with the conventional solid-state processing of ceramic electrolytes. Therefore, taking into account the inherent properties of solid electrolytes (inorganic and polymers), it becomes evident that finding a single material that meets all criteria of Table 1.7 is a grand scientific challenge.

A promising route to tackle such materials paradox, is to combine "hard" and "soft" materials (e.g., polymers or sulfides and ceramics) into a single composite electrolyte material [183–187]. This composite approach can enable achieving target performances via addition of properties as well as offering synergetic and enhanced functionalities derived from interface effects between the components. For instance, improved ion conductivity, mechanical properties and electrochemical stability, compared to the pure polymer, have been demonstrated by dispersing ceramic ion conducting ceramic particles in PEO matrices [18,21]. However, composite electrolytes are still far from practical applications in solid-state batteries, especially regarding ion conductivity and R_{ct} at the electrode–electrolyte interface.

In this section, we will describe the most relevant composite electrolytes focusing solely on those constituted by solid materials. Other composite electrolytes containing

Table 1.7 Requirements that a solid electrolyte must fulfill for application in bulk-type solid-state batteries

	Material's property/characteristic	Effect on solid-state battery performance
1	High ionic conductivity, at least 10^{-3} S/cm at room temperature	High current density and high power
2	Negligible electronic conductivity, Li transference number $t_{Li+} \sim 1$	Minimizes self-discharge rate and maximizes usable voltage (energy density)
3	Chemically inert against electrode materials and air atmosphere	Long-term stability, lifetime
4	Wide electrochemical stability window; >4.5 V for Li anode and high voltage cathodes	Higher cell voltage and energy density; long-term stability, lifetime
5	Fully dense and mechanically tough; shear modulus $>10^9$ Pa	Avoid Li dendrite growth for Li metal anodes, maximizing safety
6	Good wettability and intimate interface contact with electrodes	Low charge-transfer resistance (R_{ct}); high current, high power, and long-term stability
7	Dimensional and thermal stability	Ensures physical barrier between cathode and anode; long-term stability
8	Solution-based processing	Industrially scalable
9	Mechanically flexible and robust, allowing self-standing or electrode-supported membranes of thickness 10–30 μm	Facilitates lamination process during cell assembly; avoids fractures during cell assembly and operation
10	Abundant and nontoxic constituents	Reduces manufacturing costs and minimizes safety issues

liquid additives, for example, organic solvents or ionic liquids to plasticize the polymer component, will not be covered here as these polymer gels reintroduce the detrimental effects related to liquid electrolytes.

It is worth clarifying a conceptual confusion that is often found in the literature; that is, the difference between composites and hybrid materials. Composites are physical mixtures of two constituents with no interfacial bonding. By contrast, hybrid materials include two moieties bonded on the molecular scale [188]. Some reports use the term "hybrid" to denote the organic–inorganic nature of the solid electrolyte, although they do not provide any proof of the interfacial blending between the components. The correct term should be "composite" as these materials are physical mixtures obtained by mechanically mixing of their constituents.

3.3.1 Sulfide–oxide composite inorganic electrolytes

There are only a few examples of composite electrolytes using the combination of sulfide–oxide systems [189,190], though they are quite promising concerning the enhancement of properties (Table 1.8). The key characteristics of sulfide–oxide composites are the following:

Table 1.8 Summary of composite electrolytes containing ion-conducting fillers within an ion-conducting matrix. The listed concentrations are the best performing for each composition.

Materials						
Matrix	**Filler**					
Intrinsic σ_i at RT (mS/cm)	**Intrinsic σ_i at RT (mS/cm)**	**Particle size**	**Content (wt.%)**	**Processing route**	**σ_i composite at RT (mS/cm)**	**References**
β-Li_3PS_4 (LPS) $\sigma_i = 0.16$	Al-doped $Li_7La_3Zr_2O_{12}$ (LLZO, cubic phase) $\sigma_i = 0.4$	<1 μm	30	Dry ball milling of powders	$\sigma_i = 0.536$	[190]
	$Li_6ZnNb_4O_{14}$ $\sigma_i = 0.032$	<2 μm	10		$\sigma_i = 0.244$	[189]
PEO + LiTFSI $\sigma_i = 4 \cdot 10^{-3}$	$Li_{0.33}La_{0.55}TiO_3$ (LLTO) $\sigma_i \sim 1$	<1 μm 250 μm (fiber diameter)	20 20	Ultrasonic mixing of composite dispersion containing commercial LLTO fibers	$\sigma_i = 0.2$ $\sigma_i = 0.5$	[192]
PEO–poly(propylene oxide) copolymer + LiTFSI *σ_i not reported*	$Li_{1+x+y}Al_xTi_{2-x}Si_yP_{3-y}O_{12}$ Glass-ceramic $\sigma_i = 1.4$	1.5 μm 5 μm	75 75	Slurry, doctor blade	$\sigma_i \sim 7 \cdot 10^{-3}$ $\sigma_i \sim 8 \cdot 10^{-2}$	[193]
PEO-$LiClO_4$ $\sigma_i = 5 \cdot 10^{-5}$	$Li_7La_3Zr_2O_{12}$ (tetragonal phase) $\sigma_i = 6.2 \cdot 10^{-4}$	<1 μm	52.5	Slurry, doctor blade	$\sigma_i = 0.01$	[186]
Polyacrylonitrile-$LiClO_4$ $\sigma_i = 2.1 \cdot 10^{-4}$	$Li_{0.33}La_{0.55}TiO_3$ (LLTO) $\sigma_i \sim 1$	140 nm (fiber diameter)	15	Electrospinning + slurry and doctor blade	$\sigma_i = 0.24$	[187]

- The composites were prepared by ball-milling the powders, yielding physical mixtures of the oxide fillers and Li_3PS_4 (LPS). The particle size of the fillers as well as the surface chemistry, acid-base character and ionic conductivity play a big role in the final performances of the composite electrolyte.
- *Improvement of mechanical properties.* Combining the sticky and "soft" LPS with "hard" oxide fillers (with an LPS content as low as 10 wt.%) can result in a dense pellet free from fractures [189]. By contrast, cold-pressing a ceramic oxide pellet without addition of LPS results in poor processing and fractured/porous pellets.
- *Increase of total ion conductivity.* The addition of oxide nanofillers into LPS system has been shown to increase the conductivity of the LPS electrolyte [189,190] (Table 1.8). Recently Hood et al. have studied the so-called filler effect, by embedding ion conducting and nonconducting oxides ("active fillers" vs. "nonactive fillers") into the LPS matrix. As ion conducting filler $Li_6ZnNb_4O_{14}$ (LZNO) was chosen, with ionic conductivity of 3.28×10^{-5} S/cm^{-1} at 21.5°C, while Al_2O_3, and SiO_2 were chosen as nonconducting fillers. The results showed that as an increased amount of oxide was added to the composite, the ionic conductivity significantly decreased while the activation energy increased. This observation can be attributed to the blocking effects from the oxide filler. When 10 wt.% LZNO was added to LPS, the ionic conductivity increased to $2.44 \cdot 10^{-4}$ S/cm, practically doubling the ionic conductivity of the parent electrolyte (Table 1.8). Similarly, LPS filled with 2 wt.% Al_2O_3 (with particle size $\leq$1–2 μm) increased the conductivity to $2.28 \cdot 10^{-4}$ S/cm, demonstrating that ion-conducting and nonconducting oxide fillers are able to enhance the ionic conductivity of LPS. The mechanism by which the ionic conductivity is enhanced is yet not well-understood but it is likely to be related to synergistic interface effects between LPS and the oxide fillers.
- *Lower activation energy.* When compared to the nonconductors, the ion-conductors fillers, such LLZO and LZNO significantly decreased the thermal activation of ion hopping for composites with enhanced ionic conductivity. The activation energy of pristine LZNO is 0.39 eV [191]; however, in the best conducting LZNO–LPS composite, the activation energy decreased to 0.31 eV, which is lower than that of LPS (0.35 eV) and LZNO (0.39 eV). This lower activation energy can be also rationalized by synergistic interface effects between LPS and the oxide fillers.
- *There is an optimum volume fraction that enhances ionic conductivity.* Relatively small volume fraction of oxide filler cause an enhancement of the ionic conductivity while higher concentrations have been shown to be detrimental, causing a decrease of the ionic conductivity (Fig. 1.4). In the case of 10:90 weight ratio of LZNO:LPS, the ionic conductivity was higher than the conductivity of the pristine parent compounds (Table 1.8). This trend was previously observed in the case of LPS filled with $Li_7La_3Zr_2O_{12}$ (LLZO), where additions of 30 wt.% LLZO increased the conductivity by a factor of 5 [190] (Table 1.8). This enhancement in conductivity is general for

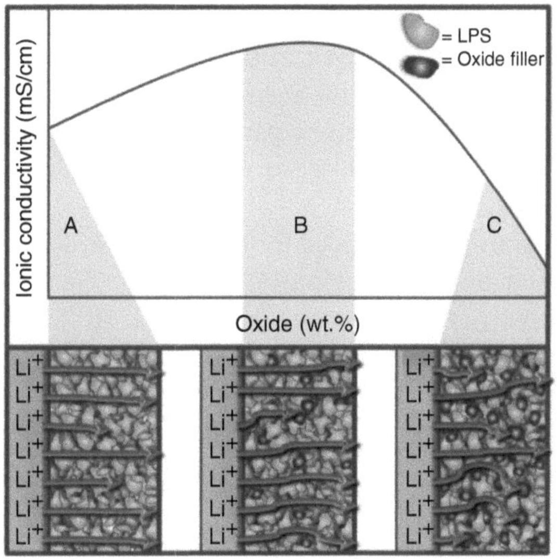

Figure 1.4 ***Ionic conductivity trend for composite solid electrolytes as a function of the filler volume fraction.*** The figure shows the particular case of LPS-oxide systems, although it can be generally applied to all composite systems. "A" represents the zone of no presence of oxide filler, "B" represents the space-charge effect, and "C" shows the blocking effect of the oxide in filler. *(Reproduced from Z.D. Hood, H. Wang, Y. Li, A.S. Pandian, M. Parans Paranthaman, C. Liang, The "filler effect": a study of solid oxide fillers with Li_3PS_4 for lithium conducting electrolytes, Solid State Ion. 283 (2015) 75–80 [189], with copyright permission from Elsevier).*

this type of composites and is attributed to the interface between LPS and the fillers. Nyquist plots showed both higher and lower conducting processes, evidencing the presence of an interfacial layer between the LPS and the oxide filler. This interface effect has been only observed for ion-conducting fillers, evidencing the important role of conducting surfaces. In contrast, the nonconducting fillers did not show two processes in the Nyquist plot, which confirms that there is competition between the higher conducting parent electrolyte and lower conducting oxide filler throughout the composite.

- *Better interfacial resistance with Li metal anode.* The cyclability and long-term stability of the sulfide–oxide composite electrolytes is typically tested by assembling symmetric Li/composite electrolyte/Li cells. Negligible interfacial resistance was observed when the composite electrolytes were cycled with metallic lithium (even after 400 cycles), supporting the compatibility of the composite electrolytes with metallic lithium.

3.3.2 Organic–inorganic composite electrolytes

Similar to the sulfide–oxide system, a relevant characteristic of ceramic–polymer composites is that, by combining dissimilar components, the overall ionic conductivity can be enhanced. In addition, more suitable mechanical properties and better interface contact with the electrodes can be achieved in this type of composites, which is key for their integration into solid-state batteries. Self-standing flexible membranes can be easily obtained by dispersing ceramic electrolyte particles into polymeric solutions. Some of the most promising ceramic–polymeric electrolytes are shown in Table 1.8, and their key characteristics summarized here:

- *Increase of ionic conductivity.* As described for the sulfide–oxide system, there is an optimum volume fraction that enhances ionic conductivity (Fig. 1.4). At ceramic loadings lower than 15–20 wt.%, the total ionic conductivity of the composite is enhanced respect to the pure polymer phase. The reason for the enhancement can be understood as a sum of various effects. First, a purely structural reconstruction effect, that is able to lower the crystalline degree of the polymer around the filler particles, thus improving ion hopping (as it was described in the polymer electrolyte, Section 3.1). Second, space-charge and percolative effects at the interface between dissimilar materials can also play a key role in enhancing the ionic conductivity. Although space-charge effects have not yet been confirmed and thoroughly investigated in ceramic–polymeric composites, they have been reported to increase of ion diffusion kinetics at solid–solid interfaces in other materials not necessarily related to Li-ion solid electrolytes [194,195]. What has been investigated in certain depth is the influence of the filler morphology and particle size on the ion conductivity [192,187].
- *Role of filler morphology and particle size in the conductivity enhancement.* Ceramic inclusions with fiber-like morphology are more beneficial than isolated particles because fibers can provide long-range diffusion pathways within the composite, especially if they have long-aspect ratio. For instance, Wang et al. demonstrated that when introducing the same loading of LLTO fillers into PEO matrix the conductivity could be enhanced by more of a factor 2, when fibers were used instead of particles [192]. Similarly, Liu et al. recently showed that the conductivity enhancement was significantly higher when using LLTO nanowires instead of nanoparticles [187]. Specifically, when filling polyacrylonitrile matrix with LLTO nanowires of 140 nm diameters, the enhancement was by 3 orders of magnitude (Table 1.8). In contrast, the improvement was limited to 1 order of magnitude, when filling the polymer matrix with LLTO nanoparticles with the same loading (15 wt.%) and similar grain size (~50 nm diameter). Nanoparticle aggregation was also pointed by the authors to contribute to the lower ionic conductivity since the aggregation would further reduce probability of forming a continuous ionic percolation path. The detailed mechanism for the enhancement of conductivity in the composite electrolyte, however, needs further investigation.

  Size effects for nonfiber like fillers have also been investigated. However, there is yet not a clear consensus on the influence of particle size and interface area density in the final transport properties of the solid electrolyte. The general trend points to that lower interfacial area densities, that is, larger particles, are desirable for larger conductivity enhancements. For instance, when embedding $Li_{1+x+y}Al_xTi_{2-x}Si_yP_{3-y}O_{12}$ glass-ceramic particles of 1.5 versus 5 μm into PEO matrices, the larger particles showed 1 order of magnitude larger enhancement (Table 1.8, [193]). However, it is not clear what would be the effect for other compositions or for nanosized fillers. The complexity of synthesizing submicron-sized ion conducting fillers and the lack of structural control during the composite processing is limiting the possibility to perform systematic studies that could shed light into the role of nanosize effects in these composites.

3.3.3 Conclusions

Combining dissimilar solid electrolytes into a single composite material is a promising strategy to achieve better performances via additive properties and synergetic interfacial effects. We have described how this composite approach leads to improved ionic conductivity, mechanical properties and electrochemical stability, compared to the individual components (Table 1.8), as well as demonstrating their integration into solid-state cells. However, composite electrolytes are still far from real application in solid-state batteries, especially regarding the low total ionic conductivity and the high R_{ct} at the electrode–electrolyte interface.

Scientific efforts to develop high performing composite electrolytes are currently carried out by several groups but publications are still scarce. The ionic conductivity values of the reported composite electrolytes are still too low for practical applications (<0.1 mS/cm) and the reason is twofold: (1) the intrinsically low ionic conductivity of the individual components and (2) experimental limitations related to the processing of the composite material. These limitations need to be overcome to push up the performances of composites. The fact that none of the composite electrolytes reported so far, show superionic conductivity, is a major challenge that could be achieved as new superionic materials are discovered and integrated into composites. Indeed in recent years there has been a renaissance of this field with promising results. However, the future successful commercial development of optimized composites will not depend only on the materials used but also on the processing methodology. It is of extreme importance that the materials processing enables fine tuning of the structure and morphology of composites at various length scales, which will give a handle to tune and enhance the resultant electrolyte performances. In addition, the processing route should be cost-efficient and industrially scalable. All in all, there is plenty of room for research in the field of composite electrolytes, from both materials science and technological point of views.

3.4 Integration of solid electrolytes into all-solid-state battery devices

From an application point of view, the biggest challenge of solid electrolytes is the integration into a full device (Fig. 1.5). The poor wettability of solid materials leads to

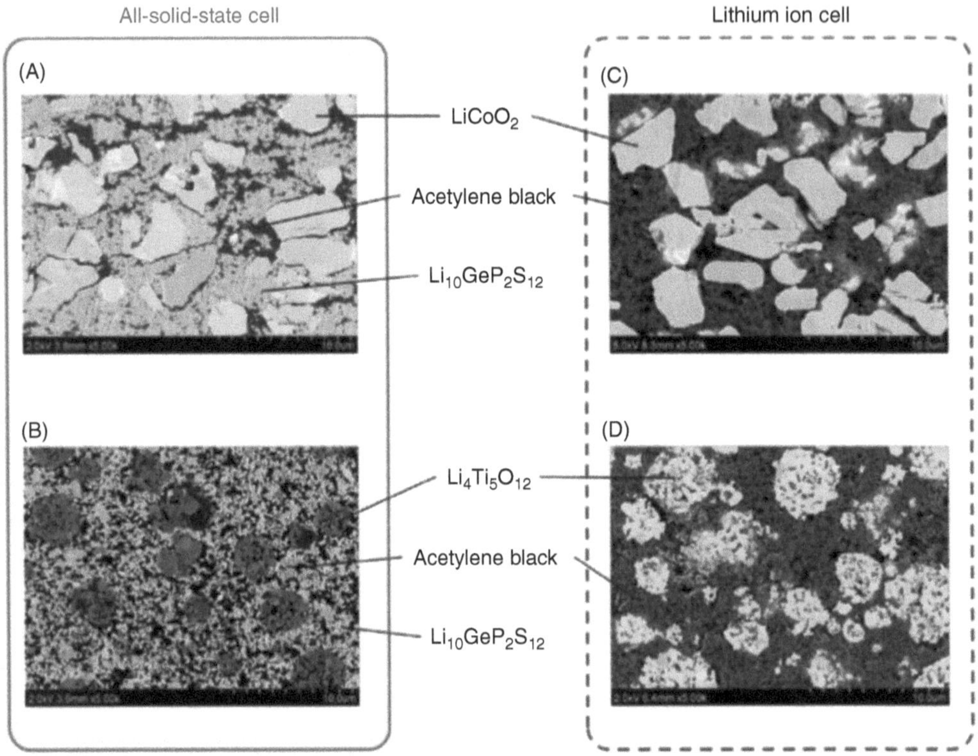

Figure 1.5 Comparison of composite electrodes in a solid-state electrochemical cell (left) and a conventional liquid-based cell (right). Cross-sectional SEM images of composite electrode layers: (A–B) cathode and anode layers of the all-solid-state cell, respectively; (C–D) cathode and anode layers of the liquid-based Li-ion cell, respectively. The cathode of the all-solid-state cell consisted of $LiNbO_3$-coated $LiCoO_2$ powder, a solid electrolyte powder, and acetylene black powder. The anode consisted of $Li_4Ti_5O_{12}$, a solid electrolyte, and acetylene black powder. *(Reproduced from Y. Kato, S. Hori, T. Saito, K. Suzuki, M. Hirayama, A. Mitsui, M. Yonemura, H. Iba, R. Kanno, High-power all-solid-state batteries using sulfide superionic conductors, Nat. Energy 1 (2016) 16030 [171], with copyright permission from Nature Publishing Group).*

nonintimate contact at the electrolyte-electrode, causing inhomogeneous current density distribution, high-charge transfer resistance and mechanical failure. Other limitations include the poor chemical stability in contact with Li metal and poor electrochemical stability under charge–discharge conditions [167,172,181]. These limitations have strongly hampered the integration of thio-LISICON electrolytes in full cell devices with metallic Li as anode. To overcome this challenge, the research community is investigating several approaches, one of the most promising is using protective coatings to avoid incompatibility with metallic Li. For instance, the surface of the solid electrolyte $Li_{3.83}Sn_{0.83}As_{0.16}S_4$ was passivated with a Li-compatible composite, $3LiBH_4{\cdot}LiI$, by simply dip-coating the ceramic pellet in a 5 wt.% solution in THF. THF solvent was removed by heating, yielding to a thin conducting $3LiBH_4 \cdot LiI$ material that prevented any interfacial reaction with metallic Li [172].

Another approach to avoid the incompatibility with metallic Li is to use other anode materials like graphite or $Li_4Ti_5O_{12}$ (LTO), which have higher reduction potential than metallic Li. This approach has typically compromised the energy density and rate capability of the cell owing to the lower capacity of the anode and narrower electrochemical voltage of the cells. However, a recent report by Kato et al. has shown impressive performances using LTO- and superionic-conducting sulfide electrolytes [171]. High specific power, superior to that of conventional cells with liquid electrolytes, was obtained using $Li_{9.54}Si_{1.74}P_{1.44}S_{11.7}Cl_{0.3}$ electrolyte, which has an ionic conductivity of 25 mS/cm. This superionic behavior together with an optimized electrode–electrolyte processing allowed a solid-state battery that delivered 70% of its maximum capacity in just 1 min at room temperature, with specific energy of ~320 Wh/kg (based on mass of active material) [171]. Analogously, the same authors prepared a solid-state cell using graphite as anode, which allowed wider electrochemical cell voltage. In that case, they use a different electrolyte, $Li_{9.6}P_3S_{12}$, which has fair stability at low voltages, though in detrimental of the ionic conductivity (1.2 mS/cm). In such high-voltage type of cells the specific energy raised to ~431 Wh/Kg, well above the values of conventional Li-ion cells, in the range of ~300 Wh/Kg.

Another important challenge for the successful integration of solid electrolytes into full devices lies in the cathode side. The cathode is a porous mixture of particles of active materials (e.g., $LiCoO_2$), conductive additives (e.g., carbon), and polymeric binders. And the question that arises is the following: how a solid material can access the pores between the particles of active material throughout the whole cathode volume? Solids, unlike liquids, have very poor wetting properties and therefore they cannot infiltrate the pores as liquids do and deliver the mobile Li-ions. The solution to this problem is to form a composite cathode structure, in which the solid electrolyte is mixed, typically by mechanical milling to achieve an intimate contact with the active material in the cathode, that is, forming a so-called catholyte (cathode material mixed with electrolyte material within the positive electrode layer).

A second problem to overcome is the reactivity of sulfides or oxides when they are in direct contact with the active material in the cathode. As a result, a highly resistive layer typically forms at the interface between sulfides and cathodes materials, such as $LiCoO_2$, limiting the current drain. In order to prevent the direct contact, the surface of the active particles can be coated with a nm-thick conductive buffer layer material, such as $LiNbO_3$ [196]. Despite the promise of this approach is worth mentioning that the surface coating adds complexity to the processing, which can be eventually limiting the large-scale production. In addition, the ionic conductivity of the buffer layer must be high enough not to limit the ion transfer from the electrolyte to the cathode particle.

3.5 The promise of nanostructured electrolytes

Nanotechnology is expected to have a key role in the development of novel and high-performing solid electrolytes. Creating nanostructures of existing solid electrolytes is a way to enhance ionic conductivity by several orders of magnitude with the advantage of not scarifying chemical stability [179]. A clear example is the case of nanocrystalline Li_3PS_4 (LPS), which is the most stable compound versus metallic Li among all Li thiophosphates. The ionic conductivity of bulk crystalline LPS at room temperature is $3 \cdot 10^{-7}$ S/cm, too low for practical application. In contrast, the ionic conductivity of a nanostructured version, known as nanoporous β-Li_3PS_4 (β-LPS) is enhanced by 3 orders of magnitude ($1.6 \cdot 10^{-4}$ S/cm) [179]. The causes of the enhancement are related to a nanosize structural effect that stabilizes the high-conduction beta phase that otherwise occurs at elevated temperatures. In addition, the high surface-to-bulk ratio of nanoporous β-LPS can promote surface conduction.

The case of nanocrystalline LPS might open the door to the exploration of size-effects in other electrolyte systems. Despite the promise of improving the ion conducting properties of inorganic solid electrolytes by exploiting nanosized effects, from a device point of view, a priori they are not that promising. A solid electrolyte must be highly dense to ensure practical application, but nanoparticulate materials would lead to high porosity. However "nanosolid electrolytes" can become an interesting functional component when embedded in a matrix material to form composite or glass-ceramic electrolytes. Ideally, the high conducting component would be the nanoelectrolyte forming a percolative ceramic network embedded in polymer or glassy matrix. An open question that arises is the high interface resistance between the nanoparticles and the matrix components, which would limit the performance of this nanocomposites. In addition, experimentally achieving this nanocomposite concept, is a challenging task since it requires accurate structural control. This high structural control is not possible with the current experimental techniques, which are mostly based in conventional sol-gel or solid-state reactions.

Despite this scientific challenge, there is plenty of room for research and optimization in the field of nanocomposite solid electrolytes. Materials scientist efforts will be needed to investigate novel preparation techniques that give high structural control at various length scales.

4 CONCLUSIONS

In the past decades, the requirements for energy storage systems kept increasing with a continuous demand for less toxic and safer solutions with higher energy densities. Despite the great importance of the active materials in an energy storage system, the electrolytes play a major role in performances and in cell aging. Fundamental research revealed that cell performances are strongly dependent on the chemical composition of the electrolyte leading to large variations in cycling efficiency, rate capability, capacity retention at various temperatures, and tolerance against abuses. Additionally, the electrochemical stability window of the electrolyte should be compatible with the selected materials used at the positive and negative electrodes. Thus the state-of-art of electrolyte systems in EES is custom-made for the cell chemistry in which the various components of electrolyte are designed for electrode surface and specific requirements of the technologies.

Electrolyte compositions in traditional batteries are not expected to have a radical modification as long as the cell chemistry remains the same. The study of electrolyte/electrode interfaces is of prime interest as local decomposition of both can have either a positive or negative impact on cell performance. Specific reactions can be driven by the careful selection of solvents and salts combined with the use of additives for liquid electrolyte and coating method and functionalized polymer for solid electrolyte. Innovations will continue to be carried out regarding rate performances, operating temperature ranges, improved safety, and "solidification" of the electrolyte.

In lithium-ion batteries, numerous efforts are focused on increasing the electrode energy density or providing safer cathode materials, such as novel >4.5 V class mixed-metal oxide or nanometric olivine iron phosphate family. On the anode side metallic lithium remains the most attractive material for high-energy density performances. The electrolyte formulation has been mainly used for controlling the undesired dendrite lithium formation during long term cycling [197,198]. The development of new cell chemistries, such as Na-ion, or the use of nanotechnology would require reformulation of the electrolytes.

Solid-based electrolytes are also proposed to be adequate in monitoring the surface morphology of metallic lithium. In regard to the successful answer that solid electrolyte gave to the safety issues of the lithium electrode, the renewal of lithium cell is considered, along with new electrolyte formulation that correspond to the new cell chemistries, such as for Li-S.

To satisfy the ever-growing demand for safer and higher energy density devices, the strategy in terms of electrolyte is to combine the benefit of liquid electrolytes for conduction with the rigidity of a solid. In principle, it is simple to combine the advantages from solid electrolytes and liquid electrolytes in one material. The final electrolyte has solid-like mechanical properties for blocking the dendrites and liquid-like ionic conductivities for Li^+ transport [199–201]. Polymer-based nanocomposites are used as the host of liquid electrolytes. The electrolyte demonstrates good mechanical properties that overcome the dendrite formation and provide an interesting approach.

The ongoing research is focused on further optimization of a solid-like electrolyte system for the assembly of lithium metal battery. It includes chemical functionalization of silica/PEO shell to prevent the diffusion of polysulfide species in Li-S battery or nano-engineering to improve the electrochemical properties in all solid batteries. Furthermore the high flexibility and compatibility enable a large combination of liquid, such as liquid electrolyte containing Na^+, Mg^{2+}, or Al^{3+} as conductive media. These properties provide a promising alternative for the development of next generation rechargeable batteries [202,203].

However, from a safety and performance point of view, true solid materials (e.g., inorganic electrolytes) are the most promising for solid-state batteries technologies. These inorganic electrolytes include metal oxides and sulfides that have superionic conducing properties and are able to withstand wider voltage window and temperatures. When integrated in a full electrochemical device (e.g., solid-state battery with Li metal anode), inorganic electrolytes still need to overcome key challenges regarding improvement of interface stability and resistance at the solid–solid interfaces with the electrodes, as well as avoiding Li dendrites growth and enhancing rate capability. Great scientific efforts are being currently devoted to face these challenges. The approaches include the development of new electrolyte materials with enhanced properties, for example, composite electrolytes and nanostructured solid electrolytes. Although there is still a long way to go before these materials can be integrated in a full device, the basic knowledge that is being generated is of great impact and will definitely accelerate the realization of a novel solid-state battery technology.

GLOSSARY

σ_i	Ionic conductivity
DEC	Diethyl carbonate
DMC	Dimethyl carbonate
EC	Ethyl carbonate
GPE	Gel polymer electrolyte
LAGP	$Li_{1+x}Al_xGe_{2-x}(PO_4)_3$
LATP	$Li_{1+x}Al_xTi_{2-x}(PO_4)_3$
LGPS	$Li_{10}GeP_2S_{12}$
LIB	Lithium-ion batteries
LISICON	Lithium Super Ionic Conductor
LLT	$La_{2/3-x}Li_{3x}\square_{1/3-2x}TiO_3$
LLZ	$Li_7La_3Zr_2O_{12}$
LPS	Li_3PS_4
LTO	$Li_4Ti_5O_{12}$
MW	Molecular weight
NASICON	Sodium Super Ion Conductor
PAN	Poly(acrylonitrile)
PC	Propyl carbonate
PEO	Poly(ethylene oxide)
PMMA	Poly(methyl methacrylate)
PPO	Poly(propylene oxide)
PVC	Poly(vinyl chloride)

PVDF	Poly(vinylidene fluoride)
PVDF–HFP	Poly(vinylidene fluoride–hexafluoropropylene)
R_{ct}	Charge transfer resistance
RT	Room temperature
SEI	Solid electrolyte interphase
SIB	Sodium-ion batteries
SPE	Solid polymer electrolyte
T_g	Glass transition of a polymer
Thio-LISICON	Thiolated versions of a LISICON-oxide type structure
T_m	Melting temperature

REFERENCES

[1] D. Linden (Ed.), Handbook of Batteries, third ed., McGraw-Hill, New York, NY, 2001.

[2] K. Xu, Nonaqueous liquid electrolytes for lithium-based rechargeable batteries, Chem. Rev. 104 (10) (2004) 4303–4418.

[3] C. Daniel, Materials and processing for lithium-ion batteries, JOM 60 (9) (2008) 43–48.

[4] E. Peled, D. Golodnitsky, G. Ardel, Advanced model for solid electrolyte interphase electrodes in liquid and polymer electrolytes, J. Electrochem. Soc. 144 (1997) L208–L210.

[5] X. Yao, B. Huang, J. Yin, G. Peng, Z. Huang, C. Gao, D. Liu, X. Xu, All-solid-state lithium batteries with inorganic solid electrolytes: review of fundamental science, Chin. Phys. B 25 (1) (2016) 018802.

[6] J. Li, C. Ma, M. Chi, C. Liang, N.J. Dudney, Solid electrolyte: the key for high-voltage lithium batteries, Adv. Energy Mater. 5 (4) (2015) 1401408.

[7] J. Motavalli, Technology: a solid future, Nature 526 (2015) S96–S97.

[8] N.A.W. Holzwarth, Solid electrolytes for battery applications: a theoretical perspective, Department of Physics, Wake Forest University, Winston-Salem, NC, USA. Available from: http://users.wfu.edu/natalie/presentations/Louisville2012.pdf

[9] K. Bullis, Solid-state battery: high-energy cells for cheaper electric cars, MIT technical review, 2011. Available from: http://www2.technologyreview.com/news/423685/solid-state-batteries/

[10] J.B. Goodenough, Y. Kim, Challenges for rechargeable Li batteries, Chem. Mater. 22 (3) (2009) 587–603.

[11] J. Scheers, S. Fantini, P. Johansson, A review of electrolytes for lithium-sulphur batteries, J. Power Sources 255 (2014) 204–218.

[12] J. Gao, M.A. Lowe, Y. Kiya, H.D. Abruña, Effects of liquid electrolytes on the charge–discharge performance of rechargeable lithium/sulphur batteries: electrochemical and in-situ X-ray absorption spectroscopic studies, J. Phys. Chem. C 115 (2011) 25132–25137.

[13] S.S. Zhang, Liquid electrolyte lithium/sulfur battery: fundamental chemistry, problems, and solutions, J. Power Sources 231 (2013) 153–162.

[14] E. Zinigrad, L. Larush-Asraf, J.S. Gnanaraj, M. Sprecher, D. Aurbach, On the thermal stability of $LiPF_6$, Thermochim. Acta 438 (2005) 184–191.

[15] A. Ponrouch, D. Monti, A. Boschin, B. Steen, P. Johansson, M.R. Palacın, Non-aqueous electrolytes for sodium-ion batteries, J. Mater. Chem. A 3 (2015) 22–42.

[16] T. Kawamura, A. Kimura, M. Egashira, S. Okada, J. Yamaki, Thermal stability of alkyl carbonate mixed-solvent electrolytes for lithium ion cells, J. Power Sources 104 (2002) 260–264.

[17] A. Manthiram, S.-H. Chung, C. Zu, Lithium-sulfur batteries: progress and prospects, Adv. Mater. 27 (12) (2015) 1980–2006.

[18] J.-Q. Huang, Q. Zhang, H.-J. Peng, X.-Y. Liu, W.-Z. Qian, F. Wei, Ionic shield for polysulfides towards highly-stable lithium–sulfur batteries, Energy Environ. Sci. 7 (2014) 347–353.

[19] V. Palomares, P. Serras, I. Villaluenga, K.B. Hucso, J. Carretero-Gonzalez, T. Rojo, Na-ion batteries, recent advances and present challenges to become low cost energy storage systems, Energy Environ. Sci. 5 (2012) 5884–5901.

[20] S.E. Sloop, J.K. Pugh, S. Wang, J.B. Kerr, K. Kinoshita, Chemical reactivity of PF 5 and $LiPF_6$ in ethylene carbonate/dimethyl carbonate solutions, Electrochem. Solid State Lett. 4 (2001) A42.

[21] A.M. Andersson, K. Edstrom, Chemical composition and morphology of the elevated temperature SEI on graphite, J. Electrochem. Soc. 148 (10) (2001) A1100–A1109.

[22] Y. Xia, Y. Zhou, M. Yoshio, Capacity fading on cycling of 4 V Li/$LiMn_2O_4$ cells, J. Electrochem. Soc. 144 (1997) 2593.

[23] S. Matsuta, Y. Kato, T. Ota, H. Kurokawa, S. Yoshimura, S.J. Fujitani, Electron-spin-resonance study of the reaction of electrolytic solutions on the positive electrode for lithium-ion secondary batteries, J. Electrochem. Soc. 148 (1) (2001) A7–A10.

[24] T.M. Bandhauer, S. Garimella, T.F. Fuller, A critical review of thermal issues in lithium-ion batteries, J. Electrochem. Soc. 158 (3) (2011) R1–R25.

[25] S. Jin An, J. Li, C. Daniel, D. Mohanty, S. Nagpure, D.L. Wood III, The state of understanding of the lithium-ion-battery graphite solid electrolyte interphase (SEI) and its relationship to formation cycling, Carbon 105 (2016) 52–76.

[26] L. Zhang, K. Amine, Z. Zhang, Redox Shuttle Additives for Lithium-Ion Battery, INTECH Open Access Publisher, (2012). http://cdn.intechopen.com/pdfs-wm/29292.pdf.

[27] Z. Zhang, Separator for a high energy rechargeable lithium battery, US Patent 6,432,586 B1, 2002.

[28] W.N. Smith, J.R. McCloskey, Non-aqueous electrolyte, US Patent 8,128,833 B1, 2006.

[29] S. Giri, S. Behera, P. Jena, Superhalogens as building blocks of halogen-free electrolytes in lithium-ion batteries, Angew. Chem. Int. Ed. 53 (50) (2014) 13916–13919.

[30] A. Manthiram, in: G.A. Nazri, G. Pistoia (Eds.), Lithium Batteries, Science and Technology, Springer Science + Business Media, New York, NY, 2009, pp. 3–37.

[31] E. Peled, C. Menachem, D. Bar-Tow, A. Melman, Improved graphite anode for lithium-ion batteries: chemically bonded solid electrolyte interface and nanochannel formation, J. Electrochem. Soc. 143 (1) (1996) L4–L7.

[32] D. Aurbach, B. Markovsky, A. Shechter, Y. Ein-Eli, H. Cohen, A comparative study of synthetic graphite and Li electrodes in electrolyte solutions based on ethylene carbonate–dimethyl carbonate mixtures, J. Electrochem. Soc. 143 (12) (1996) 3809–3820.

[33] S. Matsuta, Y. Kato, T. Ota, H. Kurokawa, S. Yoshimura, S.J. Fujitani, Electron-spin-resonance study of the reaction of electrolytic solutions on the positive electrode for lithium-ion secondary batteries, J. Electrochem. Soc. 148 (1) (2001) A7–A10.

[34] S.S. Zhang, A review on electrolyte additives for lithium-ion batteries, J. Power Sources 162 (2) (2006) 1379–1394.

[35] E. Quartarone, P. Mustarelli, Electrolytes for solid-state lithium rechargeable batteries: recent advances and perspectives, Chem. Soc. Rev. 40 (2011) 2525–2540.

[36] P. Verma, P. Maire, P. Novák, A review of the features and analyses of the solid electrolyte interphase in Li-ion batteries, Electrochim. Acta 55 (2010) 6332–6341.

[37] J. Salminen, N. Papaiconomou, A. Kumar, J.-M. Lee, J. Kerr, J.M. Prausnitz, J. Newman, Physical properties and toxicity and of selected piperadinium and pyrrolidinium cation ionic liquids, Fluid Phase Equilib. 261 (1–2) (2007) 421–426.

[38] D. Bansal, F. Croce, J. Swank, M. Solomon, in: M. Gaune-Escard, K.R. Seddon (Eds.), Molten Salts and Ionic Liquids: Newer the Twain? Wiley, Chichester, UK, 2010 (Chapter 2).

[39] N. Terasawa, S. Tsuzuki, T. Umecky, Y. Saito, H. Matsumoto, Alkoxy chains in ionic liquid anions; effect of introducing ether oxygen into perfluoroalkylborate on physical and thermal properties, Chem. Commun. 46 (2010) 1730–1732.

[40] Z. Zhang, S.S. Zhang (Eds.), Rechargeable Batteries: Materials, Technologies and New Trends, Springer International Publishing, Switzerland, 2015 (Chapter 8).

[41] X. Sun, H.S. Lee, X.Q. Yang, J. McBreen, The compatibility of a boron-based anion receptor with the carbon anode in lithium-ion batteries, Electrochem. Solid State Lett. 6 (2) (2003) A43–A46.

[42] S. Komaba, T. Itabashi, B. Kaplan, H. Groult, N. Kumagai, Enhancement of Li-ion battery performance of graphite anode by sodium ion as an electrolyte additive, Electrochem. Commun. 5 (2003) 962–966.

[43] S. Komaba, M. Watanabe, H. Groult, N. Kumagai, K. Okaharad, Impact of sodium salt coating on a graphite negative electrode for lithium-ion batteries, Electrochem. Solid State Lett. 9 (2006) A130–A133.

[44] S. Komaba, T. Itabashi, H. Kimura, H. Groult, N. Kumagai, Opposite influences of K+ versus Na+ ions as electrolyte additives on graphite electrode performance, J. Power Sources 146 (2005) 166–170.

[45] P. Bruce, B. Scrosati, J.M. Tarascon, Nanomaterials for rechargeable lithium batteries, Angew. Chem. Int. Ed. 47 (2008) 2930–2946.
[46] S.R. Narayanan, S. Surampudi, A.I. Attia, J. Bankston, Analysis of redox additive-based overcharge protection for rechargeable lithium batteries, J. Electrochem. Soc. 138 (8) (1991) 2224–2229.
[47] C. Buhrmester, J. Chen, L. Moshurchak, J. Jiang, R.L. Wang, J.R. Dahn, Studies of aromatic redox shuttle additives for $LiFePO_4$-based Li-ion cells, J. Electrochem. Soc. 152 (2005) A2390–A2399.
[48] G. Dantsin, K. Jambunathan, S.V. Ivanov, W.J. Casteel, K. Amine, J. Liu, A.N. Jansen, Z. Chen, Advanced Electrolyte Salts with Inherent Overcharge Protection for Lithium Ion Batteries, 208th ECS meeting Abstracts, 2005. Available from: http://ma.ecsdl.org/content/MA2005-02/4/223.abstract
[49] S.Y. Chen, Z.X. Wang, H.L. Zhao, L.Q. Chen, Safety-enhancing additives for lithium ion batteries, Prog. Chem. 21 (4) (2009) 629–636.
[50] G. Park, H. Nakamura, Y. Lee, M. Yoshio, The important role of additives for improved lithium ion battery safety, J. Power Sources 189 (1) (2009) 602–606.
[51] M. Galinski, A. Lewandowski, I. Stepniak, Ionic liquids as electrolytes, Electrochim. Acta 51 (2006) 5567–5580.
[52] M. Kar, T.J. Simons, M. Forsyth, D.R. Macfarlane, Ionic liquid electrolytes as a platform for rechargeable metal-air batteries: a perspective, Phys. Chem. Chem. Phys. 16 (2014) 18658–18674.
[53] M. Ishikawa, T. Sugimoto, M. Kikuta, E. Ishiko, M. Kono, Pure ionic liquid electrolytes compatible with a graphitized carbon negative electrode in rechargeable lithium-ion batteries, J. Power Sources 162 (2006) 658–662.
[54] J. Salminen, T. Kallio, N. Omar, P.V.D. Bossche, J.V. Mierlo, H. Gualous, Transport energy—lithium ion batteries, Future Energy: Improved, Sustainable and Clean Options for our Planet, second ed., Elsevier, London, UK, 2013, pp. 292–307.
[55] B. Scrosati, Recent advances in lithium ion battery materials, Electrochim. Acta 45 (2000) 2461–2466.
[56] N.J. Dudney, Lithium batteries, in: G.A. Nazri, G. Pistoia (Eds.), Lithium Batteries, Science and Technology, Springer Science + Business Media, LLC, New York, NY, 2003, pp. 623–642.
[57] N. Kamaya, K. Homma, Y. Yamakawa, M. Hirayama, R. Kanno, M. Yonemura, T. Kamiyama, Y. Kato, S. Hama, K. Kawamoto, A. Mitsui, A lithium superionic conductor, Nat. Mater. 10 (9) (2011) 682–686.
[58] C. Berthier, W. Gorecki, M. Minier, M.B. Armand, J.M. Chabagno, P. Rigaud, Microscopic investigation of ionic conductivity in alkali metal salts-poly(ethylene oxide) adducts, Solid State Ion. 11 (1) (1983) 91–95.
[59] W.H. Meyer, Polymer electrolytes for lithium-ion batteries, Adv. Mater. 10 (6) (1998) 439–448.
[60] S. Cheng, D.M. Smith, Q. Pan, S. Wanga, C.Y. Li, Anisotropic ion transport in nanostructured solid polymer electrolytes, RSC Adv. 5 (60) (2015) 48793–48810.
[61] A.M. Stephan, Review on gel polymer electrolytes for lithium batteries, Eur. Polym. J. 42 (1) (2006) 21–42.
[62] D. Golodnitsky, E. Strauss, E. Peled, S. Greenbaum, Review—on order and disorder in polymer electrolytes, J. Electrochem. Soc. 162 (14) (2015) A2551–A2566.
[63] C.D. Robitaille, D. Fauteux, Phase diagrams and conductivity characterization of some PEO-LiX electrolytes, J. Electrochem. Soc. 133 (2) (1986) 315–325.
[64] M. Marzantowicza, J.R. Dygas, F. Krok, J.L. Nowin´ski, A. Tomaszewska, Z. Florjan´czyk, E. Zygadło-Monikowska, Crystalline phases, morphology and conductivity of PEO:LiTFSI electrolytes in the eutectic region, J. Power Sources 59 (1) (2006) 420–430.
[65] S.K. Fullerton-Shirey, J.K. Maranas, Effect of $LiClO_4$ on the structure and mobility of PEO-based solid polymer electrolytes, Macromolecules 42 (6) (2009) 2142–2156.
[66] N.A. Stolwijk, M. Wiencierz, C. Heddier, J. Kösters, What can we learn from ionic conductivity measurements in polymer electrolytes? A case study on poly(ethylene oxide) (PEO)–NaI and PEO–LiTFSI, J. Phys. Chem. B 116 (10) (2012) 3065–3074.
[67] D. Devaux, R. Bouchet, D. Glé, R. Denoyel, Mechanism of ion transport in PEO/LiTFSI complexes: effect of temperature, molecular weight and end groups, Solid State Ion. 227 (2012) 119–127.
[68] J. Ni, Y. Lin, F. Wang, Synthesis of a novel polysiloxane-based polymer electrolyte and its ionic conductivity, Polym. Adv. Technol. 4 (2–3) (1993) 80–84.
[69] M. Watanabe, T. Endo, A. Nishimoto, K. Miura, M. Yanagida, High ionic conductivity and electrode interface properties of polymer electrolytes based on high molecular weight branched polyether, J. Power Sources 81–82 (1999) 786–789.

[70] S. York, E.C. Kellam III, H.R. Allcock, R. Frech, A vibrational spectroscopic study of lithium triflate in polyphosphazenes with linear oligoethyleneoxy side-chains of different lengths, Electrochim. Acta 46 (10–11) (2001) 1553–1557.
[71] P. Jannasch, Phase behavior and ion conductivity of electrolytes based on aggregating combshaped polyethers, Electrochim. Acta 46 (10–11) (2001) 1641–1649.
[72] J.F. Le Nest, A. Gandini, H. Cheradame, Crosslinked polyethers as media for ionic conduction, Polym. Int. 20 (3) (1988) 253–268.
[73] M. Kono, K. Furuta, S. Mori, M. Watanabe, N. Ogata, Synthesis of polymer electrolytes based on poly[2-(2-methoxyethoxy)ethyl glycidyl ether] and their high ionic conductivity, Polym. Adv. Technol. 4 (2–3) (1993) 85–91.
[74] S. Matsui, T. Muranaga, H. Higobashi, S. Inoue, T. Sakai, Liquid-free rechargeable Li polymer battery, J. Power Sources 97–98 (2001) 772–774.
[75] T. Niitani, M. Shimada, K. Kawamura, K. Dokko, Y.H. Rho, K. Kanamura, Synthesis of Li^+ ion conductive PEO-PSt block copolymer electrolyte with microphase separation structure, Electrochem. Solid State Lett. 8 (8) (2005) A385–A388.
[76] M. Singh, O. Odusanya, G.M. Wilmes, H.B. Eitouni, E.D. Gomez, A.J. Patel, V.L. Chen, M.J. Park, P. Fragouli, H. Iatrou, N. Hadjichristidis, D. Cookson, N.P. Balsara, Effect of molecular weight on the mechanical and electrical properties of block copolymer electrolytes, Macromolecules 40 (13) (2007) 4578–4585.
[77] E.D. Gomez, A. Panday, E.H. Feng, V. Chen, G.M. Stone, A.M. Minor, C. Kisielowski, K.H. Downing, O. Borodin, G.D. Smith, N.P. Balsara, Effect of ion distribution on conductivity of block copolymer electrolytes, Nano Lett. 9 (3) (2009) 1212–1216.
[78] A. Panday, S. Mullin, E.D. Gomez, N. Wanakule, V.L. Chen, A. Hexemer, J. Pople, N.P. Balsara, Effect of molecular weight and salt concentration on conductivity of block copolymer electrolytes, Macromolecules 42 (13) (2009) 4632–4637.
[79] W.S. Young, W.F. Kuan, T.H. Epps, Block copolymer electrolytes for rechargeable lithium batteries, J. Polym. Sci. B 52 (1) (2014) 1–16.
[80] J.E. Weston, B.C.H. Steele, Effects of inert fillers on the mechanical and electrochemical properties of lithium salt-poly(ethylene oxide) polymer electrolytes, Solid State Ion. 7 (1) (1982) 75–79.
[81] F. Croce, G.B. Appetecchi, L. Persi, B. Scrosati, Nanocomposite polymer electrolytes for lithium batteries, Nature 394 (1998) 456–458.
[82] S.H. Chung, Y. Wang, L. Persi, F. Croce, S.G. Greenbaum, B. Scrosati, E. Plicht, Enhancement of ion transport in polymer electrolytes by addition of nanoscale inorganic oxides, J. Power Sources 97–98 (2001) 644–648.
[83] J. Zhou, P.S. Fedkiw, Ionic conductivity of composite electrolytes based on oligo(ethylene oxide) and fumed oxides, Solid State Ion. 166 (3–4) (2004) 275–293.
[84] J.K. Lee, Y.J. Lee, W.S. Chae, Y.M. Sung, Enhanced ionic conductivity in PEO-$LiClO_4$ hybrid electrolytes by structural modification, J. Electroceram. 17 (2) (2006) 941–944.
[85] Z. Gadjourova, Y.G. Andreev, D.P. Tunstall, P.G. Bruce, Ionic conductivity in crystalline polymer electrolytes, Nature 412 (2001) 520–523.
[86] H. Luo, X. Liang, L. Wang, A. Zheng, C. Liu, J. Feng, Highly mobile segments in crystalline poly(ethylene oxide)$_8$: $NaPF_6$ electrolytes studied by solid-state NMR spectroscopy, J. Chem. Phys. 140 (7) (2014) 074901.
[87] R. Mezianea, J.P. Bonnet, M. Courty, K. Djellab, M. Armand, Single-ion polymer electrolytes based on a delocalized polyanion for lithium batteries, Electrochim. Acta 57 (2011) 14–19.
[88] S. Feng, D. Shi, F. Liu, L. Zheng, J. Nie, W. Feng, X. Huang, M. Armand, Z. Zhou, Single lithium-ion conducting polymer electrolytes based on poly[(4-styrenesulfonyl)(trifluoromethanesulfonyl)imide] anions, Electrochim. Acta 93 (2013) 254–263.
[89] R. Bouchet, S. Maria, R. Meziane, A. Aboulaich, L. Lienafa, J.P. Bonnet, T.N.T. Phan, D. Bertin, D. Gigmes, D. Devaux, R. Denoyel, M. Armand, Single-ion BAB triblock copolymers as highly efficient electrolytes for lithium-metal batteries, Nat. Mater. 12 (2013) 452–457.
[90] Q. Ma, H. Zhang, C. Zhou, L. Zheng, P. Cheng, J. Nie, W. Feng, Y.S. Hu, H. Li, X. Huang, L. Chen, M. Armand, Z. Zhou, Single lithium-ion conducting polymer electrolytes based on a super-delocalized polyanion, Angew. Chem. Int. Ed. 55 (2016) 2521.

[91] M. Watanabe, M. Kanba, K. Nagaoka, I. Shinohara, Ionic conductivity of hybrid films based on polyacrylonitrile and their battery application, J. Appl. Polym. Sci. 27 (1982) 4191–4198.
[92] H. Hong, C. Liquan, H. Xuejie, X. Rongjian, Studies on PAN-based lithium salt complex, Electrochim. Acta 37 (9) (1992) 1671–1673.
[93] G.B. Appetecchi, F. Croce, P. Romagnoli, B. Scrosati, U. Heider, R. Oesten, High-performance gel-type lithium electrolyte membranes, Electrochem. Commun. 1 (2) (1999) 83–86.
[94] O. Bohnke, G. Frand, M. Rezrazi, C. Rousselot, C. Truche, Fast ion transport in new lithium electrolytes gelled with PMMA. 1. Influence of polymer concentration, Solid State Ion. 66 (1–2) (1993) 97–104.
[95] O. Bohnke, G. Frand, M. Rezrazi, C. Rousselot, C. Truche, Fast ion transport in new lithium electrolytes gelled with PMMA. 2. Influence of lithium salt concentration, Solid State Ion. 66 (1–2) (1993) 105–112.
[96] G.B. Appetecchi, F. Croce, B. Scrosati, Kinetics and stability of the lithium electrode in poly(methylmethacrylate)-based gel electrolytes, Electrochim. Acta 40 (8) (1995) 991–997.
[97] M. Alamgir, K.M. Abraham, Li Ion conductive electrolytes based on poly(vinyl chloride), J. Electrochem. Soc. 140 (6) (1993) L96–L97.
[98] M. Watanabe, M. Kanba, H. Matsuda, K. Tsunemi, K. Mizoguchi, E. Tsuchida, I. Shinohara, High lithium ionic conductivity of polymeric solid electrolytes, Makromol. Rapid Commun. 2 (12) (1981) 741–744.
[99] H.S. Choe, J. Giaccai, M. Alamgir, K.M. Abraham, Preparation and characterization of poly(vinyl sulfone)- and poly(vinylidene fluoride)-based electrolytes, Electrochim. Acta 40 (13–14) (1995) 2289–2293.
[100] Z. Jiang, B. Carroll, K.M. Abraham, Studies of some poly(vinylidene fluoride) electrolytes, Electrochim. Acta 42 (17) (1997) 2667–2677.
[101] A.M. Voice, G.R. Davies, I.M. Ward, Structure of poly(vinylidene fluoride) gel electrolytes, Polym. Gels Networks 5 (2) (1997) 123–144.
[102] I.M. Ward, H.V.A. St. Hubbard, S.C. Wellings, Gl.P. Thompson, J. Kaschmitter, H.P. Wang, Separator-free rechargeable lithium ion cells produced by the extrusion lamination of polymer gel electrolytes, J. Power Sources 162 (2) (2006) 818–822.
[103] I.M. Ward, H.V.A. St. Hubbard, Polymer gel electrolytes: conduction mechanism and battery applications, in: A.P. Alberto Ciferri, A. Perico (Eds.), Ionic Interactions in Natural and Synthetic Macromolecules, John Wiley & Sons, Hoboken, NJ, 2012, pp. 817–840.
[104] C. Capiglia, Y. Saito, H. Kataoka, T. Kodama, E. Quartarone, P. Mustarelli, Structure and transport properties of polymer gel electrolytes based on PVdF–HFP and $LiN(C_2F_5SO_2)_2$, Solid State Ion. 131 (3–4) (2000) 291–299.
[105] D. Saikia, A. Kumar, Ionic conduction in P(VDF-HFP)/PVDF–(PC+DEC)–$LiClO_4$ polymer gel electrolytes, Electrochim. Acta 49 (16) (2004) 2581–2589.
[106] H.S. Kim, P. Periasamy, S.I. Moon, Electrochemical properties of the Li-ion polymer batteries with P(VdF-*co*-HFP)-based gel polymer electrolyte, J. Power Sources 141 (2) (2005) 293–297.
[107] C.W. Huang, C.A. Wu, S.S. Hou, P.L. Kuo, C.T. Hsieh, H. Teng, Gel electrolyte derived from poly(ethylene glycol) blending poly(acrylonitrile) applicable to roll-to-roll assembly of electric double-layer capacitors, Adv. Funct. Mater. 22 (2012) 4677–4685.
[108] S.H. Wang, P.L. Kuo, C.T. Hsieh, H. Teng, Design of poly(acrylonitrile)-based gel electrolytes for high-performance lithium ion batteries, ACS Appl. Mater. Interfaces 6 (21) (2014) 19360–19370.
[109] P.L. Kuo, C.A. Wu, C.Y. Lu, C.H. Tsao, C.H. Hsu, S.S. Hou, High performance of transferring lithium ion for polyacrylonitrile-interpenetrating cross-linked polyoxyethylene network as gel polymer electrolyte, ACS Appl. Mater. Interfaces 6 (5) (2014) 3156–3162.
[110] C.M. Costa, L.C. Rodrigues, V. Sencadas, M.M. Silva, J.G. Rocha, S. Lanceros-Méndez, Effect of degree of porosity on the properties of poly(vinylidene fluoride–trifluorethylene) for Li-ion battery separators, J. Membr. Sci. 407–408 (2012) 193–201.
[111] N.H. Idris, M. Md. Rahmana, J.Z. Wanga, H.K. Liu, Microporous gel polymer electrolytes for lithium rechargeable battery application, J. Power Sources 201 (2012) 294–300.
[112] F. Croce, M.L. Focarete, J. Hassoun, I. Meschini, B. Scrosati, A safe, high-rate and high-energy polymer lithium-ion battery based on gelled membranes prepared by electrospinning, Energy Environ. Sci. 4 (2011) 921–927.

[113] P. Carola, P. Ramakrishnanb, B. Johna, G. Cheruvally, Preparation and characterization of electrospun poly(acrylonitrile) fibrous membrane based gel polymer electrolytes for lithium-ion batteries, J. Power Sources 196 (23) (2011) 10156–10162.

[114] R. Prasantha, V. Aravindanb, M. Srinivasan, Novel polymer electrolyte based on cob-web electrospun multi component polymer blend of polyacrylonitrile/poly(methyl methacrylate)/polystyrene for lithium ion batteries—preparation and electrochemical characterization, J. Power Sources 202 (2012) 299–307.

[115] H. Bi, G. Sui, X. Yang, Studies on polymer nanofibre membranes with optimized core–shell structure as outstanding performance skeleton materials in gel polymer electrolytes, J. Power Sources 267 (2014) 309–315.

[116] J. Adebahr, N. Byrne, M. Forsyth, D.R. MacFarlane, P. Jacobsson, Enhancement of ion dynamics in PMMA-based gels with addition of TiO_2 nano-particles, Electrochim. Acta 48 (14–16) (2003) 2099–2103.

[117] M. Stolarska, L. Niedzicki, R. Borkowska, A. Zalewska, W. Wieczorek, Structure, transport properties and interfacial stability of PVdF/HFP electrolytes containing modified inorganic filler, Electrochim. Acta 53 (4) (2007) 1512–1517.

[118] O. Krejza, J. Velická, M. Sedlarˇíková, J. Vondrák, The presence of nanostructured Al_2O_3 in PMMA-based gel electrolytes, J. Power Sources 178 (2) (2008) 774–778.

[119] S.M. Eo, E. Cha, D.W. Kim, Effect of an inorganic additive on the cycling performances of lithium-ion polymer cells assembled with polymer-coated separators, J. Power Sources 198 (1) (2009) 766–770.

[120] A. Zalewska, M. Walkowiak, L. Niedzicki, T. Jesionowski, N. Langwald, Study of the interfacial stability of PVdF/HFP gel electrolytes with sub-micro- and nano-sized surface-modified silicas, Electrochim. Acta 55 (4) (2010) 1308–1313.

[121] Y.T. Chen, Y.C. Chuang, J.H. Su, H.C. Yu, Y.W. Chen-Yang, High discharge capacity solid composite polymer electrolyte lithium battery, J. Power Sources 196 (5) (2011) 2802–2809.

[122] B. Kurc, Gel electrolytes based on poly(acrylonitrile)/sulpholane with hybrid TiO_2/SiO_2 filler for advanced lithium polymer batteries, Electrochim. Acta 125 (2014) 415–420.

[123] S. Ferraria, E. Quartarone, P. Mustarelli, A. Magistris, M. Fagnoni, S. Protti, C. Gerbaldi, A. Spinella, Lithium ion conducting PVdF-HFP composite gel electrolytes based on *N*-methoxyethyl-*N*-methylpyrrolidinium bis(trifluoromethanesulfonyl)-imide ionic liquid, J. Power Sources 195 (2) (2010) 559–566.

[124] S. Ramesh, C.W. Liew, K. Ramesh, Evaluation and investigation on the effect of ionic liquid onto PMMA-PVC gel polymer blend electrolytes, J. Non Cryst. Solids 357 (10) (2011) 2132–2138.

[125] M. Rao, X. Geng, Y. Liao, S. Hu, W. Li, Preparation and performance of gel polymer electrolyte based on electrospun polymer membrane and ionic liquid for lithium ion battery, J. Membr. Sci. 399–400 (2012) 37–42.

[126] R. Che, W. Yang, J. Wang, J. Lei, Electrolyte-based antistatic plasticizer for soft poly(vinyl chloride) composites, J. Appl. Polym. Sci. 116 (2010) 1718–1724.

[127] D. Kumar, S.A. Hashmi, Ion transport and ion–filler-polymer interaction in poly(methyl methacrylate)-based, sodium ion conducting, gel polymer electrolytes dispersed with silica nanoparticles, J. Power Sources 195 (15) (2010) 5101–5108.

[128] D. Kumar, S.A. Hashmi, Ionic liquid based sodium ion conducting gel polymer electrolytes, Solid State Ion. 181 (8–10) (2010) 416–423.

[129] D. Kumar, M. Suleman, S.A. Hashmi, Studies on poly(vinylidene fluoride-*co*-hexafluoropropylene) based gel electrolyte nanocomposite for sodium–sulfur batteries, Solid State Ion. 202 (1) (2011) 45–53.

[130] H. Gao, B. Guo, J. Song, K. Park, J.B. Goodenough, A composite gel–polymer/glass–fiber electrolyte for sodium-ion batteries, Adv. Energy Mater. 5 (2015) 1402235.

[131] Y.Q. Yang, Z. Chang, M.X. Li, X.W. Wang, Y.P. Wu, A sodium ion conducting gel polymer electrolyte, Solid State Ion. 269 (2015) 1–7.

[132] A.S. Gozdz, C.N. Schmutz, J.M. Tarascon, P.C. Warren, Lithium secondary battery extraction method, US Patent No. 5,540,741, 1997.

[133] J.L. Souquet, Ionic transport in amorphous solid electrolytes, Annu. Rev. Mater. Sci. 11 (1981) 211–231.

[134] K. Hayamizu, Y. Aihara, T. Watanabe, T. Yamada, S. Ito, N. Machida, NMR studies on lithium ion migration in sulfide-based conductors, amorphous and crystalline Li_3PS_4, Solid State Ion. 285 (2016) 51–58.
[135] Y.F.Y. Yao, J.T. Kummer, Ion exchange properties of and rates of ionic diffusion in beta-alumina, J. Inorg. Nucl. Chem. 29 (1967) 2453–2475.
[136] J.C. Bachman, S. Muy, A. Grimaud, H.H. Chang, N. Pour, S.F. Lux, O. Paschos, F. Maglia, S. Lupart, P. Lamp, L. Giordano, Y. Shao-Horn, Inorganic solid-state electrolytes for lithium batteries: mechanisms and properties governing ion conduction, Chem. Rev. 116 (1) (2016) 140–162.
[137] M.S. Whittingham, R.A. Huggins, Measurement of sodium ion transport in beta alumina using reversible solid electrodes, J. Chem. Phys. 54 (1) (1971) 414–416.
[138] J.L. Briant, G.C. Farrington, Ionic conductivity in Na^+, K^+, and Ag^+ β''-alumina, J. Solid State Chem. 33 (3) (1980) 385–390.
[139] X. Lu, G. Xia, J.P. Lemmon, Z. Yang, J. Power Sources 195 (9) (2010) 2431–2442.
[140] J. Fergus, Mechanism of ion transport in PEO/LiTFSI complexes: effect of temperature, molecular weight and end groups, Solid State Ion. 227 (2012) 102–112.
[141] J.B. Goodenough, H.Y.P. Hong, J.A. Kafalas, Fast Na^+-Ion transport in skeleton structures, Mater. Res. Bull. 11 (2) (1976) 203–220.
[142] N. Anantharamulu, K.K. Rao, G. Rambabu, B.V. Kumar, V. Radha, M. Vithal, A wide-ranging review on Nasicon type materials, J. Mater. Sci. 46 (9) (2011) 2821–3283.
[143] M. Cretin, P. Fabry, Comparative study of lithium ion conductors in the system $Li_{1+x}Al_xA^{IV}_{2-x}(PO_4)_3$ with A^{IV} = Ti or Ge and $0 \le x \le 0.7$ for use as Li^+ sensitive membranes, J. Eur. Ceram. Soc. 19 (16) (1999) 2931–2940.
[144] K. Arbi, et al., Non-Arrhenius conductivity in the fast lithium conductor $Li_{1.2}Ti_{1.8}Al_{0.2}(PO_4)_3$: a ^7Li NMR and electric impedance study, Phys. Rev. B 72 (2005) 094302.
[145] H. Aono, E. Sugimoto, Y. Sadaoka, N. Imanaka, G.Y. Adachi, Ionic conductivity of solid electrolytes based on lithium titanium phosphate, J. Electrochem. Soc. 137 (4) (1990) 1023–1027.
[146] P. Hartmann, T. Leichtweiss, M.R. Busche, M. Schneider, M. Reich, J. Sann, P. Adelhelm, J. Janek, Degradation of NASICON-type materials in contact with lithium metal: formation of mixed conducting interphases (MCI) on solid electrolytes, J. Phys. Chem. C 117 (41) (2013) 21064–21074.
[147] M. Yashima, M. Itoh, Y. Inaguma, Y. Morii, Crystal structure and diffusion path in the fast lithium-ion conductor $La_{0.62}Li_{0.16}TiO_3$, J. Am. Chem. Soc. 127 (10) (2005) 3491–3495.
[148] Y. Inaguma, C. Liquan, M. Itoh, T. Nakamura, T. Uchida, H. Ikuta, M. Wakihara, High ionic conductivity in lithium lanthanum titanate, Solid State Commun. 86 (50) (1993) 689–693.
[149] F. Aguesse, J.M. López del Amo, V. Roddatis, A. Aguadero, J.A. Kilner, Enhancement of the grain boundary conductivity in ceramic $Li_{0.34}La_{0.55}TiO_3$ electrolytes in a moisture-free processing environment, Adv. Mater. Interfaces 1 (7) (2014) 1300143.
[150] C.W. Ban, G.M. Choi, The effect of sintering on the grain boundary conductivity of lithium lanthanum titanates, Solid State Ion. 140 (3–4) (2001) 285–292.
[151] K.Y. Yang, J.W. Wang, K.Z. Fung, Roles of lithium ions and La/Li-site vacancies in sinterability and total ionic conduction properties of polycrystalline $Li_{3x}La_{2/3-x}TiO_3$ solid electrolytes ($0.21 \le 3x \le 0.50$), J. Alloys Compd. 458 (1–2) (2008) 415–424.
[152] A. Mei, Q.H. Jiang, Y.H. Lin, C.W. Nan, Lithium lanthanum titanium oxide solid-state electrolyte by spark plasma sintering, J. Alloys Compd. 486 (1–2) (2009) 871–875.
[153] C. Cao, Z.B. Li, X.L. Wang, X.B. Zhao, W.Q. Han, Recent advances in inorganic solid electrolytes for lithium batteries, Front. Energy Res. 2 (2014) 25.
[154] M.P. O'Callaghan, D.R. Lynham, E.J. Cussen, G.Z. Chen, Structure and ionic-transport properties of lithium-containing garnets $Li_3Ln_3Te_2O_{12}$ (Ln = Y, Pr, Nd, Sm–Lu), Chem. Mater. 18 (19) (2006) 4681–4689.
[155] E.J. Cussen, T.W.S. Yip, G. O'Neill, M.P. O'Callaghan, A comparison of the transport properties of lithium-stuffed garnets and the conventional phases $Li_3Ln_3Te_2O_{12}$, J. Solid State Chem. 184 (2) (2011) 470–475.
[156] V. Thangadurai, S. Narayanana, D. Pinzarua, Garnet-type solid-state fast Li ion conductors for Li batteries: critical review, Chem. Soc. Rev. 43 (13) (2014) 4714–4727.
[157] W.G. Zeier, Structural limitations for optimizing garnet-type solid electrolytes: a perspective, Dalton Trans. 43 (2014) 16133–16138.

[158] C. Bernuy-Lopez, W. Manalastas, J.M. Lopez del Amo, A. Aguadero, F. Aguesse, J.A. Kilner, Atmosphere controlled processing of Ga-substituted garnets for high Li-ion conductivity ceramics, Chem. Mater. 26 (12) (2014) 3610–3617.
[159] C. Galven, J.L. Fourquet, M.P. Crosnier-Lopez, F. Le Berre, Instability of the lithium garnet $Li_7La_3Sn_2O_{12}$: Li^+/H^+ exchange and structural study, Chem. Mater. 23 (7) (2011) 1892–1900.
[160] G. Larraz, A. Orera, M.L. Sanjuán, Cubic phases of garnet-type $Li_7La_3Zr_2O_{12}$: the role of hydration, J. Mater. Chem. A 1 (37) (2013) 11419–11428.
[161] M. Kotobuki, H. Munakata, K. Kanamura, Y. Satob, T. Yoshida, Compatibility of $Li_7La_3Zr_2O_{12}$ solid electrolyte to all-solid-state battery using Li metal anode, J. Electrochem. Soc. 157 (10) (2010) A1076–A1079.
[162] Y. Ren, Y. Shen, Y. Lin, C.W. Nan, Direct observation of lithium dendrites inside garnet-type lithium-ion solid electrolyte, Electrochem. Commun. 57 (2015) 27–30.
[163] C.L. Tsai, V. Roddatis, C.V. Chandran, Q. Ma, S. Uhlenbruck, M. Bram, P. Heitjans, O. Guillon, $Li_7La_3Zr_2O_{12}$ interface modification for Li dendrite prevention, ACS Appl. Mater. Interfaces 8 (16) (2016) 10617–10626.
[164] R. Kanno, M. Murayama, Lithium ionic conductor thio-LISICON: the Li_2S-GeS_2-P_2S_5 system, J. Electrochem. Soc. 148 (7) (2001) A742–A746.
[165] A. Sakuda, A. Hayashi, M. Tatsumisago, Sulfide solid electrolyte with favorable mechanical property for all-solid-state lithium battery, Sci. Rep. 3 (2013) 2261.
[166] A. Hayashi, K. Noi, A. Sakuda, M. Tatsumisago, Superionic glass-ceramic electrolytes for room-temperature rechargeable sodium batteries, Nat. Commun. 3 (2012) 856.
[167] N. Ohta, K. Takada, I. Sakaguchi, L. Zhang, R. Ma, K. Fukuda, M. Osada, T. Sasaki, $LiNbO_3$-coated $LiCoO_2$ as cathode material for all solid-state lithium secondary batteries, Electrochem. Commun. 9 (7) (2007) 1486–1490.
[168] S. Wenzel, S. Randau, T. Leichtweiß, D.A. Weber, J. Sann, W.G. Zeier, J. Janek, Direct observation of the interfacial instability of the fast ionic conductor $Li_{10}GeP_2S_{12}$ at the lithium metal anode, Chem. Mater. 28 (7) (2016) 2400–2407.
[169] J.M. Whiteley, J.H. Woo, E. Hu, K.-W. Nam, S.-H. Lee, Empowering the lithium metal battery through a silicon-based superionic conductor, J. Electrochem. Soc. 161 (12) (2014) A1812–A1817.
[170] P. Bron, S. Johansson, K. Zick, J.S. auf der Günne, S. Dehnen, B. Roling, $Li_{10}SnP_2S_{12}$: an affordable lithium superionic conductor, J. Am. Chem. Soc. 135 (42) (2013) 15694–15697.
[171] Y. Kato, S. Hori, T. Saito, K. Suzuki, M. Hirayama, A. Mitsui, M. Yonemura, H. Iba, R. Kanno, High-power all-solid-state batteries using sulfide superionic conductors, Nat. Energy 1 (2016) 16030.
[172] G. Sahu, Z. Lin, J. Li, Z. Liu, N. Dudney, C. Liang, Air-stable, high-conduction solid electrolytes of arsenic-substituted Li_4SnS_4, Energy Environ. Sci. 7 (3) (2014) 1053–1058.
[173] R. Kanno, T. Hata, Y. Kawamoto, M. Irie, Solid State Ion. 130 (2000) 97.
[174] R. Kanno, M. Maruyama, Lithium ionic conductor thio-LISICON: the Li_2S-GeS_2-P_2S_5 system, J. Electrochem. Soc. 148 (7) (2001) A742–A746.
[175] M. Murayama, R. Kanno, Y. Kawamoto, T. Kamiyama, Structure of the thio-LISICON, Li_4GeS_4, Solid State Ion. 154–155 (2002) 789–794.
[176] H. Yamane, et al. Crystal structure of a superionic conductor, $Li_7P_3S_{11}$, Solid State Ion. 178 (2007) 1163–1167.
[177] F. Mizuno, A. Hayashi, K. Tadanaga, M. Tatsumisago, New, highly ion-conductive crystals precipitated from Li_2S–P_2S_5 glasses, Adv. Mater. 17 (7) (2005) 918–921.
[178] Y. Seino, T. Ota, K. Takada, A. Hayashi, M. Tatsumisago, A sulphide lithium super ion conductor is superior to liquid ion conductors for use in rechargeable batteries, Energy Environ. Sci. 7 (2) (2014) 627–631.
[179] Z. Liu, W. Fu, E.A. Payzant, X. Yu, Z. Wu, N.J. Dudney, J. Kiggans, K. Hong, A.J. Rondinone, C. Liang, Anomalous high ionic conductivity of nanoporous beta-Li_3PS_4, J. Am. Chem. Soc. 135 (3) (2013) 975–978.
[180] A. Kuhn, V. Duppel, B.V. Lotsch, Tetragonal $Li_{10}GeP_2S_{12}$ and Li_7GePS_8: exploring the Li ion dynamics in LGPS Li electrolytes, Energy Environ. Sci. 6 (12) (2013) 3548–3552.
[181] S.P. Ong, Y. Mo, W.D. Richards, L. Miara, H.S. Lee, G. Ceder, Phase stability{,} electrochemical stability and ionic conductivity of the $Li_{10+/-1}MP_2X_{12}$ (M = Ge, Si, Sn, Al or P and X = O, S or Se family of superionic conductors, Energy Environ. Sci. 6 (1) (2013) 148–156.

[182] Y. Wang, W.D. Richards, S.P. Ong, L.J. Miara, J.C. Kim, Y. Mo, G. Ceder, Design principles for solid-state lithium superionic conductors, Nat. Mater. 14 (10) (2015) 1026–1031.
[183] S. Skaarup, K. West, B. Zachauchristiansen, Mixed phase solid electrolytes, Solid State Ion. 28 (1988) 975–978.
[184] C.S. Wang, X.W. Zhang, A.J. Appleby, Solvent-free composite PEO-ceramic fiber/mat electrolytes for lithium secondary cells, J. Electrochem. Soc. 152 (2005) A205–A209.
[185] Y. Inda, T. Katoh, M. Baba, Development of all-solid lithium-ion battery using Li-ion conducting glass-ceramics, J. Power Sources 174 (2007) 741–744.
[186] J.-H. Choi, C.-H. Lee, J.-H. Yu, C.-H. Doh, S.-M. Lee, Enhancement of ionic conductivity of composite membranes for all-solid-state lithium rechargeable batteries incorporating tetragonal Li7La3Zr2O12 into a polyethylene oxide matrix, J. Power Sources 274 (2015) 458–463.
[187] W. Liu, N. Liu, J. Sun, P.-C. Hsu, Y. Li, H.-W. Lee, Y. Cui, Ionic conductivity enhancement of polymer electrolytes with ceramic nanowire fillers, Nano Lett. 15 (2015) 2740–2745.
[188] G. Kickelbick (Ed.), Hybrid Materials, Wiley-VCH Verlag GmbH & Co. KGaA, Weinheim, 2007.
[189] Z.D. Hood, H. Wang, Y. Li, A.S. Pandian, M. Parans Paranthaman, C. Liang, The "filler effect": a study of solid oxide fillers with Li_3PS_4 for lithium conducting electrolytes, Solid State Ion. 283 (2015) 75–80.
[190] E. Rangasamy, G. Sahu, J.K. Keum, A.J. Rondinone, N.J. Dudney, C. Liang, A high conductivity oxide-sulfide composite lithium superionic conductor, J. Mater. Chem. A 2 (2014) 4111–4116.
[191] J. Maier, Space charge regions in solid two-phase systems and their conduction contribution—I. Conductance enhancement in the system ionic conductor—'inert' phase and application on AgC1:Al_2O_3 and AgC1:SiO_2, J. Phys. Chem. Solids 46 (1985) 309–320.
[192] C.S. Wang, X.W. Zhang, A.J. Appleby, Solvent-free composite PEO-ceramic fiber/mat electrolytes for lithium secondary cells, J. Electrochem. Soc. 152 (2005) A205–A209.
[193] Y. Inda, T. Katoh, M. Baba, Development of all-solid lithium-ion battery using Li-ion conducting glass-ceramics, J. Power Sources 174 (2007) 741–744.
[194] N. Sata, K. Eberman, K. Eberl, J. Maier, Mesoscopic fast ion conduction in nanometre-scale planar heterostructures, Nature 408 (2000) 946–949.
[195] A. Schirmeisen, A. Taskiran, H. Fuchs, H. Bracht, S. Murugavel, B. Roling, Fast interfacial ionic conduction in nanostructured glass ceramics, Phys. Rev. Lett. 98 (2007) 225901.
[196] N. Ohta, K. Takada, I. Sakaguchi, L. Zhang, R. Ma, K. Fukuda, M. Osada, T. Sasaki, $LiNbO_3$-coated $LiCoO_2$ as cathode material for all solid-state lithium secondary batteries, Electrochem. Commun. 9 (7) (2007) 1486–1490.
[197] D. Aurbach, Y. Talyosef, B. Markovsky, E. Markevich, E. Zinigrad, L. Asraf, J.S. Gnanaraj, H.-J. Kim, Design of electrolyte solutions for Li and Li-ion batteries: a review, Electrochim. Acta 50 (2004) 247.
[198] A.S. Aricò, P. Bruce, B. Scrosati, J.-M. Tarascon, W. van Schalkwijk, Nanostructured materials for advanced energy conversion and storage devices, Nat. Mater. 4 (2005) 366.
[199] Z. Tu, Y. Kambe, Y. Lu, L.A. Archer, Nanoporous polymer–ceramic composite electrolytes for lithium metal batteries, Adv. Energy Mater. 4 (2014) 1300654.
[200] K. Park, J.H. Cho, K. Shanmuganathan, J. Song, J. Peng, M. Gobet, S. Greenbaum, C.J. Ellison, J.B. Goodenough, New battery strategies with a polymer/Al_2O_3 separator, J. Power Sources 263 (2014) 52–58.
[201] Y. Ansari, B. Guo, J.H. Cho, K. Park, J. Song, C.J. Ellison, J.B. Goodenough, Low-cost, dendrite-blocking polymer-Sb_2O_3 separators for lithium and sodium batteries, J. Electrochem. Soc. 161 (10) (2014) A1655–A1661.
[202] M. Armand, J.-M. Tarascon, Building better batteries, Nature 451 (2008) 652–657.
[203] N. Yabuuchi, K. Kubota, M. Dahbi, S. Komaba, Research development on sodium-ion batteries, Chem. Rev. 114 (23) (2014) 11636–11682.

CHAPTER TWO

Review of Nanotechnology for Anode Materials in Batteries

S. Goutam*, N. Omar*, P. Van Den Bossche*', J. Van Mierlo***
*Mobility, Logistic and Automotive Technology Research Center (MOBI), Vrije Universiteit Brussel, Brussels, Belgium
**Vrije Universiteit Brussel, Anderlecht, Belgium

Contents

1 A HIGH-PERFORMANCE ANODE

In terms of performance of an electrochemical energy storage system, such as a battery cell, the anode is one of the three most crucial components apart from the electrolyte and the cathode. An anode is generally identified as the negative electrode of a rechargeable battery cell. Performance indexes of a battery cell, for example, energy density, power density, cycle life, are defined by the combination of the physiochemical properties of all three components. Recent progression in the use of electric vehicle traction batteries require much higher energy density (compared to ~240 Wh/kg offered by available commercial cells), higher power density and long cycle life [1–3]. In case of an anode in a battery cell, specific capacity, potential, cycle life, and coulombic efficiency are the most common performance defining indexes. An anode with high specific capacity signifies how the anode can reversely accommodate the working cations quantitatively and efficiently. Here, "working cations" implies, Li^+ for a lithium-based battery and Na^+ for a sodium-ion-based battery. The energy density of a complete cell is generally determined by the specific capacity of the electrodes and working voltage of the cell (where the working voltage of a cell is the difference between the potential of cathode and anode). For example, in a typical lithium–sulfur cell (Li–S), theoretical capacity of lithium (as anode) and sulfur (as cathode) are 3.861 and 1.672 Ah/g, respectively. The theoretical capacity of the cell is 1.167 Ah/g and average voltage during discharge is 2.15 V. Thus, the theoretical gravimetric energy density of a Li–S cell is 2.51 Wh/g [4]. Hence, a high-energy density anode should possess a high specific capacity and low working voltage. But it is important to mention that the electrochemical potential of an anode should have lower energy than the lower unoccupied molecular orbital (LUMO) of the electrolyte. Otherwise the electrolyte will be reduced forming a passivating solid-electrolyte interface (SEI) film on anode. Careful control of this passivation and SEI formation mechanism is needed to achieve high stability and performance [5,6]. Volumetric energy density (Wh/cm^3) is another important index, especially for contemporary applications. Higher volumetric energy density implies higher energy with a cell of lower volume of the electrodes or the complete cell. Cycle life of a cell signifies the ability of a cell to deliver the demanded output or a reasonable percentage for a duration. Cycle life is another crucial performance index of a cell; for example, traction battery for automotive applications demands a life from 500 to over thousands of cycles. Repeated charge/discharge process results in degradation of cell performance, such as capacity fade, increment of internal resistance, etc. Anode with high cycle life signifies its ability to repeated reversible accommodation of the working cation without significant change of it structural integrity. For instance, during charging and discharging process in a typical Li-ion battery cell the Li^+ insertion/extraction accompanies volume change of the electrodes and change of the crystal structure. If an anode is not stable enough to accommodate these changes, SEI layer will be nonuniform and can act as a defect sites for the unwanted interaction between the electrolyte and the anode causing performance degradation/decay. Power density specified by the unit W/g implies the peak power that a battery cell can provide. In the case of anode, how fast the diffusion process of the working cation is, will determine the power density. Current density (A/g) is also commonly used in this regard to characterize peak power capability of a battery.

2 BENEFITS OF A NANOSTRUCTURED ANODE

The majority of the performance indexes discussed in the previous section are defined by the material properties of the core element of an anode. These material properties include crystal structure; physical properties, such as electronic conductivity, ionic diffusivity, and mechanical stability; chemical properties, such as redox potential, reversibility, and so on. Choice of material also includes properties, such as nonexpensive and nontoxic properties. Over the past decades, a significant number of studies have been focused on finding suitable anode materials [1–3,6–28]. Among them, carbon-based

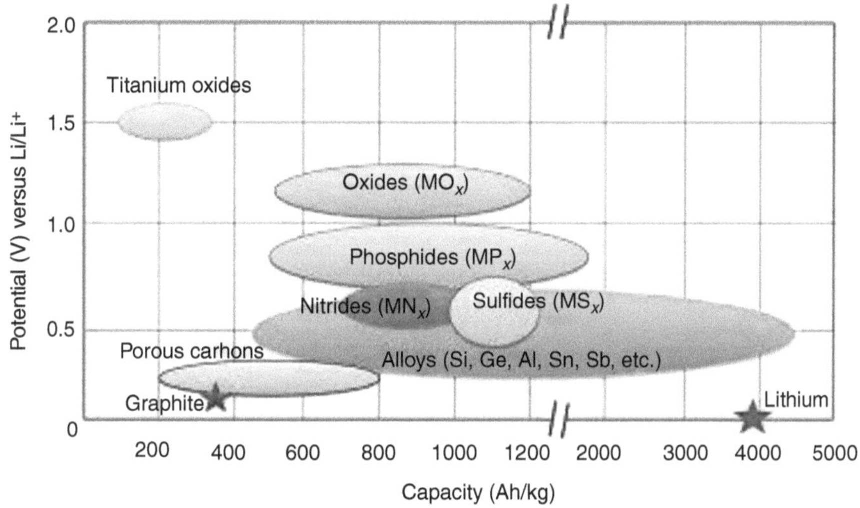

Figure 2.1 ***Potential versus Li/Li$^+$ and the corresponding capacity density of active anode materials for the next generation of Li-ion batteries.*** *(Taken from S. Goriparti, E. Miele, F. De Angelis, E. Di Fabrizio, R. Proietti Zaccaria, C. Capiglia, J. Power Sources 257 (2014) 421–443 [1]; under a creative common license).*

materials have been widely explored and led to successful implementation in commercial battery cells in numerous applications, such as portable electronics, electric vehicles, and so on [29]. Other than carbon-based material, several elements and their compounds have been extensively explored. Due to their suitable properties for anode, elements of IA (Li), IIA (Mg), IIIA (Al, Ga, In), IVA (Si, Ge, Sn, Pb), VA (P, As, Sb, Bi), IB (Ag, Au), and IIB (Zn, Cd) have been implemented in Li-ion, Na-ion, and Li-S battery systems. Fig. 2.1 shows an overview of potential anode materials for Li-ion batteries. Potential of these elements was explored based on one or more suitable properties. But unfortunately, in their pure or bulk form they do not satisfy all of the requirements. For example, graphite, a carbon-based material, has good electronic conductivity, low potential against Li (~0.05 V), and an inexpensive manufacturing process but due to its structural limitation (six C-atoms per Li atom in LiC_6) the specific capacity is low (~372 mAh/g). Whereas, silicon-based material can provide a specific capacity more than 3500 mAh/g and suitable potential against Li/Li$^+$ (~0.5 V) [19]. However, during insertion/deinsertion of Li$^+$, silicon-based materials showed huge volume expansion (~400%) which leads to loss of proper contact between the anode and the current collector and finally results in poor performance of the battery cell.

To achieve high performance of an anode, designing of material properties at nanoscale has gathered significant attention. Among the numerous benefits of nanostructured materials, most significant is the improvement in ion and electron transport properties by providing shorter and efficient diffusion paths. In a typical Li-ion battery system, during a charging process, simultaneous insertion of Li^+ and e^- take place. Energy and power performance of the system can be significantly improved by shortening the diffusion length of Li^+ in the anode during insertion/extraction. Diffusivity or diffusion time, τ is related to diffusion length as:

$$\tau = L_{\text{ion}}^2 / D_{\text{ion}} \tag{2.1}$$

where, L_{ion} is the diffusion length and D_{ion} is the diffusion coefficient of the respective ion. Moreover, the transition from macro to nanostructured materials increases the accessible active sites for working cations to be incorporated; ultimately, gaining high energy and high power by synergistic effect of both. For example, diffusion of Li^+ in rutile TiO_2 is anisotropic and slow at low temperature ($\sim 10^{-15}$ cm^2/s), which prevents reaching the favorable lattice site and resulting in inferior performance [30]. On the contrary, nanocrystalline rutile TiO_2 of particle size 15 nm improves the reversible capacity of the cell up to 263 mAh/g [30]. Here, a comparatively higher amount of Li^+ storage at suitable lattice sites resulted in a higher capacity anode. In regards to the improvement of electronic conductivity, similar to shortening of ion diffusion paths, nanostructured materials also provide shorter transport paths for electrons. Another very effective method for improving electronic conductivity is to add conductive coating on nanostructures [13]. Conductive coating provides effective paths through grain boundaries, reducing the contact resistance. Details of the different conductive coatings will be discussed in later sections.

Structural integrity and mechanical stability can be boosted by nanostructuring. During insertion/extraction of ions, the anodes encounter volume expansion/reduction respectively. Strain generated from this volume change can cause irreversible damages, such as pulverization, poor contact with current collector, and unstable SEI layer. These are some of the causes of capacity fade and ultimately poor cycle life. Owing to high surface area, nanostructures can provide more spaces for volume expansion and can absorb the strain preserving the structural integrity and keeping the connection with current collector intact. Silicon-based anode material is a perfect example to illustrate the volume change issue. As stated in an earlier section, silicon-based anode suffers huge volume expansion. Si nanowire–based anode showed high performance presenting a discharge capacity of 75% of the theoretical capacity with little capacity fading because of its capability to better accommodate volume changes [31]. Furthermore, nanostructured material can prevent agglomeration of active material, which may cause inferior performance [32]. Another benefit of nanostructured anode is its flexibility and robustness. Modern devices can benefit from flexibility in the shape of the battery. To achieve that, carefully designed packing density of nanoanode material can play a vital role [32]. Benefits of nanostructure will be described in depth respective to different materials and architecture in later sections.

3 GEOMETRICAL ASPECTS AND DESIGN OF NANOSTRUCTURED ANODES

Proper sizing and architecture is very important for nanostructured anodes. It has been shown by several studies that the synergistic effect of material properties and nano-architectures can provide desirable high performance anodes. Over the past decades, revealing different mechanical and chemical synthesis processes has opened the opportunity to create alloys and composites with unique nanostructures, which have been showing satisfactory improvement in anode performance. A good overview of the synthesis process can be found here [33]. The following section is a brief overview of the nanostructured anodes according to their geometrical aspects.

3.1 Low-dimensional nanostructures

Low-dimensional structures [e.g., 0D nanoparticles (Fig. 2.2A), 1D nanotube, 2D nanosheets, and so on] possess numerous unique favorable properties, mostly related to transport kinetics and mechanical stability.

1D structures include nanotubes, nanowires, nanorods, and nanobelts, where the aspect ratio of length to diameter of the structure is high. Nanotubes are hollow inside (Fig. 2.2B) and nanorods and nanowires are solid (Fig. 2.2B). Nanorods are smaller in length and larger in diameter compared to nanowires, giving more flexibility to the later. Nanobelts are simply elongated rectangular structures. The long length ensures

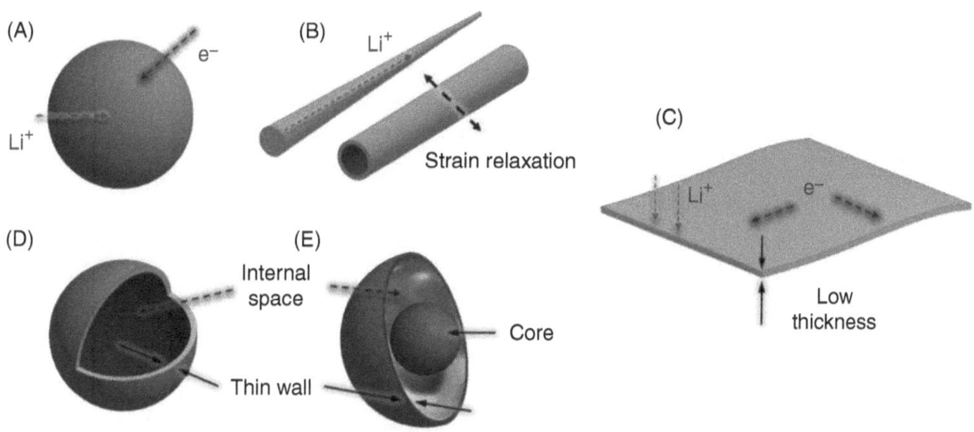

Figure 2.2 ***Schematic of 0D, 1D, 2D, and 3D nanostructure unit.*** (A) A 0D nanoparticle with Li^+ and e^- transport path, (B) a 1D nanowire and nanotube with Li^+ transport path and strain relaxation properties, (C) a 2D thin nanosheet with Li^+ and e^- transport path, (D) a 3D hollow sphere with illustration of thin walled shell and internal empty space, (E) a 3D core–shell structure with illustration of the core and thin walled shell.

good electrical contact and shorter diameter promotes the ion diffusion. Evidence of benefits of 1D structure was shown by Tang et al. who synthesized elongated (length up to 30 μm) TiO_2(B) via mechanical force-driven growth in hydrothermal process [34]. A high capacity of 114 mAh/g at 8.4 A/g current rate for an impressive 10,000 cycles was achieved. It was concluded by the authors that the capacitor-like rate performance and battery-like high capacity for long-time cycling was the result of the pseudocapacitive charge storage process, short diffusion length, and large surface area. Besides the favorable transfer and diffusion kinetics, 1D materials offer facile strain relaxation, maintaining structural integrity during repeated charge/discharge process. Further, a demonstration has been presented by Chan et al. where Si nanowire showed intact structure after a few cycles at high current rate despite suffering a ~400% volume change [31]. The conclusions made from this work were that the generated strain was well accommodated due to the small diameter of the nanowires while maintaining good contact with current collector resulted in a high capacity (~3500 mAh/g) anode. And the achieved capacity was maintained with little insignificant capacity fade until 10 cycles. Even at the very high current rate of 1 C the stable capacity was over 2100 mAh/g.

2D nanostructures are usually very thin sheet-like structures. The thickness of these sheets ranges from a few to several tenths of nanometers. Thin sheets and 2D directional kinetics make nanosheets a potential nanostructure for anodes. Fig. 2.2C shows schematic of a nanosheet along with its geometric and kinetic properties. Exceptional rate capability can be achieved by 2D nanosheets due to its facile electron transport along 2D plane and fast reaction kinetics [32,35,36]. A crystalline SnS_2 nanosheets–based nanoanode was prepared via a one-step solvothermal process by Kim et al. [36]. These nanosheets exhibited 95% charge capacity retention at a high current rate of 323 mA/g after 50 cycles. In the case of TiO_2(B), nanosheets-based structure showed better rate capability compared to nanoparticles and nanotubes [32,35]. Additionally, 2D structures can also provide strain relaxation generated from volume change and can offer flexibility to the anode [32]. Due to these properties, the anode can exhibit long cycle life [37] as discussed for 1D structures. Furthermore, 2D structures create the opportunity of selectively exposing crystallographic facets, which boost the storage capacity of an anode due to abundance of the exposed surface [32].

3.2 High-dimensional nanostructure

High-dimensional nanostructures are generally 3D structures that consist of a combination of 1D and/or 2D structures along with interconnection and designed void spaces between them. Most of the recent nanostructures are designed based on this principle. These structures offer all the advantages of low-dimensional structures. Additionally, as they are designed to act to rectify the limitations of low-dimensional structure and material properties, the resulted anodes exhibit improved performance. Examples of high-dimensional nanostructures are porous, hollow, core shell, hybrid, and heterogeneous structures. Different structures offer material specific improvement to the anodes that are presented in this section.

Porous structures are the most basic form of all high-dimensional nanostructures. A porous structure is a 3D assembly of 0D/1D/2D structures, such as nanoparticles, nanowires, and so on. Fig. 2.3A shows a schematic of a porous structure with a network of interconnected pores. 3D porous structures offer superior performance as evident in the work of Ge et al. who synthesized porous boron-doped silicon nanowires (SiNWs) by chemical etching process [38]. Due to the high porosity of such structures, the anode showed a capacity of 2000, 1600, and 1100 mAh/g at high current rates of 2, 4, and 18 A/g, respectively, after 250 cycles. Important to mention here, improved capacity also partially resulted from the boron doping; doping of boron increased the electronic conductivity of Si. Similar performance was also observed for other materials, such as Germanium (Ge) [39]. Fig. 2.3B shows an SEM image of a 3D porous Ge nanoparticle assembly. In general, there are three categories based on the size of the pores: micro (<2 nm), meso (2–50 nm), and macro (>50 nm). Anodes can be composed of one size of pores or a combination of different sizes. For instance, an hierarchical arrangement of pores offers on one hand shorter diffusion length, that is, faster diffusion kinetics; on the other hand, it offers larger surface area and higher electrolyte penetration. Hence, it is possible to achieve high capacity at high current rate by implementing hierarchical structure. Moreover, void spaces enhance the anode's ability to better manage the strain originated from volume change. Recently, Hasegawa et al. has demonstrated the high rate performance of a crystalline $Li_4Ti_5O_{12}$ hierarchical nanostructured anode for both Li-ion and Na-ion batteries [40]. Without implementing conductive coating, the battery exhibited a capacity of 146 and 105 mAh/g at 10 and 30 C current rate, respectively. For Na-ion battery, however, further improvement is needed for this structure. Porous structured anode can also be designed by random or ordered arrays of 1D structures, such as nanotubes/nanowires, and so on. Ordered array design provide facile transport path, the void spaces sustain structural deviation, and strong adhesion with the substrates provide electrical and mechanical stability (Fig. 2.3C). For instance, mesoporous Co_3O_4 arrays on Ti foil showed excellent capacity retention and rate capability; 700 mAh/g capacity at 1 C and up to 50% retention at 50 C [41]. Fig. 2.3D shows a SEM image of the Co_3O_4 nanowire arrays. Another design method of porous structures is to create a network of cross-linked 1D structures. Fig. 2.3E shows a schematic of such structure. Lee et al. reported fabrication of TiO_2 nanowebs via electrospun of nanofibers directly on the current collector [42]. Fig. 2.3F shows a SEM image of the fabricated nanowebs. A stable cycle performance of 100 cycles with 60% retention at 10 C current rate was achieved with this nanostructure anode. Another example of such design is carbon cloth–based anode. A variety of nanostructured anodes can be fabricated on carbon cloth. $ZnCO_2O_4$-urchins on carbon cloth exhibited a capacity of 750 mAh/g at a high current rate of 18 A/g [43]. These web- or net-like structures offer high electrolyte penetration,

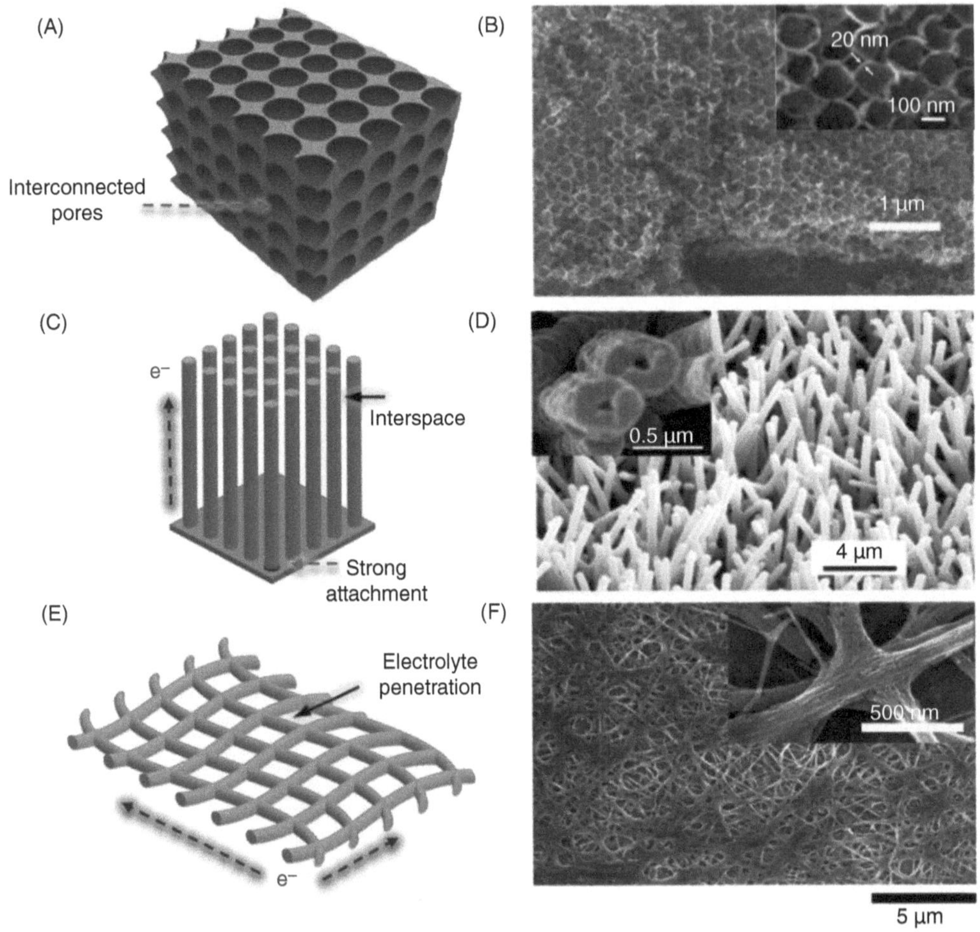

Figure 2.3 ***Illustration of 3D designed anode nanostructures.*** Porous design: (A) schematic of a 3D porous structure unit; (B) SEM image of a 3D porous Ge nanoparticle assembly, with enlarged portion shown in inset. Ordered array design: (C) a schematic of an ordered nanowire array with illustrations of interspatial property and strong attachment with substrate; (D) SEM image of Co_3O_4 ordered nanowire arrays grown on Ti foil, with inset showing a few enlarged nanowires. Cross-linked nanoweb design: (E) a schematic of cross-linked nanoweb with illustration of 2D e^- transport path and increased area for electrolyte penetration; (F) SEM image of TiO_2 nanoweb, with inset showing an enlargement of interlinked nanoweb. *(Part A–B, Reprinted from M.-H. Park, K. Kim, J. Kim, J. Cho, Adv. Mater. 22 (2010) 415–418 [39], with permission from John Wiley & Sons; part C–D, reprinted with permission from Y. Li, B. Tan, Y. Wu, Nano Lett. 8 (2008) 265–270 [41]. Copyright © 2008 American Chemical Society; part E–F, reprinted with permission from S. Lee, J. Ha, J. Choi, T. Song, J.W. Lee, U. Paik, ACS Appl. Mater. Interfaces 5 (2013) 11525–11529 [42]. Copyright © 2013 American Chemical Society).*

easy diffusion path, and flexibility and thus display stable performance. Porous structures fabricated by 2D nanosheets and nanoflakes also showed superior performance [32].

Hollow structures with very thin shell have the potential to provide better ion diffusion kinetics and better structural stability against volume change. Fig. 2.2D shows a schematic of a single hollow structure unit. An example of high performance of hollow nanostructure was presented by Koo et al. where anode based on iron oxide hollow nanoparticle exhibited a capacity of 133 mA/g at a current rate 3 A/g with negligible capacity fading up to 500 cycles [44]. Similarly, 3D structures based on interconnected network of hollow nanostructures showed superior performance. The outer shell of the hollow structure can act as a base for 1D, 2D, or porous subunits [32,45]. Zhong et al. prepared a rambutan-like $FeCO_3$ hollow microsphere–based anode via hydrothermal method (Fig. 2.4A–B [45]). In this unique structure, the subunit is the assembly of numerous 1D nanofibers anchored on a submicrometer hollow sphere and exhibited a capacity of more than 700 mA/g at high current rate (200 mA/g) after 200 cycles. Besides creating complex structures on the thin shell surface, performance of the anode can be further refined by tuning the hollow structure, such as fabricating multiple shells, increasing porosity, and tuning shell thickness. For instance, it was demonstrated by Wang et al. that the porous triple-shell structure provides maximum Li storage and displays superior cycling behavior in the case of α-Fe_2O_3 hollow microsphere–based structure [46]. Here, the porous thin shell facilitated rapid electrochemical kinetics and buffered stresses of volume change. As a result, a reversible capacity of 1702 mAh/g at 50 mA/g current rate was achieved [46]. Yolk–shell structures are modified hollow structures with a configuration of a core covered by a shell and void in between. In addition to offering all the advantage of hollow structure this kind of structure offers a window for further tuning, such as tuning of core, shell, and void spaces. A comparison of performance of hollow and yolk structure was presented by Choi et al. [47]. The authors prepared $ZnCo_2O_4$ yolk–shell, hollow, and single crystalline powder. The demonstrated discharge capacities were 753, 538, and 206 mAh/g, respectively, after 200 cycles at a current rate of 3 A/g. Fig. 2.4C shows the cycling performance along with the TEM images of the respective unit structures. Apparently, the existence of the core component and smart utilization of void spaces in case of volume expansion made the difference. Core–shell structures possess similar geometrical configuration, but here the core and the shell consist of different elements, making them composite nanostructures. Fig. 2.2E shows a schematic of core–shell structure unit. The synergistic effect of different materials in a single structure unit is highly beneficial. For instance, carbon-based materials in the form of nanotubes or graphene nanosheets possess high electron conductivity and mechanical robustness properties but inferior storage properties (capacity). Incorporating a carbon-based shell on an active material with superior storage properties in a core–shell configuration can omit or reduce the limitations of homogenous material–based nanostructured anodes. A good example of such a structure was reported by Liu et al. where

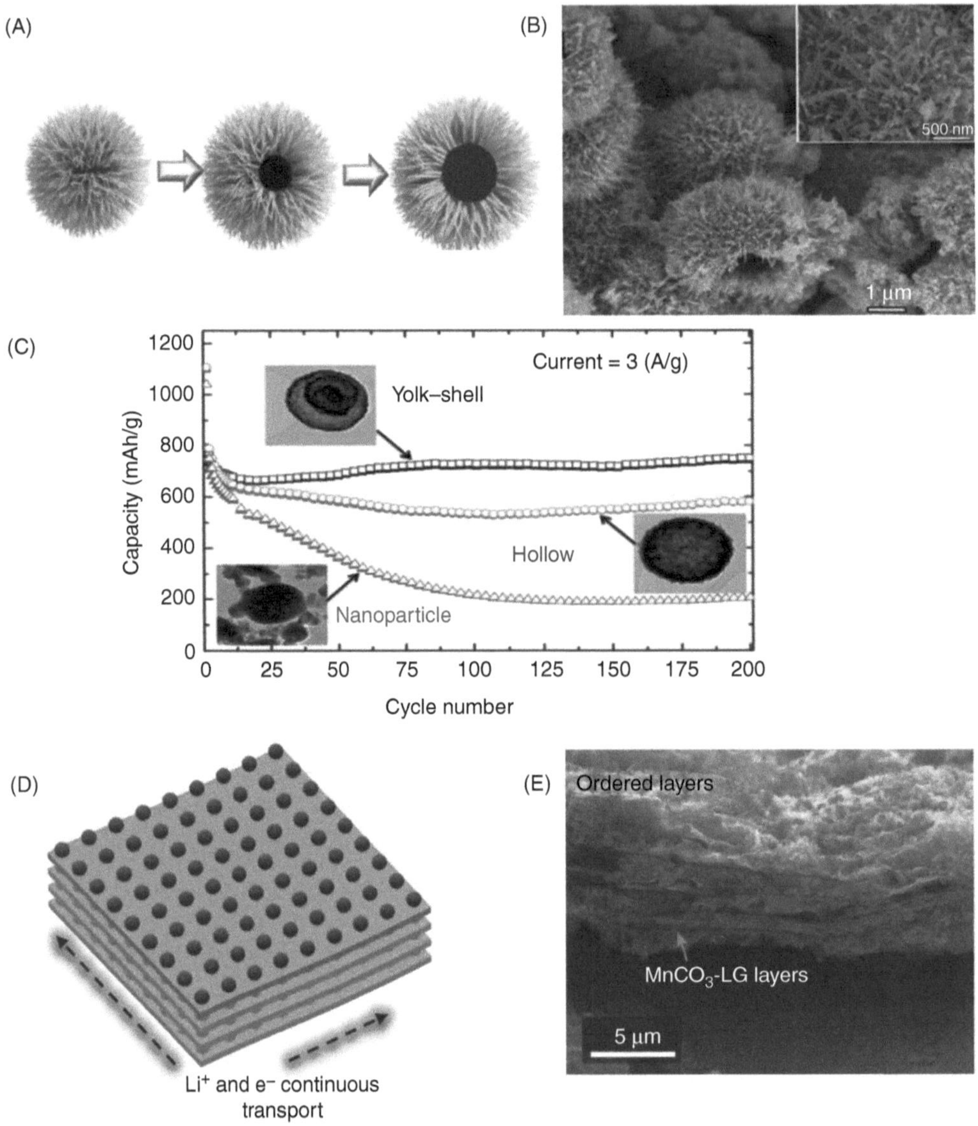

Figure 2.4 (A) Schematic process of rambutan-like structure formation; (B) SEM image of $FeCO_3$ hollow microspheres, with inset showing enlarged image of the surface of a microsphere. (C) Cycling performance of $ZnCo_2O_4$ yolk–shell, hollow, and nanoparticle-based anodes with their TEM images. (D) Schematic of layered or sandwich structure. (E) SEM image of a cross section of an ordered structure $MnCO_3$ nanoparticles arrays on long graphene sheet. *(Part A–B, Reprinted with permission from Y. Zhong, L. Su, M. Yang, J. Wei, Z. Zhou, ACS Appl. Mater. Interfaces 5 (2013) 11212–11217 [45]. Copyright © 2013 American Chemical Society; part C, reprinted from S.H. Choi, Y.C. Kang, ChemSusChem 6 (2013) 2111–2116 [47], with permission from John Wiley & Sons; part E, reprinted from Y. Zhong, M. Yang, X. Zhou, Y. Luo, J. Wei, Z. Zhou, Adv. Mater. 27 (2015) 806–812 [49], with permission from John Wiley & Sons).*

a pomegranate-inspired Si core carbon shell structure was fabricated. The reported system achieved 97% capacity retention after 1000 cycles [48]. The authors highlighted two independent properties of such a nanostructure: (1) the accommodation of volume expansion of Si and (2) the maintenance of the integrity of the stable SEI layer. Benefits of different elements and their composites will be discussed in the following sections.

Hybrid structures are a combination of 1D/2D/3D structures. By inheriting all the favorable features of the 1D/2D/3D structures, hybrid structures can provide outstanding electrochemical and mechanical properties. 2D-reduced graphene oxide (RGO) supported active nanostructure (such as nanoparticles, nanospheres, core–shell structures), that is, hybrid structures are some excellent examples. Such 2D sheets can host versatile structures on it and can form layered structure by face-to-face stacking and can be termed as alternating layered or sandwich structure. Fig. 2.4D shows a schematic of layered structure and Fig. 2.4E shows a SEM image of layered structure. These structures can boost the stability of the support between the sheet and the hosted structure on each layer. For instance, Zhong et al. presented a layered composite structure comprised of uniformly distributed $MnCO_3$ nanoparticles on long and continuous graphene sheets. The fabricated anode exhibited high capacity, power capability, and cycling stability (capacity over 1000 mAh/g at 2 A/g for 1100 cycles) [49].

As described previously, structural aspects of nanostructured anode are of paramount importance. With the progression of time, novel structures are being reported with unique advantages. But the performances of an anode are not solely dependent on the structural aspects. One kind of structure can be suitable for a particular material or material group but may not be suitable for other materials. Numerous combinations of materials and structures is possible, offering a wide window of new possibilities. Hybrid structures can meet multiple requirements simultaneously. Thus more and more works are being published concentrating on hybrid structure. However, synthesizing hybrid structures with precise control is generally complex. Therefore, balancing between high performance anode and complexity of the synthesis process and finally upscaling of the complete system is a great challenge. Nevertheless, it is safe to say that the future direction of nanostructured anodes will be toward hybrid structures. Readers are encouraged to follow the work written on the design of nanostructured anodes [32]. The later sections will be focused on different types of material for anodes.

4 CARBON-BASED ANODES

Carbon-based materials represented the primary choice for anodes and still are commercially the most successful material for Li-ion batteries because of their chemical stability, low redox potential, ease of manufacturing methods, and reasonably high specific charge. Among the different bulk forms of carbon material, graphite or graphitized, nongraphitized hard and soft carbons have been used as anode materials. There are several review articles published on the use of carbon material as anodes in electrochemical energy storage devices [50–54]. Theoretically, highly crystalline graphitic carbon possesses a specific capacity of 372 mAh/g and nongraphitized hard carbon up to 600 mAh/g for Li-ion batteries [3]. Owing to relatively high irreversible capacity and poor power capability, great effort was focused toward nanostructured carbon materials in the recent years.

As stated earlier, the 1D structure can improve the ion kinetics, and carbon nanotubes (CNTs) are considered as a potential candidate for anodes. Single-walled CNTs (SWCNT) and multiple-walled CNTs (MWCNT) are the two 1D morphologies of CNTs and are generally prepared by rolling up graphene layers via different chemical and mechanical routes, such as acid oxidation, ball milling, chemical vapor deposition, and so forth [3,55]. High reversible capacity of 1116 mAh/g was achieved by using CNT-based anodes. Reducing the diameter of CNTs, strain is generated causing delocalization of electrons and thus making them more electronegative [3]. As a consequence, the capacity increases. However, in the case of MWCNTs, the existence of multiple layers buffer the lithium extraction process, thus showing inferior coulombic efficiency [3,56]. Carbon nanofibers (CNFs) were also used as anodes in Li-ion batteries and showed longer cycle life but inferior coulombic efficiency remained [57]. A hybrid structure consisting of nitrogen-doped hollow CNT–CNF hybrid nanostructured anode was reported by Chen et al. and showed high capacity of ~1150 mAh/g at 0.1 A/g current rate after 70 cycles [58]. Due to N-doping and improved kinetics of the hybrid structure, the capacity fade was less than 20% after 3500 cycles at 8 A/g (21.5 C). 2D RGO nanosheets have been significantly investigated as anode material, either solely or in hybrid forms. The most attractive features of graphene nanosheets are one atomic layer thick aromatic carbon network with high conductivity, and great physical and chemical stability. Additionally, the presence of defect sites has a beneficial influence on electrochemical performance [3,5,59,60]. As the nanosheets can provide storing area on both surfaces, the specific capacity increases. For instance, it was reported that the graphene anode showed a high charge capacity of more than 1200 mAh/g [10]. However, discharge capacity fell to less than 700 mAh/g with a coulombic efficiency loss of 55% and maintained a reversible capacity of ~500 mAh/g for 30 cycles. Along with the beneficiary effect of having high surface area, detrimental effects are also associated, such as increased undesirable reactions, undesirable electrolyte integration with Li^+, unstable SEI layer formation on defect sites, and so forth [3,10,12]. Apparently, these are the root causes of low efficiency of the graphene anode. In this regard, 3D porous carbon structure can overcome the limitation of low efficiency. A reversible capacity of ~1000 mAh/g was achieved by a mesoporous carbon anode [3]. Very high reversible capacity of 2640 mAh/g at a current rate of 0.186 A/g was reported by Cohn et al. with a freestanding, flexible graphene and CNT hybrid foam [61]. At a very high current rate of 27.9 A/g, the system showed a capacity of 236 mAh/g, demonstrating high rate capability.

Table 2.1 A summary of recent development of nanostructured carbon anode for Na-ion battery

Nanostructure	Synthesis process	Performances: Capacity (mAh/g)	Current rate (mA/g)	Stable cycles	References
Hollow carbon nanosphere	Hydrothermal	~200	50	~100	[63]
Hollow carbon nanowire	Pyrolysis	~250	50	400	[64]
N-doped interconnected nanofibers	Pyrolysis	~135	200	200	[65]
Reduced graphene oxide nanosheets	Modified Hummers' method	~140	40	1000	[66]
N-doped porous carbon fibers	Pyrolysis	~300	50	—	[67]
CNFs	Electrospinning	~230	50	200	[68]
S-doped mesoporous carbon	Thermal decomposition and subsequent leaching	~310	20	50	[69]

CNFs, Carbon nanofiber.

Unlike Li insertion, Na insertion in graphite anode is not thermodynamically favorable, thus showing very high irreversible capacity loss along with poor cycle life and power capability [27,28,62]. It is believed that the major source of this high irreversible capacity is the surface passivation on the graphite anode by electrolyte decomposition [62]. Obvious shift of searching for better carbon anodes is directed toward disordered carbon nanostructures [27]. Table 2.1 shows a summary of some remarkable achievements by using carbon nanostructures. From the summary, it is clear that nanostructuring along with doping of a foreign element, such as nitrogen showed promising improvement. Doping provides an additional defect site to boost the Na insertion in the carbon host structure.

Although carbon-based nanostructured anode offered suitable working potential, low cost production, and reasonable safety, further developments are expected because of its low columbic efficiency and irreversible capacity. Hybridization with other materials has proven to be a viable solution. Carbon-based nanostructure can act as a base for other nanomaterials and provides synergetic effect on the electrochemical performance anode.

5 SILICON-BASED ANODES

As stated in the earlier sections, silicon (Si) is a promising material due to its attractive theoretical specific capacity (>3500 mAh for Li-ion system) and low discharge potential (~370 mV vs. Li/Li^+) [3]. Application of silicon nanostructures as Li-ion anode is so versatile that entire reviews have been published on it [70,71]. But one of the major challenges of silicon-based anode is the severe volume expansion/contraction during insertion/deinsertion (~400%). Another challenge for silicon anode is its low electric conductivity. It has already been mentioned in the previous section that one of the advantages of nanostructure is to provide space to relax the stress and strain during volume change. Incorporation of nanowires, nanorods, nanotubes, nanosheets, and porous silicon nanostructures have improved the anode performance significantly [70]. Moreover, composites with carbon coatings, graphene, metal nanoparticles (e.g., Cu, Ag, and so on), and polymer [e.g., polypyrrole (PPy)] have been used to improve the conductivity and overall electrochemical performance. Among 1D nanostructures SiNWs are one of the most explored nanostructures. Most recent studies on Li-ion anode showed SiNWs can provide an initial capacity of over 3000 mAh/g, ~3000 mAh/g after 50 cycles at 0.5 C, and 1300 mAh/g after 100 cycles at 0.2 C current rate [72–74]. Chemical vapor deposition (CVD) is the most common synthesis process of SiNWs [70]. Very recently Toan et al. synthesized SiNW anode by Al-catalyzed plasma enhanced CVD, which showed high initial capacity and maintained a capacity of ~1400 mAh/g after 150 cycles [75]. A shorter form of SiNWs is silicon nanorods (SiNRs), also being explored as anodes. SiNWs were prepared by chemical etching of Si thin films. Recent progress showed up to ~1000 mAh/g after 170 cycles at high current rate 0.2 A/g [76,77]. 2D silicon nanosheets (SiNSs), especially thin sheets (e.g., 5-nm thickness) showed improved number of cycles by better managing volume expansion [78]. Hollow interior silicon nanotubes (SiNTs) exhibited better performance compared to SiNWs [70,79]. Double-walled Si–SiO_x nanotubes prevented the formation of unstable SEI layer, prepared by using electrospun nanofiber template. This particular nanoanode demonstrated excellent performance in terms of capacity, power capability, and cycle life. High charge capacity was achieved of ~2970 mAh/g at C/5 rate, 1000 mAh/g at 12 C rate, and 6000 cycles with 88% capacity retention at 12 C [80]. The outer oxide coating allowed the formation of stable SEI layer, thus preserving the integrity of the NTs. Hollow spheres synthesized by CVD process also showed improved performances [70]. 3D porous silicon is another explored structure for Li-ion anode. There are usually two methods to synthesize porous Si structure: (1) chemical etching without using sacrificial template and (2) using a hard template (e.g., SiO_2) and depositing Si on the template from a Si source by CVD [70,71,81]. Although these nanostructures showed improved performance, there are limitations in terms of performance. For example, anodes based on these structures possess low volumetric density, which results in inferior volumetric capacity and energy. Composite and hybrid nanostructured silicon anodes have been designed to overcome this limitation.

Carbon-based materials are by far the most used materials for making silicon composites [82]. Besides using bulk carbon composite, nanocarbon composite in the form of coatings, nanofibers, nanotubes, and nanosheets have been widely explored. Carbon

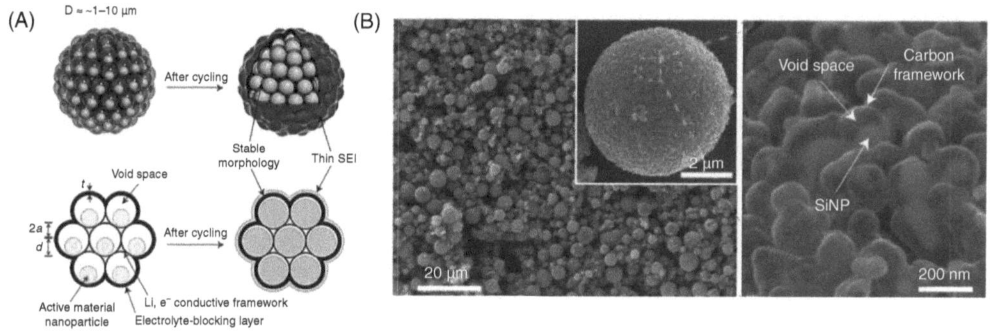

Figure 2.5 (A) Schematics of the pomegranate-inspired SiNPs/C composite anode design and its unique properties and simplified two-dimensional cross-section view. (B) SEM images of silicon pomegranates showing the micrometer-sized and spherical morphology and magnified SEM image showing the local structure of silicon nanoparticles and the conductive carbon framework with well-defined void space between. *(Reprinted by permission from Macmillan Publishers: N. Liu, Z. Lu, J. Zhao, M.T. McDowell, H.-W. Lee, W. Zhao, Y. Cui, Nat. Nanotechnol. 9 (2014) 187–192 [48] Copyright © 2014).*

composites improve the conductivity, restrict the silicon anode volume expansion, and help to form stable SEI layer. To fabricate these nanostructures, several processes, such as CVD and pyrolysis were used [70]. A unique yolk–shell designed, Si nanoparticle (SiNPs)/C composite anode was reported by Liu et al. [48]. In this pomegranate-like design, silicon nanoparticles were encapsulated by a conductive carbon layer providing ample room for expansion and contraction. An ensemble of these hybrid nanoparticles was then encapsulated by a thicker carbon layer in micrometer-size pouches to act as an electrolyte barrier (Fig. 2.5). This hierarchical arrangement resulted in a stable SEI layer and finally a superior cyclability: 97% capacity retention after 1000 cycles. Similar yolk–shell nanocomposite with high performance was also reported recently by Su et al. [83]. SiNPs/1D carbon (e.g., Nanofibers, Nanotubes, and so on) have also exhibited improved performance, as evident in several recent reports [84–88]. 2D carbon structure, graphene nanosheets do not offer high reversible capacity, but the synergistic effect of SiNPs/graphene composite has proven to be attractive for anode application [89–95]. While SiNPs prove high capacity and graphene sheets maintain contact with current collector, buffer contact between SiNPs and the electrolyte provide space during volume variation. Reported anode showed up to >1100 mAh/g capacity after 100 cycles [70].

Metal nanoparticles, such as Cu, Ag, and so forth have been implemented as surface modification of Si nanostructure to improve overall performance, particularly coulombic efficiency and power capability. For instance, Yoo et al. synthesized electrically conductive Ag-coated Si particles via a simple chemical reduction process and showed improved performance, such as a high reversible capacity (1550 mAh/g at 0.1 C rate), a stable cycling performance (a specific capacity of 1320 mAh/g at 0.2 C rate after 100 cycles), and excellent rate capability (a charge capacity of 640 mAh/g at 5 C rate) [96].

Conductive polymer, such as PPy, polyaniline (PANi) have also been widely employed as composite coating [97].

Unlike Li insertion, silicon does not intercalate sodium at room temperature. Moreover, considering ~400% volume expansion of silicon on Li insertion and bigger size of Na ion, designing a silicon anode for Na ion battery is pragmatically very challenging [98].

Silicon has been employed as anode material in Li-S cells. $Li_{15}Si_4$ anode possesses a room temperature theoretical capacity of ~3579 mAh/g [4]. Li is generally introduced in the anode before the cell assembly either by prelithiated anode or as in Si–Li_2S system. For Li–sulfur cells, there are few reports regarding nanostructured silicon anode. SiNWs-based anode for Si–Li_2S cell was reported by Yang et al. [99]. The reported performance indicated a theoretical specific energy of 1550 Wh/kg but with inferior cycle life. Later, Bogart et al. reported Si nanowires with a conductive carbon skin–based anodes. The Si nanowire electrodes exhibited capacities of over 2000 mAh/g for 100 cycles at C/10 rate and over 1200 mAh/g at 1 C rate [100].

As a summary, Si-based anodes are most promising because of high theoretical capacity but volume expansion problem makes it challenging for practical implementation (e.g., inferior cycle life, specific power). Nanostructuring significantly overcame this challenge; however, further effort is needed to implement silicon nanoanode in a practical application (e.g., automotive traction battery). Better understanding of the degradation mechanism in the case of nanoscale anode is one of the ways to further improve the performance. Another important factor, all the reported results are based on lab scale small cells (e.g., coin cell); the performance of full-scale cells is yet to be widely investigated.

6 METAL ALLOY ANODES

The concept of using binary metal alloys as electrode was in present since mid-1990s. Since then, a significant number of reports have been presented as promising electrode materials, particularly Li-based alloys [1,3,24,101]. An excellent review on metal alloy anodes can be found here [102]. Initially, Li_xM alloys were exploited where M is metal, electrochemically active or inactive. The role of M here is to support the Li ion generated during de/intercalation and to improve the electrical conductivity. However, due to the presence of electrochemically inactive material, a decrease in the reversible capacity was observed. Binary metal alloys with nanostructured architecture have been found to be a potential choice. Si, Sn, Al, Ga Pb, Sb, Ag, and Ge-based Li–M alloys have been extensively studied for the application in Li-ion system during the past decade [1,3,101]. Si alloy is the most studied among them, forming several alloys, such as $Li1_2Si_{17}$, Li_7Si_{17}, $Li1_3Si_4$, and $Li_{22}Si_5$ in several steps [3,103]. As discussed in the previous section, like Si metal, Si alloys show high capacity but suffer from a high volume expansion problem. Similar to Si, Sn metal also forms Li–Sn alloys; reversible capacity, however, is less compared to Li–Si alloy and effect of inhomogeneous volume expansion is still a challenge for such alloys [104,105]. Li metal is also known to form alloys with other group V metal, such as Al (Li_xAl), P (Li_3P), As (Li_3As), Sb (Li_3Sb), and Bi (Li_3Bi) with high capacity but a major challenge of volume expansion was inevitable. In the Li–S battery system, the Li–M alloy system was explored and found to significantly reduce dendrite formation [26]. However, the challenge of low capacity compared to lithium metal anode hindered further progress.

Another potential method of metal alloy as anode is the fabrication of intermetallic ternary Li–M1–M2 system, where M1 and M2 are metals, which are either electrochemically active–active or active–inactive. Li–M1–M2 alloy system showed improved performance in terms of reversible capacity. Si–Mg (or Al) systems are the example of Si-based intermetallic anode (e.g., Li_xMg_2Si). Si–M where M represents metals, such as Ni, Fe, Co, Mn, Cr, and so on are other known anode materials. Si–M where M stands for electrochemically inactive metals (such as Fe, Ni, and so on) showed capacity value in the range of 60–1000 mAh/g [3,101,106,107]. Sn-based all intermetallic alloys have also been investigated as anode materials. In this regard, an excellent comparative study was reported by Wang et al. who prepared $FeSn_2$, Cu_6Sn_5, $CoSn_3$, and Ni_3Sn_4 nanospheres of particle size ~40 nm [108]. Fig. 2.6 shows the SEM image of the nanospheres. The

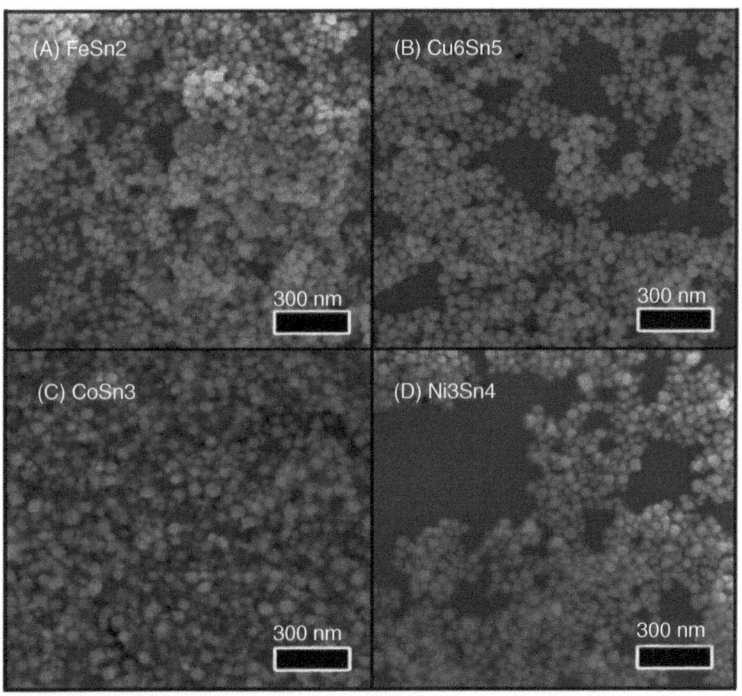

Figure 2.6 SEM images of (A) $FeSn_2$, (B) Cu_6Sn_5, (C) $CoSn_3$, and (D) Ni_3Sn_4 nanospheres. The scale bars are 300 nm. *(Adapted with permission from X.-L. Wang, W.-Q. Han, J. Chen, J. Graetz, ACS Appl. Mater. Interfaces 2 (2010) 1548–1551 [108]. Copyright © 2010 American Chemical Society).*

capacity values were in the following order: $FeSn_2$ > Cu_6Sn_5 ≈ $CoSn_3$ > Ni_3Sn_4, with highest capacity for $FeSn_2$ of 500 mAh/g. However, increased capacity value of more than 700 mAh/g was reported recently by employing Sn–Fe–C [109] and Cu_6Sn_5/Sn [110] composite anodes. Electrochemically, active M–Sn intermetallic compounds, such as SbSn [111–114], Ag_3Sn [115], and Mg_2Sn [116] and their composites were also reported as anodes. Anode consisting of other binary and ternary alloy were also reported, such as Cu_2Sb–TiC–C alloy composites [117] and Cd–Sn–SnO_2@C hetero-composites [118].

Although owing lower specific capacity (1623 mAh/g) and being more expensive than Si, Ge-based anodes were investigated because of its 10^4 times higher electrical conductivity, 15 and 400 times faster Li diffusion at 360°C and at room temperature compared to Si, respectively, higher capacity than graphite anode and a narrow band gap (0.67 eV) [1]. Different nanostructured anodes based on Ge, GeO, and their alloys/composites were reported and showed improvement in specific capacity and specially the power capability [119–122]. Recently Han et al. presented that by inserting ultrathin amorphous germanium (a-Ge) (~5 nm) at the interface of $NiSi_x$/amorphous Si (a-Si), the rate capacity was substantially enhanced, 477 mAh/g at a high rate of 40 A/g [123]. In addition, batteries containing the $NiSi_x$/GeSi anodes cycled over 1000 times at 10 A/g while the capacity retaining more than 877 mAh/g.

Metal alloys employed as anode materials for Na-ion battery showed attractive specific capacity [124]. Composite electrodes based on Sn, Sb, and Ge are among the most promising nanostructured anodes [125]. Liu et al. designed a high-capacity alloy anode, a forest of Sn nanorods with a unique core–shell structure [126]. The resultant 3D nanoforest of C/Sn/Ni/TMV1cys binder-free composite electrode exhibited supreme initial capacity [722 mAh/(g Sn)] along with high-retained capacity [405 mAh/(g Sn)] after 150 deep cycles. Although the Sn-based alloy electrode showed promising capacity, its cycling life was not remarkable. Several strategies were reported to improve the cycle life [125,127]. Several reports on Sb-based nanostructured anode showed a stable capacity above 600 mAh/g and high rate capability up to 20 C [127–130].

Due to the high number of alloyable materials and their combination in binary and ternary forms, a significant number of metal alloys anode has been already investigated. Additionally, topology of designing nanostructures further broadens the choice of anode. It is difficult to extract a summary of comparison between them. In general, a few key material parameters, such as theoretical capacity of alloying reaction and abundance in environment can act as selection criteria. In terms of gravimetric capacity of allowing reaction the order of alloys is Si > Ge ≈ As ≈ Sn ≈ Al > Ga >Sb ≈ Pb > Bi. And for volumetric capacity the order is Si ≈ Ge > Sn > Pb ≈ As > Ga ≈ Sb > Bi > Al. Al and Si are the most abundant among all the materials, but Al possesses inferior volumetric capacity. Sn has lower gravimetric capacity than Si but the volumetric capacity is similar. Thus for alloy anodes Si and Sn are better choices. As discussed previously hybrid nanostructure and carbon–metal alloys composite nanostructures showed better performance, and the future focus will be more concentrated on these anodes.

7 METAL OXIDE-BASED ANODES

Metal oxide–based anodes are the most versatile and widely studied group of anodes [131]. Increasing interest of metal oxide anodes is mainly due to their high structural stability and electronic conductivity. The methods of Li/Na hosting vary for different metal oxides anode as insertion and conversion. Titanium-based anodes, such as lithium titanate ($Li_4Ti_5O_{12}$) and titanium dioxide are alternatives to conventional graphite insertion anodes. Ti-based anode offers very high cycle life, high power, and improved safety, and is already commercialized, known as lithium titanate oxide (LTO) battery. However, low specific capacity in the range of 170–330 mAh/g and low electronic conductivity are still major challenges. In this regard, nanostructured Ti-based anode were investigated and found to provide better performance in terms of specific capacity [132]. LTO nanowires grown directly on Ti substrate showed high rate capability (121 mAh/g at 30 C) [133]. Titanium dioxide or titania is an attractive anode material [134–138]. In case of Na-ion battery, TiO_2, spinel $Li_4Ti_5O_{12}$, layered $Na_2Ti_3O_7$, and $Na_{0.66}[Li_{0.22}Ti_{0.78}]O_2$ are also known to be used as anodes. TiO_2-based nanotubes showed an important relation between the performances of the anode and diameter of the nanotubes. With nanotubes with an internal diameter of 45 nm, the electrochemical profile remained inactive, compared to nanotubes with an internal diameter of 80 nm, where 150 mAh/g of capacity was achieved after 15 cycles [139] and another report showed that with a diameter of 70–130 nm, 300 cycles were achieved [140]. Composite of TiO_2 and doped anodes (e.g., N, Fe as dopant) were also reported with improved performance [141–144]. Spinel $Li_4Ti_5O_{12}$ showed promising Na storage behavior with a reversible capacity of ~150 mAh/g, and a first coulombic efficiency of up to 81% [145]. Zhou et al. reported recently that $Li_4Ti_5O_{12}$ nanorods (Fig. 2.7) showed a discharge capacity of ~100 mAh/g after 1000 cycles at 20 C in Li-ion system and 131.6 mAh/g after 100 cycles at 0.1 C rate in a Na-ion system [146]. This result implies that implementing nanostructured titanate improved the rate capacity with reasonable capacity retention over long cycle life. In particular, for Na-ion system, Ti-based nanoanode partially overcame the limitation of low reversible capacity and poor power capability. Higher surface area for Na^+ insertion and shortened diffusion length is the root cause of the performance improvement. Moreover, nanostructured $Li_4Ti_5O_{12}$ composited with carbon showed even greater rate capability because of increased electronic conductivity [147,148]. Kim et al. achieved a capacity of 168 mAh/g at a rate of 0.2 C (35 mA/g), 117 mAh/g at a rate of 10 C, 88 mAh/g at a rate of 30 C, 67 mAh/g at a rate of 50 C, and 38 mAh/g at a rate of 100 C [147,148] by an anode composed of carbon-coated $Li_4Ti_5O_{12}$ nanowires. Na titanate anodes also exhibited promising performance in terms of rate capability and reversible capacity [149–152].

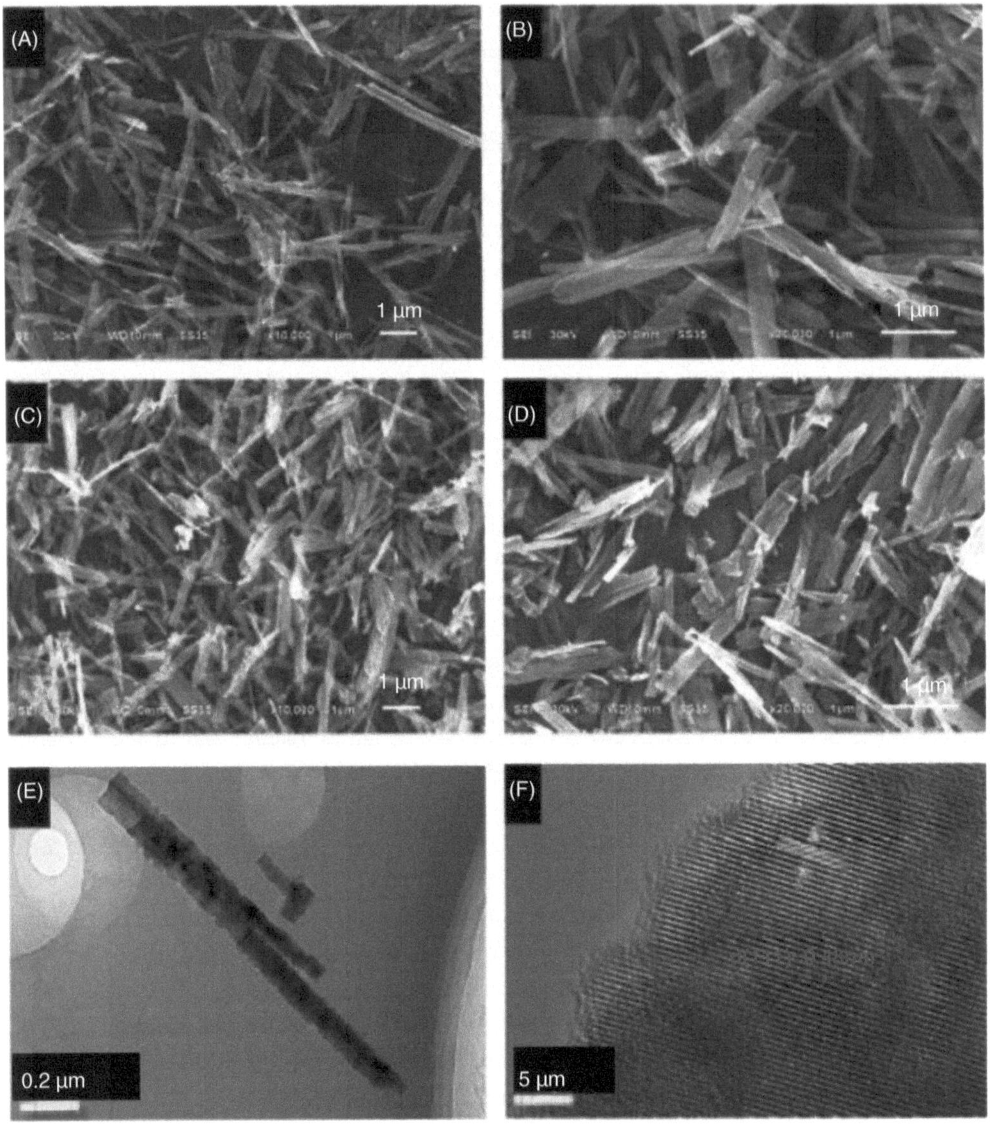

Figure 2.7 (A–B) SEM images of TiO_2–B; (C–D) SEM image; (E) HRTEM; (F) images of $Li_4Ti_5O_{12}$ nanorods. *(Reprinted from Q. Zhou, L. Liu, J. Tan, Z. Yan, Z. Huang, X. Wang, J. Power Sources 283 (2015) 243–250 [146]. Copyright © 2015, with permission from Elsevier).*

Other transition metal oxides host Li/Na ion by conversion mechanism according to the following general reaction:

$$MO + Z^+ + e^- \leftrightarrow Z_2O + M$$

where M = Fe, Co, Ni, Cu metal, and $Z^+ = Li^+/Na^+$.

Although the reaction in the forward direction is thermodynamically favorable, the backward direction is not. The reversibility of the backward direction can be enhanced by nanoscale design mostly because of shorter diffusion length and high surface area. Low conductivity, thick SEI layer are the other limitations of metal oxide anodes, which can be mitigated by combining different metal oxide along with 3D hybrid structures. Several metal oxide nanoanodes have been reported; Roy et al. gathered an extensive list of nanostructured metal oxide anodes published until the year 2014 [3]. In addition to the summary of that list, the most recent reports on metal oxide anodes for Li and Na battery systems will be discussed here.

SnO_2 is one of the choices of oxide anodes. SnO_2 hollow nanospheres, nanobox, nanowire anodes were able to achieve >500 mAh/g for more than 50 cycles, whereas SnO_2/C nanostructures delivered a capacity of more than 800 mAh/g for ~100 cycles [3]. Among the recent reports, SnO_2/C, SnO_2/N-doped carbon nanostructures displayed the most promising performance (>1400 mAh/g initial capacity and ~900 mAh/g stable capacity after 100 cycles) [153–157] as Li-ion anode. SnO_2 anodes were also reported for Na-ion battery, showing a high capacity of >300 mAh/g over 200 cycles at 0.2 C rate [158,159].

Theoretical capacity values of manganese oxides (MnO_x) are on the high end and vary depending on their different forms and phases of oxides, for example, MnO_2 (1230 mAh/g), MnO (755 mAh/g), and Mn_3O_4 (936 mAh/g) for Li-ion system. However, pristine MnO_x suffers from poor conductivity and rapid capacity fading, and thus focus was given to C-based composite anode. MnO_x nanorod, nanoribbon, nanoflake on C structures (e.g., graphene, carbon nanohorns) anodes were reported previously and found to show capacity in the range of 500–900 mAh/g with a cycle life in the range of 10–250 [3] in Li-ion system. Recent studies suggest that MnO_x–graphene hybrid structure and surface modification can further extend the value up to >1000 mAh/g with a stable capacity of more than 500 mAh/g after 100 cycles [160–166]. In the case of Na-ion system, MnO_2 nanocrystalite anode was recently presented for the first time as Na conversion anode by Wu et al. [167]. The reported anode exhibited a reversible sodiation capacity of 567 mAh/g and 70% retention after 500 cycles, which is by far the best performance as Na-ion anode.

Among different iron oxides (α-Fe_2O_3, Fe_3O_4, FeO), α-Fe_2O_3 is thermodynamically the most stable with high theoretical capacity (1007 mAh/g) in Li-ion system. J.M. Tarascon et al. showed that nanostructuring can significantly increase the Li ion hosting capacity of α-Fe_2O_3 [168]. Different Fe oxides as different nanostructure, such as nanoparticles, nanorods, nanoflakes, nanowires and their composites with carbon nanostructure showed excellent performance (up to >1200 mAh/g capacity) for more than 1000 cycles [3]. Recent studies showed further improvement of energy and power capability [169–177]. For instance, Zheng et al. prepared ultrasmall Fe_2O_3 nanoparticles embedded in nitrogen-doped hollow carbon sphere shells (Fe_2O_3@N-C) by direct pyrolysis of Fe-based zeolitic imidazolate frameworks (Fe-ZIF) at 620°C in air. The reported Li ion system displayed a capacity retention of 1573 mAh/g after 50 cycles at a current density of 0.1 C (1 C = 1000 mA/g) and delivered a remarkable reversible and stable capacity of 1142 mAh/g after 100 cycles at a current density of 1 C [175]. In the case of Na-ion system, iron oxide–based anode also showed promising performance (capacity up to 400 mAh/g with 98% retention after 200 cycles) [178–181].

Co_3O_4 is a mixed valence oxide with spinel structure containing both Co^{2+} as well as Co^{3+}. It can uptake 8 mole of Li per mole of Co_3O_4 in a reversible reaction showing the theoretical capacity 890 mAh/g. Co oxide nanorods, nanowire, porous, nanosphere, yolk–shell designed anode, and hybrid and composite anodes have been greatly investigated. Nanostructuring of Co oxide provided improved capacity up to ~1300 mAh/g with near 100% capacity retention after 50 cycles [3] in a Li-ion system. Further improvement in the performance have also been reported very recently [182–188]. Among other oxides, hybrid structures consisting of more than one oxide material became an anode material of interest recently. Examples of such materials are $A_2Mo_3O_8$ (A = Co, Mn and Zn), $CoMn_2O_4$, ZnO–$ZnCo_2O_4$, $MgCo_2O_4$, and so on. Synergistic effects of two or more oxides delivered improved performance compared to the anode based on single oxides [189–194]. Fig. 2.8 shows SEM images of few mixed oxide hybrid composite anodes.

In summary, with metal oxide–based anodes a specific capacity up to ~1300 mAh/g for Li-ion and >500 mAh/g for Na-ion system can be achieved, which is considerably higher than graphite anode. Thanks to nanotechnology, by nanosizing of these metal-oxides, reversible storage of Li^+/Na^+ with high capacity, particularly Na^+ was possible to achieve. Additionally, transition metals are usually abundant, which makes them anodes of interest. But challenges still remain in the life cycle performance. Better understanding of the working mechanism will play a vital role to further improve the performance.

8 METAL PHOSPHIDE AND SULFIDE ANODES

Metal phosphide anode materials are very interesting by offering cation hosting both through intercalation and conversion mechanisms [1]. Storage by intercalation mechanism occurs according to the following reactions:

$$M_xP_y + ZLi^+ + Ze^- \leftrightarrow Li_zM_{x-z}P_y$$

and the conversion mechanism according to the following reaction:

$$M_xP_y + ZLi^+ + Ze^- \leftrightarrow Li_zP_y + xM$$

Phosphates of Cu, Co, Fe, Sn, and Ge are usually the investigated anode materials. The attractive features of these phosphide materials are (1) high capacity, theoretically

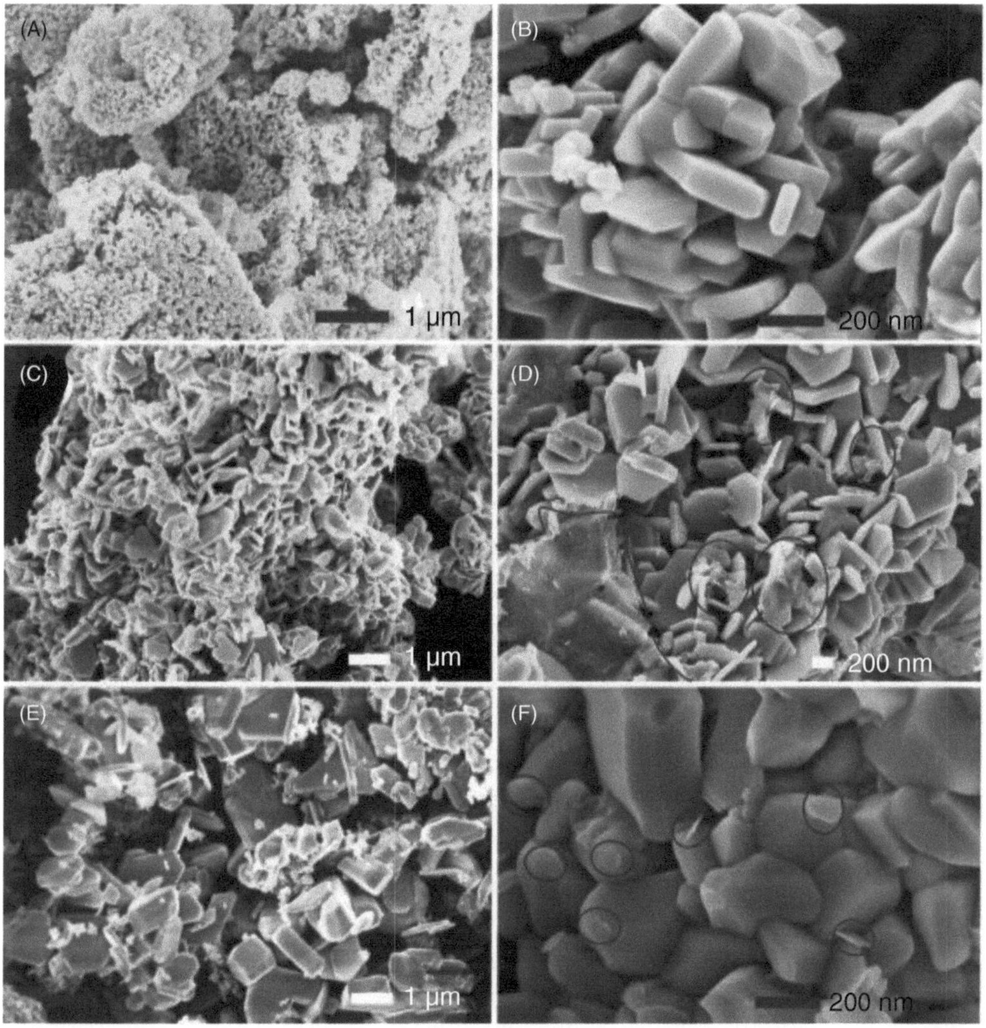

Figure 2.8 ***Secondary electron micrographs.*** (A–B) Graphene–$Co_2Mo_3O_8$; (C–D) graphene–$Mn_2Mo_3O_8$; (E–F) graphene–$Zn_2Mo_3O_8$ composites, respectively. The *black circles* represent the particles covered with transparent FLG (few layered graphene). *(Reprinted from S. Petnikota, S.K. Marka, V.V.S.S. Srikanth, M.V. Reddy, B.V.R. Chowdari, Electrochim. Acta 178 (2015) 699–708 [193]. Copyright © 2015, with permission from Elsevier).*

500 to >2000 mAh/g, (2) high degree of electron delocalization, (3) lower insertion potential compared to oxides. However, poor electrical conductivity and volume change are still the major challenges for these materials. In previous reviews [18,195], past progress on nanophosphate anode can be found. Following are some recent progress on phosphate anode for Li-ion systems (Table 2.2).

Table 2.2 Recent progress on nanostructured phosphide anode for Li-ion batteries

Nanostructure	Synthesis process	Performances: Capacity (mAh/g)	Current rate (mA/g)	Stable cycles	References
Diphosphide/carbon tube (FeP_2/C) nanohybrids	Pyrolysis	435	137	100	[196]
Fe_2P nanoparticles in graphited carbon sheets	Hydrothermal	602	100	200	[197]
GeP_5/C nanocomposite	Mechanical milling	2127	5000	40	[198]
Silicon doped MoP_2 nanocluster	Mechanical milling	783	0.3 mA/cm^2	100	[199]
Sn_4P_3 hollow spherical structures	Facile solvothermal	261	500	100	[200]
3D hierarchical cobalt orthophosphate	Facile wet chemical route	260	~40	100	[201]
CoP nanowires–reduced graphene oxide composite	Facile hydrothermal	960	200	200	[202]
C@NiCoP peapods	Anion exchange reaction	670	200	350	[203]
Ni_3P–Ni array	Electroless deposition	360	0.02 mA/cm^2	100	[204]

Metal phosphide anodes are also promising for the Na-ion system. Walter et al. prepared nanocrystals of FeP, CoP, NiP_2, and CuP_2 using colloidal synthesis methods and compared their performances [205]. Sn and Cu phosphide showed improved capacity of over 800 mAh/g when composited with carbon nanostructures [195,206–209].

Metal sulfide anodes have attracted attention because of their storage capacity in their layered crystal structure. Sulfides of Mo, W, Sn, Nb, and Ta are known to be anode materials for Li ion batteries. MoS_2 undergo lithiation at relatively high potential value ~ 2 V and the theoretical capacity value for MoS_2 is >670 mAh/g [3]. Nanosheets are the most common nanoforms used for MoS_2; other structures include hollow structures, nanoparticles, nanotubes, and so forth, also combined with carbon as composites. Nanostructuring and compositing increased the capacity of the anode in up to 1200 mAh/g with >90% retention [210–216]. Other nanostructured sulfide materials, such as WS_2 [217–220], SnS_2 [221–223] have proven to be promising anodes. Similar to Li-ion system metal sulfide anodes showed excellent capacity up to ~700 mAh/g for Na-ion system. Since the last couple of years a significant number of nanostructured metal sulfide anodes, such as MoS_2 nanoflakes, CoS–graphene composite, Co_9S_8–C composite, Cu_3S_4 nanosheet, SnS_2 nanosheet, SnS–C composite, BiS_3 nanorods, $NaBiS_2$, NiS–graphene, and so on were reported [224–232]

9 SUMMARY AND CONCLUSIONS

Undeniably nanostructured anodes opened a new paradigm in the development of rechargeable battery cells in terms of performance. In principal, nanostructured anodes: (1) facilitate cation (e.g., Li^+/Na^+) and electron transport by providing shorter diffusion paths, (2) provide larger area to host cations, (3) provide spaces to accommodate expansion during cation hosting and thus maintain structural integrity, and (4) allow effective hybridizing/compositing of different materials. There are two major interdependent aspects of nanostructure anodes: the geometrical aspect and the material element. 0D (such as nanoparticles), 1D (such as nanotubes, nanorods), and 2D (such as nanosheets) each has unique properties in terms of ion transport, stability, and so forth. Each of them solely cannot effectively satisfy all the demands for overall high performance robust anodes. Designing 3D hybrid structures (such as yolk–shell structure) by combining low dimensional structures can satisfy most of the requirements. The possibility of designing such structures has created the vast opportunity of exploring different nanomaterials as anodes, including some materials that were once thought to be unsuitable for the anode application. Research has come a long way since commercial graphite was first introduced as anode. Many different anode materials are now being implemented, and many more combinations of these materials are yet to be discovered in the future. Carbon-based nanomaterials showed considerable higher capacity than graphite, but their inferior stability issue drove the search into other materials. Silicon-based nanoanodes are the most promising nanomaterials for Li-ion systems, although extensive effort is needed to overcome the volume expansion challenge. Other metal alloys possess similar issue. Nanostructuring of metal oxide anodes have significantly accelerated the possibility of their implementation into practical electrodes. Metal oxides can be a potential successor of graphite anode in terms of price and environmental friendly synthesis process. But unfortunately, poor cycle life and low efficiency are still challenges in their practical implementation. Nanotechnology is relentlessly providing improvement and will continue to do so. Titanate-based nanoanodes in Na-ion systems significantly reduced the irreversible capacity. Phosphide and sulfide-based anodes are relatively less explored but have already shown their potential. Similar to different geometrical aspects of nanoanode, synergistic effects of different materials can mitigate each other's (materials) limitations. Therefore, composite materials with hybrid 3D structure will be the future direction for nanobased anodes.

Despite the beneficial effects of nanomaterial, adopting nanomaterials also presents significant limitations. First of all, the low volumetric efficiency, owing to large surface area and porous structure nanomaterials have low packing density compared to nonnanocounterparts. Low packing densities usually result in inferior volumetric efficiencies. In applications like electric vehicle, where space is limited, volumetric efficiency is very important. Large surface area provides more sites for insertion/conversion reactions but at the same accelerates undesirable surface reactions, which often causes instability problem during operation. This phenomenon challenges researchers to understand degradation mechanisms. And without proper understanding of the degradation mechanism, it is difficult to develop paths for performance improvement. In general, reaction mechanisms at nanoscale are complex and it is difficult to have precise control without proper understanding. In many published studies, attempts were made to understand the working mechanisms for different nanoanodes. However, most of these studies were performed on lab scale cell (such as coin cell) with standard Li/Na metal cathode and standard electrolytes. Translation of the technology from lab scale cell to application scale cell is demanding and requires further focus and research effort. It is especially crucial to have unequivocal understanding of both small- and large-scale systems. Furthermore, implementing nanomaterial anodes produces thicker SEI layers. On one hand, thick SEI layer consumes active materials, and on the other hand, it increases the internal impedance of the battery. Delicacy increases to maintain stable SEI layer.

Last but not least, most of the synthesis processes of nanomaterials are very complex, sensitive, and time consuming. All of these characteristics are highly unfavorable for mass production. To improve the possibility of industrialization and to avoid nanomaterials drawbacks, a few strategies can be adopted. Designing materials with nano/micro combinations can significantly improve packing density and overall volumetric efficiency. Surface coating with materials like carbon significantly decreases the undesired reactions. Moreover, coating can play a role on the SEI layer to make it more stable and favorable. It is foreseen that by solving these issues, nanomaterial-based anodes will be commercialized in the very near future and fulfill the high requirements of modern applications.

GLOSSARY

CNF Carbon nanofiber
CNT Carbon nanotube
CVD Chemical vapor deposition
HRTEM High-resolution transmission electron microscope
LTO Lithium titanate oxide
LUMO Lower unoccupied molecular orbital
MWCNT Multiwalled carbon nanotube
PANi Polyaniline
PPy Polypyrrole
RGO Reduced graphene oxide
SEI Solid-electrolyte interface
SEM Scanning electron microscope
SiNP Silicon nanoparticle
SiNR Silicon nanoribbon
SiNT Silicon nanotube
SiNW Silicon nanowire
SWCNT Single-walled carbon nanotube
TEM Transmission electron microscope

REFERENCES

[1] S. Goriparti, E. Miele, F. De Angelis, E. Di Fabrizio, R. Proietti Zaccaria, C. Capiglia, Review on recent progress of nanostructured anode materials for Li-ion batteries, J. Power Sources 257 (2014) 421–443, doi: 10.1016/j.jpowsour.2013.11.103.

[2] N. Nitta, G. Yushin, High-capacity anode materials for lithium-ion batteries: choice of elements and structures for active particles, Part. Part. Syst. Charact. 31 (2014) 317–336, doi: 10.1002/ppsc.201300231.

[3] P. Roy, S.K. Srivastava, Nanostructured anode materials for lithium ion batteries, J. Mater. Chem. A. 3 (2015) 2454–2484, doi: 10.1039/C4TA04980B.

[4] A. Manthiram, Y. Fu, S.-H. Chung, C. Zu, Y.-S. Su, Rechargeable lithium–sulfur batteries, Chem. Rev. 114 (2014) 11751–11787, doi: 10.1021/cr500062v.

[5] C. Liu, Z.G. Neale, G. Cao, Understanding electrochemical potentials of cathode materials in rechargeable batteries, Mater. Today 19 (2016) 109–123, doi: 10.1016/j.mattod.2015.10.009.

[6] J.M. Tarascon, M. Armand, Issues and challenges facing rechargeable lithium batteries, Nature 414 (2001) 359–367, doi: 10.1038/35104644.

[7] P. Poizot, S. Laruelle, S. Grugeon, L. Dupont, J.M. Tarascon, Nano-sized transition-metal oxides as negative-electrode materials for lithium-ion batteries, Nature 407 (2000) 496–499, doi: 10.1038/35035045.

[8] P.G. Bruce, B. Scrosati, J.-M. Tarascon, Nanomaterials for rechargeable lithium batteries, Angew. Chem. Int. Ed. Engl. 47 (2008) 2930–2946, doi: 10.1002/anie.200702505.

[9] Y.G. Guo, J.S. Hu, L.J. Wan, Nanostructured materials for electrochemical energy conversion and storage devices, Adv. Mater. 20 (2008) 2878–2887, doi: 10.1002/adma.200800627.

[10] P. Guo, H. Song, X. Chen, Electrochemical performance of graphene nanosheets as anode material for lithium-ion batteries, Electrochem. Commun. 11 (2009) 1320–1324, doi: 10.1016/j.elecom.2009.04.036.

[11] Y. Wang, H. Li, P. He, E. Hosono, H. Zhou, Nano active materials for lithium-ion batteries, Nanoscale 2 (2010) 1294, doi: 10.1039/c0nr00068j.

[12] L. Ji, Z. Lin, M. Alcoutlabi, X. Zhang, Recent developments in nanostructured anode materials for rechargeable lithium-ion batteries, Energy Environ. Sci. 4 (2011) 2682, doi: 10.1039/c0ee00699h.

[13] H. Li, H. Zhou, Enhancing the performances of Li-ion batteries by carbon-coating: present and future, Chem. Commun. 48 (2012) 1201, doi: 10.1039/c1cc14764a.

[14] V. Palomares, P. Serras, I. Villaluenga, K.B. Hueso, J. Carretero-González, T. Rojo, Na-ion batteries, recent advances and present challenges to become low cost energy storage systems, Energy Environ. Sci. 5 (2012) 5884, doi: 10.1039/c2ee02781j.

[15] M.D. Slater, D. Kim, E. Lee, C.S. Johnson, Sodium-ion batteries, Adv. Funct. Mater. 23 (2013) 947–958, doi: 10.1002/adfm.201200691.

[16] B.C. Melot, J.M. Tarascon, Design and preparation of materials for advanced electrochemical storage, Acc. Chem. Res. 46 (2013) 1226–1238, doi: 10.1021/ar300088q.

[17] Q. Zhang, E. Uchaker, S.L. Candelaria, G. Cao, Nanomaterials for energy conversion and storage, Chem. Soc. Rev. 42 (2013) 3127–3171, doi: 10.1039/c3cs00009e.

[18] X. Rui, H. Tan, Q. Yan, Nanostructured metal sulfides for energy storage, Nanoscale 6 (2014) 9889–9924, doi: 10.1039/c4nr03057e.

[19] X. Su, Q. Wu, J. Li, X. Xiao, A. Lott, W. Lu, B.W. Sheldon, J. Wu, Silicon-based nanomaterials for lithium-ion batteries: a review, Adv. Energy Mater. 4 (2014) n/a–n/a10, doi: 10.1002/aenm.201300882.

[20] W. Wang, I. Ruiz, S. Guo, Z. Favors, H.H. Bay, M. Ozkan, C.S. Ozkan, Hybrid carbon nanotube and graphene nanostructures for lithium ion battery anodes, Nano Energy 3 (2014) 113–118, doi: 10.1016/j.nanoen.2013.10.005.

[21] T. Osaka, Z. Ogumi (Eds.), Nanoscale Technology for Advanced Lithium Batteries,, Springer, New York, NY, 2014doi: 10.1007/978-1-4614-8675-6.

[22] L. Ma, K.E. Hendrickson, S. Wei, L.A. Archer, Nanomaterials: science and applications in the lithium–sulfur battery, Nano Today 10 (2015) 315–338, doi: 10.1016/j.nantod.2015.04.011.

[23] M. Sawicki, L.L. Shaw, Advances and challenges of sodium ion batteries as post lithium ion batteries, RSC Adv. 5 (2015) 53129–53154, doi: 10.1039/C5RA08321D.

[24] J. Tang, A.D. Dysart, V.G. Pol, Advancement in sodium-ion rechargeable batteries, Curr. Opin. Chem. Eng. 9 (2015) 34–41, doi: 10.1016/j.coche.2015.08.007.

[25] S.Wu, R. Ge, M. Lu, R. Xu, Z. Zhang, Graphene-based nano-materials for lithium–sulfur battery and sodium-ion battery, Nano Energy 15 (2015) 379–405, doi: 10.1016/j.nanoen.2015.04.032.
[26] R. Cao, W. Xu, D. Lv, J. Xiao, J.-G. Zhang, Anodes for rechargeable lithium-sulfur batteries, Adv. Energy Mater. 5 (2015) n/a–n/a10, doi: 10.1002/aenm.201402273.
[27] M.-S. Balogun, Y. Luo, W. Qiu, P. Liu, Y. Tong, A review of carbon materials and their composites with alloy metals for sodium ion battery anodes, Carbon N.Y. 98 (2016) 162–178, doi: 10.1016/j.carbon.2015.09.091.
[28] L.P. Wang, L. Yu, X. Wang, M. Srinivasan, Z.J. Xu, Recent developments in electrode materials for sodium-ion batteries, J. Mater. Chem. A 3 (2015) 9353–9378, doi: 10.1039/C4TA06467D.
[29] N. Omar, M. Daowd, P. van den Bossche, O. Hegazy, J. Smekens, T. Coosemans, J. van Mierlo, Rechargeable energy storage systems for plug-in hybrid electric vehicles—assessment of electrical characteristics, Energies 5 (2012) 2952–2988, doi: 10.3390/en5082952.
[30] C. Jiang, I. Honma, T. Kudo, H. Zhou, Nanocrystalline rutile TiO[sub 2] electrode for high-capacity and high-rate lithium storage, Electrochem. Solid-State Lett. 10 (2007) A127, doi: 10.1149/1.2712041.
[31] C.K. Chan, H. Peng, G. Liu, K. McIlwrath, X.F. Zhang, R.a. Huggins, Y. Cui, High-performance lithium battery anodes using silicon nanowires, Nat. Nanotechnol. 3 (2008) 31–35, doi: 10.1038/nnano.2007.411.
[32] Y. Zhong, M. Yang, X. Zhou, Z. Zhou, Structural design for anodes of lithium-ion batteries: emerging horizons from materials to electrodes, Mater. Horiz. 2 (2015) 553–566, doi: 10.1039/C5MH00136F.
[33] M.A. Virji, A.B. Stefaniak, Comprehensive Materials Processing, Elsevier, (2014)doi: 10.1016/B978-0-08-096532-1.00811-6.
[34] Y. Tang, Y. Zhang, J. Deng, J. Wei, H. Le Tam, B.K. Chandran, Z. Dong, Z. Chen, X. Chen, Mechanical force-driven growth of elongated bending TiO2 -based nanotubular materials for ultrafast rechargeable lithium ion batteries, Adv. Mater. 26 (2014) 6111–6118, doi: 10.1002/adma.201402000.
[35] S. Liu, H. Jia, L. Han, J. Wang, P. Gao, D. Xu, J. Yang, S. Che, Nanosheet-constructed porous TiO2-B for advanced lithium ion batteries, Adv. Mater. 24 (2012) 3201–3204, doi: 10.1002/adma.201201036.
[36] T.-J. Kim, C. Kim, D. Son, M. Choi, B. Park, Novel SnS2-nanosheet anodes for lithium-ion batteries, J. Power Sources 167 (2007) 529–535, doi: 10.1016/j.jpowsour.2007.02.040.
[37] C. Yu, X. Li, T. Ma, J. Rong, R. Zhang, J. Shaffer, Y. An, Q. Liu, B. Wei, H. Jiang, Silicon thin films as anodes for high-performance lithium-ion batteries with effective stress relaxation, Adv. Energy Mater. 2 (2012) 68–73, doi: 10.1002/aenm.201100634.
[38] M. Ge, J. Rong, X. Fang, C. Zhou, Porous doped silicon nanowires for lithium ion battery anode with long cycle life, Nano Lett. 12 (2012) 2318–2323, doi: 10.1021/nl300206e.
[39] M.-H. Park, K. Kim, J. Kim, J. Cho, Flexible dimensional control of high-capacity Li-Ion-battery anodes: from 0D hollow to 3D porous germanium nanoparticle assemblies, Adv. Mater. 22 (2010) 415–418, doi: 10.1002/adma.200901846.
[40] G. Hasegawa, K. Kanamori, T. Kiyomura, H. Kurata, K. Nakanishi, T. Abe, Hierarchically porous Li 4 Ti 5 O 12 anode materials for Li- and Na-ion batteries: effects of nanoarchitectural design and temperature dependence of the rate capability, Adv. Energy Mater. 5 (2015) n/a–n/a10, doi: 10.1002/aenm.201400730.
[41] Y. Li, B. Tan, Y. Wu, Mesoporous Co 3 O 4 nanowire arrays for lithium ion batteries with high capacity and rate capability, Nano Lett. 8 (2008) 265–270, doi: 10.1021/nl0725906.
[42] S. Lee, J. Ha, J. Choi, T. Song, J.W. Lee, U. Paik, 3D cross-linked nanoweb architecture of binder-free TiO(2) electrodes for lithium ion batteries, ACS Appl. Mater. Interfaces 5 (2013) 11525–11529, doi: 10.1021/am404082h.
[43] B. Liu, X. Wang, B. Liu, Q. Wang, D. Tan, W. Song, X. Hou, D. Chen, G. Shen, Advanced rechargeable lithium-ion batteries based on bendable ZnCo2O4-urchins-on-carbon-fibers electrodes, Nano Res. 6 (2013) 525–534, doi: 10.1007/s12274-013-0329-3.
[44] B. Koo, H. Xiong, M.D. Slater, V.B. Prakapenka, M. Balasubramanian, P. Podsiadlo, C.S. Johnson, T. Rajh, E.V. Shevchenko, Hollow iron oxide nanoparticles for application in lithium ion batteries, Nano Lett. 12 (2012) 2429–2435, doi: 10.1021/nl3004286.
[45] Y. Zhong, L. Su, M. Yang, J. Wei, Z. Zhou, Rambutan-like FeCO3 hollow microspheres: facile preparation and superior lithium storage performances, ACS Appl. Mater. Interfaces 5 (2013) 11212–11217, doi: 10.1021/am403453r.

[46] S. Xu, C.M. Hessel, H. Ren, R. Yu, Q. Jin, M. Yang, H. Zhao, D. Wang, α-Fe 2 O 3 multi-shelled hollow microspheres for lithium ion battery anodes with superior capacity and charge retention, Energy Environ. Sci. 7 (2014) 632–637, doi: 10.1039/C3EE43319F.

[47] S.H. Choi, Y.C. Kang, Yolk-shell, hollow, and single-crystalline ZnCo 2 O 4 powders: preparation using a simple one-pot process and application in lithium-ion batteries, ChemSusChem. 6 (2013) 2111–2116, doi: 10.1002/cssc.201300300.

[48] N. Liu, Z. Lu, J. Zhao, M.T. McDowell, H.-W. Lee, W. Zhao, Y. Cui, A pomegranate-inspired nanoscale design for large-volume-change lithium battery anodes, Nat. Nanotechnol. 9 (2014) 187–192, doi: 10.1038/nnano.2014.6.

[49] Y. Zhong, M. Yang, X. Zhou, Y. Luo, J. Wei, Z. Zhou, Orderly packed anodes for high-power lithium-ion batteries with super-long cycle life: rational design of MnCO 3 /large-area graphene composites, Adv. Mater. 27 (2015) 806–812, doi: 10.1002/adma.201404611.

[50] M. Endo, C. Kim, K. Nishimura, T. Fujino, K. Miyashita, Recent development of carbon materials for Li ion batteries, Carbon N.Y. 38 (2000) 183–197, doi: 10.1016/S0008-6223(99)00141-4.

[51] S. Flandrois, B. Simon, Carbon materials for lithium-ion rechargeable batteries, Carbon N.Y. 37 (1999) 165–180, doi: 10.1016/S0008-6223(98)00290-5.

[52] M. Inagaki, H. Konno, O. Tanaike, Carbon materials for electrochemical capacitors, J. Power Sources 195 (2010) 7880–7903, doi: 10.1016/j.jpowsour.2010.06.036.

[53] H. Azuma, H. Imoto, S. Yamada, K. Sekai, Advanced carbon anode materials for lithium ion cells, J. Power Sources 81-82 (1999) 1–7, doi: 10.1016/S0378-7753(99)00122-6.

[54] Y.P. Wu, E. Rahm, R. Holze, Carbon anode materials for lithium ion batteries, J. Power Sources 114 (2003) 228–236, doi: 10.1016/S0378-7753(02)00596-7.

[55] D. Janas, K.K. Koziol, A review of production methods of carbon nanotube and graphene thin films for electrothermal applications, Nanoscale 6 (2014) 3037, doi: 10.1039/c3nr05636h.

[56] E. Frackowiak, S. Gautier, H. Gaucher, S. Bonnamy, F. Beguin, Electrochemical storage of lithium in multiwalled carbon nanotubes, Carbon N.Y. 37 (1999) 61–69, doi: 10.1016/S0008-6223(98)00187-0.

[57] D. Deng, J.Y. Lee, One-step synthesis of polycrystalline carbon nanofibers with periodic dome-shaped interiors and their reversible lithium-ion storage properties, Chem. Mater. 19 (2007) 4198–4204, doi: 10.1021/cm0707206.

[58] Y. Chen, X. Li, K. Park, J. Song, J. Hong, L. Zhou, Y.-W. Mai, H. Huang, J.B. Goodenough, Hollow carbon-nanotube/carbon-nanofiber hybrid anodes for Li-ion batteries, J. Am. Chem. Soc. 135 (2013) 16280–16283, doi: 10.1021/ja408421n.

[59] R. Raccichini, A. Varzi, S. Passerini, B. Scrosati, The role of graphene for electrochemical energy storage, Nat. Mater. 14 (2015) 271–279, doi: 10.1038/nmat4170.

[60] O.A.V. C., Á. Caballero, J. Morales, Can the performance of graphene nanosheets for lithium storage in Li-ion batteries be predicted?, Nanoscale 4 (2012) 2083–2092, doi: 10.1039/c2nr11936f.

[61] A.P. Cohn, L. Oakes, R. Carter, S. Chatterjee, A.S. Westover, K. Share, C.L. Pint, Assessing the improved performance of freestanding, flexible graphene and carbon nanotube hybrid foams for lithium ion battery anodes, Nanoscale 6 (2014) 4669–4675, doi: 10.1039/c4nr00390j.

[62] S.Y. Hong, Y. Kim, Y. Park, A. Choi, N.-S. Choi, K.T. Lee, Charge carriers in rechargeable batteries: Na ions vs. Li ions, Energy Environ. Sci. 6 (2013) 2067, doi: 10.1039/c3ee40811f.

[63] K. Tang, L. Fu, R.J. White, L. Yu, M.M. Titirici, M. Antonietti, J. Maier, Hollow carbon nanospheres with superior rate capability for sodium-based batteries, Adv. Energy Mater. 2 (2012) 873–877, doi: 10.1002/aenm.201100691.

[64] Y. Cao, L. Xiao, M.L. Sushko, W. Wang, B. Schwenzer, J. Xiao, Z. Nie, L.V. Saraf, Z. Yang, J. Liu, Sodium ion insertion in hollow carbon nanowires for battery applications, Nano Lett. 12 (2012) 3783–3787, doi: 10.1021/nl3016957.

[65] Z. Wang, L. Qie, L. Yuan, W. Zhang, X. Hu, Y. Huang, Functionalized N-doped interconnected carbon nanofibers as an anode material for sodium-ion storage with excellent performance, Carbon N.Y. 55 (2013) 328–334, doi: 10.1016/j.carbon.2012.12.072.

[66] Y.-X. Wang, S.-L. Chou, H.-K. Liu, S.-X. Dou, Reduced graphene oxide with superior cycling stability and rate capability for sodium storage, Carbon N.Y. 57 (2013) 202–208, doi: 10.1016/j.carbon.2013.01.064.

[67] L. Fu, K. Tang, K. Song, P. a van Aken, Y. Yu, J. Maier, Nitrogen doped porous carbon fibres as anode materials for sodium ion batteries with excellent rate performance, Nanoscale 6 (2014) 1384–1389, doi: 10.1039/C3NR05374A.
[68] T. Chen, Y. Liu, L. Pan, T. Lu, Y. Yao, Z. Sun, D.H.C. Chua, Q. Chen, Electrospun carbon nanofibers as anode materials for sodium ion batteries with excellent cycle performance, J. Mater. Chem. A. 2 (2014) 4117–4121, doi: 10.1039/c3ta14806h.
[69] S. Zhang, F. Yao, L. Yang, F. Zhang, S. Xu, Sulfur-doped mesoporous carbon from surfactant-intercalated layered double hydroxide precursor as high-performance anode nanomaterials for both Li-ion and Na-ion batteries, Carbon N.Y. 93 (2015) 143–150, doi: 10.1016/j.carbon.2015.04.091.
[70] F.-H. Du, K.-X. Wang, J.-S. Chen, Strategies to succeed in improving the lithium-ion storage properties of silicon nanomaterials, J. Mater. Chem. A. 4 (2016) 32–50, doi: 10.1039/C5TA06962A.
[71] M.A. Rahman, G. Song, A.I. Bhatt, Y.C. Wong, C. Wen, Nanostructured silicon anodes for high-performance lithium-ion batteries, Adv. Funct. Mater. 26 (2016) 647–678, doi: 10.1002/adfm.201502959.
[72] Z. Wen, J. Stark, R. Saha, J. Parker, P.A. Kohl, Fabrication and electrochemical performance of interconnected silicon nanowires synthesized from AlCu catalyst, J. Phys. Chem. C. 117 (2013) 8604–8610, doi: 10.1021/jp3096709.
[73] Y. Xiao, D. Hao, H. Chen, Z. Gong, Y. Yang, Economical synthesis and promotion of the electrochemical performance of silicon nanowires as anode material in Li-ion batteries, ACS Appl. Mater. Interfaces 5 (2013) 1681–1687, doi: 10.1021/am302731y.
[74] P.P. Prosini, C. Cento, F. Alessandrini, P. Gislon, A. Mancini, A. Rufoloni, F. Rondino, A. Santoni, Electrochemical characterization of silicon nanowires as an anode for lithium batteries, Solid State Ionics 260 (2014) 49–54, doi: 10.1016/j.ssi.2014.03.004.
[75] L.D. Toan, E. Moyen, M.R. Zamfir, J. Joe, Y.W. Kim, D. Pribat, Si nanowires grown by Al-catalyzed plasma-enhanced chemical vapor deposition: synthesis conditions, electrical properties and application to lithium battery anodes, Mater. Res. Express. 3 (2016) 15003, doi: 10.1088/2053-1591/3/1/015003.
[76] Y. Zhou, X. Jiang, L. Chen, J. Yue, H. Xu, J. Yang, Y. Qian, Novel mesoporous silicon nanorod as an anode material for lithium ion batteries, Electrochim. Acta. 127 (2014) 252–258, doi: 10.1016/j.electacta.2014.01.158.
[77] S. Soleimani-Amiri, S.A. Safiabadi Tali, S. Azimi, Z. Sanaee, S. Mohajerzadeh, Highly featured amorphous silicon nanorod arrays for high-performance lithium-ion batteries, Appl. Phys. Lett. 105 (2014) 193903, doi: 10.1063/1.4902068.
[78] X. Yu, F. Xue, H. Huang, C. Liu, J. Yu, Y. Sun, X. Dong, G. Cao, Y. Jung, Synthesis and electrochemical properties of silicon nanosheets by DC arc discharge for lithium-ion batteries, Nanoscale 6 (2014) 6860–6865, doi: 10.1039/c3nr06418b.
[79] J.-K. Yoo, J. Kim, Y.S. Jung, K. Kang, Scalable fabrication of silicon nanotubes and their application to energy storage, Adv. Mater. 24 (2012) 5452–5456, doi: 10.1002/adma.201201601.
[80] H. Wu, G. Chan, J.W. Choi, I. Ryu, Y. Yao, M.T. McDowell, S.W. Lee, A. Jackson, Y. Yang, L. Hu, Y. Cui, Stable cycling of double-walled silicon nanotube battery anodes through solid-electrolyte interphase control, Nat. Nanotechnol. 7 (2012) 310–315, doi: 10.1038/nnano.2012.35.
[81] M. Ge, Y. Lu, P. Ercius, J. Rong, X. Fang, M. Mecklenburg, C. Zhou, Large-scale fabrication, 3D tomography, and lithium-ion battery application of porous silicon, Nano Lett. 14 (2014) 261–268, doi: 10.1021/nl403923s.
[82] M. Zhang, T. Zhang, Y. Ma, Y. Chen, Latest development of nanostructured Si/C materials for lithium anode studies and applications, Energy Storage Mater. 4 (2016) 1–14, doi: 10.1016/j.ensm.2016.02.001.
[83] L. Su, J. Xie, Y. Xu, L. Wang, Y. Wang, M. Ren, Preparation and lithium storage performance of yolk-shell Si@void@C nanocomposites, Phys. Chem. Chem. Phys. 17 (2015) 17562–17565, doi: 10.1039/c5cp01954k.
[84] D. Nan, Z.-H. Huang, R. Lv, Y. Lin, L. Yang, X. Yu, L. Ye, W. Shen, H. Sun, F. Kang, Silicon-encapsulated hollow carbon nanofiber networks as binder-free anodes for lithium ion battery, J. Nanomater. 2014 (2014) 1–10, doi: 10.1155/2014/139639.
[85] W.-J. Yu, C. Liu, P.-X. Hou, L. Zhang, X.-Y. Shan, F. Li, H.-M. Cheng, Lithiation of silicon nanoparticles confined in carbon nanotubes, ACS Nano 9 (2015) 5063–5071, doi: 10.1021/acsnano.5b00157.

[86] Y.-S. Na, H. Yoo, T.-H. Kim, J. Choi, W.I. Lee, S. Choi, D.-W. Park, Electrochemical performance of Si-multiwall carbon nanotube nanocomposite anode synthesized by thermal plasma, Thin Solid Films 587 (2015) 14–19, doi: 10.1016/j.tsf.2014.12.038.

[87] Y.-S. Na, S. Choi, D.-W. Park, Carbon nanotube surface modification with the attachment of Si nanoparticles in a thermal plasma jet, Phys. Status Solidi 211 (2014) 2749–2755, doi: 10.1002/pssa.201431377.

[88] Z.-L. Xu, B. Zhang, S. Abouali, M. Akbari Garakani, J. Huang, J.-Q. Huang, E. Kamali Heidari, J.-K. Kim, Nanocavity-engineered Si/multi-functional carbon nanofiber composite anodes with exceptional high-rate capacities, J. Mater. Chem. A. 2 (2014) 17944–17951, doi: 10.1039/C4TA04257C.

[89] B. Xu, H. Wu, C.X. (Cynthia) Lin, B. Wang, Z. Zhang, X.S. Zhao, Stabilization of silicon nanoparticles in graphene aerogel framework for lithium ion storage, RSC Adv. 5 (2015) 30624–30630, doi: 10.1039/C5RA00566C.

[90] S. Yang, G. Li, Q. Zhu, Q. Pan, Covalent binding of Si nanoparticles to graphene sheets and its influence on lithium storage properties of Si negative electrode, J. Mater. Chem. 22 (2012) 3420, doi: 10.1039/c2jm15232k.

[91] Y.-S. Ye, X.-L. Xie, J. Rick, F.-C. Chang, B.-J. Hwang, Improved anode materials for lithium-ion batteries comprise non-covalently bonded graphene and silicon nanoparticles, J. Power Sources 247 (2014) 991–998, doi: 10.1016/j.jpowsour.2013.08.048.

[92] H. Tang, J. Zhang, Y.J. Zhang, Q.Q. Xiong, Y.Y. Tong, Y. Li, X.L. Wang, C.D. Gu, J.P. Tu, Porous reduced graphene oxide sheet wrapped silicon composite fabricated by steam etching for lithium-ion battery application, J. Power Sources 286 (2015) 431–437, doi: 10.1016/j.jpowsour.2015.03.185.

[93] F. Maroni, R. Raccichini, A. Birrozzi, G. Carbonari, R. Tossici, F. Croce, R. Marassi, F. Nobili, Graphene/silicon nanocomposite anode with enhanced electrochemical stability for lithium-ion battery applications, J. Power Sources 269 (2014) 873–882, doi: 10.1016/j.jpowsour.2014.07.064.

[94] J. Zhang, L. Zhang, P. Xue, L. Zhang, X. Zhang, W. Hao, J. Tian, M. Shen, H. Zheng, Silicon-nanoparticles isolated by in situ grown polycrystalline graphene hollow spheres for enhanced lithium-ion storage, J. Mater. Chem. A 3 (2015) 7810–7821, doi: 10.1039/C5TA00457H.

[95] V. Chabot, K. Feng, H.W. Park, F.M. Hassan, A.R. Elsayed, A. Yu, X. Xiao, Z. Chen, Graphene wrapped silicon nanocomposites for enhanced electrochemical performance in lithium ion batteries, Electrochim. Acta 130 (2014) 127–134, doi: 10.1016/j.electacta.2014.02.135.

[96] S. Yoo, J.-I. Lee, S. Ko, S. Park, Highly dispersive and electrically conductive silver-coated Si anodes synthesized via a simple chemical reduction process, Nano Energy 2 (2013) 1271–1278, doi: 10.1016/j.nanoen.2013.06.006.

[97] M. Feng, J. Tian, H. Xie, Y. Kang, Z. Shan, Nano-silicon/polyaniline composites with an enhanced reversible capacity as anode materials for lithium ion batteries, J. Solid State Electrochem. 19 (2015) 1773–1782, doi: 10.1007/s10008-015-2807-x.

[98] D. Kundu, E. Talaie, V. Duffort, L.F. Nazar, The emerging chemistry of sodium ion batteries for electrochemical energy storage, Angew. Chemie Int. Ed. 54 (2015) 3431–3448, doi: 10.1002/anie.201410376.

[99] Y. Yang, M.T. McDowell, A. Jackson, J.J. Cha, S.S. Hong, Y. Cui, New nanostructured Li2S/silicon rechargeable battery with high specific energy, Nano Lett. 10 (2010) 1486–1491, doi: 10.1021/nl100504q.

[100] T.D. Bogart, D. Oka, X. Lu, M. Gu, C. Wang, B.A. Korgel, Lithium ion battery performance of silicon nanowires with carbon skin, ACS Nano 8 (2014) 915–922, doi: 10.1021/nn405710w.

[101] T.T. Tran, M.N. Obrovac, Alloy negative electrodes for high energy density metal-ion cells, J. Electrochem. Soc. 158 (2011) A1411, doi: 10.1149/2.083112jes.

[102] M.N. Obrovac, V.L. Chevrier, Alloy negative electrodes for Li-ion batteries, Chem. Rev. 114 (2014) 11444–11502, doi: 10.1021/cr500207g.

[103] S. Iwamura, H. Nishihara, Y. Ono, H. Morito, H. Yamane, H. Nara, T. Osaka, T. Kyotani, Li-rich Li-Si alloy as a lithium-containing negative electrode material towards high energy lithium-ion batteries, Sci. Rep. 5 (2015) 8085, doi: 10.1038/srep08085.

[104] D.H. Youn, A. Heller, C.B. Mullins, Simple synthesis of nanostructured Sn/nitrogen-doped carbon composite using nitrilotriacetic acid as lithium ion battery anode, Chem. Mater. 28 (2016) 1343–1347, doi: 10.1021/acs.chemmater.5b04282.

[105] Y. Yan, L. Ben, Y. Zhan, X. Huang, Nano-Sn embedded in expanded graphite as anode for lithium ion batteries with improved low temperature electrochemical performance, Electrochim. Acta 187 (2016) 186–192, doi: 10.1016/j.electacta.2015.11.015.

[106] L. Deng, Y. Cui, J. Chen, J. Wu, A.P. Baker, Z. Li, X. Zhang, A core-shell Si@NiSi2/Ni/C nanocomposite as an anode material for lithium-ion batteries, Electrochim. Acta 192 (2016) 303–309, doi: 10.1016/j.electacta.2016.01.197.

[107] Z. Yi, X. Tian, Q. Han, Y. Cheng, J. Lian, Y. Wu, L. Wang, One-step synthesis of Ni3Sn2@reduced graphene oxide composite with enhanced electrochemical lithium storage properties, Electrochim. Acta 192 (2016) 188–195, doi: 10.1016/j.electacta.2016.01.204.

[108] X.-L. Wang, W.-Q. Han, J. Chen, J. Graetz, Single-crystal intermetallic M Sn (M = Fe, Cu, Co, Ni) nanospheres as negative electrodes for lithium-ion batteries, ACS Appl. Mater. Interfaces 2 (2010) 1548–1551, doi: 10.1021/am100218v.

[109] Z. Dong, R. Zhang, D. Ji, N.A. Chernova, K. Karki, S. Sallis, L. Piper, M.S. Whittingham, The anode challenge for lithium-ion batteries: a mechanochemically synthesized Sn-Fe-C composite anode surpasses graphitic carbon, Adv. Sci. (2016) n/a–n/a10, doi: 10.1002/advs.201500229.

[110] Q. Han, Z. Yi, Y. Cheng, Y. Wu, L. Wang, Simple preparation of Cu 6 Sn 5 /Sn composites as anode materials for lithium-ion batteries, RSC Adv. 6 (2016) 15279–15285, doi: 10.1039/C5RA23788B.

[111] Q. Ru, X. Chen, J. Li, L. Guo, S. Hu, The lamella SnSbCux/MCMB/carbon composite as high stability and durable anodes for lithium ion battery, Electrochim. Acta 193 (2016) 180–190, doi: 10.1016/j.electacta.2016.02.029.

[112] P. Nithyadharseni, M.V. Reddy, B. Nalini, B.V.R. Chowdari, Electrochemical investigation of SnSb nano particles for lithium-ion batteries, Mater. Lett. 150 (2015) 24–27, doi: 10.1016/j.matlet.2015.02.124.

[113] X. Chen, Q. Ru, D. Zhao, Y. Mo, S. Hu, Flake structured SnSbCo/MCMB/C composite as high performance anodes for lithium ion battery, J. Alloys Compd. 646 (2015) 794–802, doi: 10.1016/j.jallcom.2015.06.078.

[114] P. Nithyadharseni, M.V. Reddy, B. Nalini, M. Kalpana, B.V.R. Chowdari, Sn-based intermetallic alloy anode materials for the application of lithium ion batteries, Electrochim. Acta 161 (2015) 261–268, doi: 10.1016/j.electacta.2015.02.057.

[115] X. Wang, Z. Wen, B. Lin, J. Lin, X. Wu, X. Xu, Preparation and electrochemical characterization of tin/graphite/silver composite as anode materials for lithium-ion batteries, J. Power Sources 184 (2008) 508–512, doi: 10.1016/j.jpowsour.2008.03.025.

[116] D. Larcher, A.S. Prakash, J. Saint, M. Morcrette, J.-M. Tarascon, Electrochemical reactivity of Mg 2 Sn phases with metallic lithium, Chem. Mater. 16 (2004) 5502–5511, doi: 10.1021/cm040132h.

[117] I.T. Kim, Cu_2Sb nano-alloys dispersed in TiC-C hybrid matrix for sodium-ion batteries, Sci. Adv. Mater. 8 (2016) 80–83, doi: 10.1166/sam.2016.2605.

[118] W. Guo, F. Li, X. Duan, J. Ma, Synthesis of Cd–Sn–SnO2@C heterocomposite anode with superior electrochemical performance, Mater. Lett. 166 (2016) 210–214, doi: 10.1016/j.matlet.2015.12.081.

[119] J. Liu, K. Song, C. Zhu, C.-C. Chen, P.A. van Aken, J. Maier, Y. Yu, Ge/C nanowires as high-capacity and long-life anode materials for Li-ion batteries, ACS Nano. 8 (2014) 7051–7059, doi: 10.1021/nn501945f.

[120] C.K. Chan, X.F. Zhang, Y. Cui, High capacity Li ion battery anodes using ge nanowires, Nano Lett. 8 (2008) 307–309, doi: 10.1021/nl0727157.

[121] K.H. Seng, M. Park, Z.P. Guo, H.K. Liu, J. Cho, Catalytic role of Ge in highly reversible GeO2/Ge/C nanocomposite anode material for lithium batteries, Nano Lett. 13 (2013) 1230–1236, doi: 10.1021/nl304716e.

[122] F.-W. Yuan, H.-J. Yang, H.-Y. Tuan, Alkanethiol-passivated ge nanowires as high- performance anode materials for lithium-ion batteries: the role of chemical surface functionalization, ACS Nano 6 (2012) 9932–9942, doi: 10.1021/nn303519g.

[123] X. Han, H. Chen, X. Li, S. Lai, Y. Xu, C. Li, S. Chen, Y. Yang, NiSix/a-Si nanowires with interfacial a-Ge as anodes for high-rate lithium-ion batteries, ACS Appl. Mater. Interfaces 8 (2016) 673–679, doi: 10.1021/acsami.5b09783.

[124] W. Luo, F. Shen, C. Bommier, H. Zhu, X. Ji, L. Hu, Na-ion battery anodes: materials and electrochemistry, Acc. Chem. Res. 49 (2016) 231–240, doi: 10.1021/acs.accounts.5b00482.

[125] Z. Li, J. Ding, D. Mitlin, Tin and tin compounds for sodium ion battery anodes: phase transformations and performance, Acc. Chem. Res. 48 (2015) 1657–1665, doi: 10.1021/acs.accounts.5b00114.
[126] Y. Liu, Y. Xu, Y. Zhu, J.N. Culver, C.A. Lundgren, K. Xu, C. Wang, Tin-coated viral nanoforests as sodium-ion battery anodes, ACS Nano 7 (2013) 3627–3634, doi: 10.1021/nn400601y.
[127] Y. Zhu, X. Han, Y. Xu, Y. Liu, S. Zheng, K. Xu, L. Hu, C. Wang, Electrospun Sb/C fibers for a stable and fast sodium-ion battery anode, ACS Nano 7 (2013) 6378–6386, doi: 10.1021/nn4025674.
[128] L. Wu, H. Lu, L. Xiao, X. Ai, H. Yang, Y. Cao, Electrochemical properties and morphological evolution of pitaya-like Sb@C microspheres as high-performance anode for sodium ion batteries, J. Mater. Chem. A 3 (2015) 5708–5713, doi: 10.1039/C4TA06086E.
[129] M. He, K. Kravchyk, M. Walter, M.V. Kovalenko, Monodisperse antimony nanocrystals for high-rate Li-ion and Na-ion battery anodes: nano versus bulk, Nano Lett. 14 (2014) 1255–1262, doi: 10.1021/nl404165c.
[130] L. Wu, X. Hu, J. Qian, F. Pei, F. Wu, R. Mao, X. Ai, H. Yang, Y. Cao, Sb–C nanofibers with long cycle life as an anode material for high-performance sodium-ion batteries, Energy Environ. Sci. 7 (2014) 323–328, doi: 10.1039/C3EE42944J.
[131] Y. Zhao, X. Li, B. Yan, D. Xiong, D. Li, S. Lawes, X. Sun, Recent developments and understanding of novel mixed transition-metal oxides as anodes in lithium ion batteries, Adv. Energy Mater. (2016) n/a–n/a10, doi: 10.1002/aenm.201502175.
[132] Y. Ren, Z. Liu, F. Pourpoint, A.R. Armstrong, C.P. Grey, P.G. Bruce, Nanoparticulate TiO2(B): an anode for lithium-ion batteries, Angew. Chem. Int. Ed. Engl. 51 (2012) 2164–2167, doi: 10.1002/anie.201108300.
[133] L. Shen, E. Uchaker, X. Zhang, G. Cao, Hydrogenated Li(4)Ti(5)O(12) nanowire arrays for high rate lithium ion batteries, Adv. Mater. 24 (2012) 6502–6506, doi: 10.1002/adma.201203151.
[134] J. Wang, Y. Zhou, B. Xiong, Y. Zhao, X. Huang, Z. Shao, Fast lithium-ion insertion of TiO2 nanotube and graphene composites, Electrochim. Acta 88 (2013) 847–857, doi: 10.1016/j.electacta.2012.10.010.
[135] C. Zhao, L. Liu, Q. Zhang, J. Rogers, H. Zhao, Y. Li, Synthesis of carbon-TiO2 nanocomposites with enhanced reversible capacity and cyclic performance as anodes for lithium-ion batteries, Electrochim. Acta. 155 (2015) 288–296, doi: 10.1016/j.electacta.2014.12.167.
[136] Y. Tang, L. Hong, Q. Wu, J. Li, G. Hou, H. Cao, L. Wu, G. Zheng, TiO2(B) nanowire arrays on Ti foil substrate as three-dimensional anode for lithium-ion batteries, Electrochim. Acta 195 (2016) 27–33, doi: 10.1016/j.electacta.2016.01.235.
[137] Y. Liu, J.-L. Lan, Q. Cai, Y.-H. Yu, Y.-H. Lin, X.-P. Yang, Encapsulating tin dioxide@porous carbon in carbon tubes: a fiber-in-tube hierarchical nanostructure for superior capacity and long-life lithium storage, Part. Part. Syst. Charact. 32 (2015) 952–961, doi: 10.1002/ppsc.201500073.
[138] H. Liu, Z. Bi, X.-G. Sun, R.R. Unocic, M.P. Paranthaman, S. Dai, G.M. Brown, Mesoporous TiO2-B microspheres with superior rate performance for lithium ion batteries, Adv. Mater. 23 (2011) 3450–3454, doi: 10.1002/adma.201100599.
[139] H. Xiong, M.D. Slater, M. Balasubramanian, C.S. Johnson, T. Rajh, Amorphous TiO 2 nanotube anode for rechargeable sodium ion batteries, J. Phys. Chem. Lett. 2 (2011) 2560–2565, doi: 10.1021/jz2012066.
[140] D. Prutsch, M. Wilkening, I. Hanzu, Long-cycle-life Na-ion anodes based on amorphous titania nanotubes-interfaces and diffusion, ACS Appl. Mater. Interfaces 7 (2015) 25757–25769, doi: 10.1021/acsami.5b07508.
[141] Y. Lai, W. Liu, J. Li, K. Zhang, F. Qin, M. Wang, J. Fang, High performance sodium storage of Fe-doped mesoporous anatase TiO2/amorphous carbon composite, J. Alloys Compd. 666 (2016) 254–261, doi: 10.1016/j.jallcom.2016.01.101.
[142] J.-Y. Hwang, S.-T. Myung, J.-H. Lee, A. Abouimrane, I. Belharouak, Y.-K. Sun, Ultrafast sodium storage in anatase TiO2 nanoparticles embedded on carbon nanotubes, Nano Energy 16 (2015) 218–226, doi: 10.1016/j.nanoen.2015.06.017.
[143] J.-M. Feng, L. Dong, Y. Han, X.-F. Li, D.-J. Li, Facile synthesis of graphene-titanium dioxide nanocomposites as anode materials for Na-ion batteries, Int. J. Hydrogen Energy 41 (2016) 355–360, doi: 10.1016/j.ijhydene.2015.10.137.
[144] H.A. Cha, H.M. Jeong, J.K. Kang, Nitrogen-doped open pore channeled graphene facilitating electrochemical performance of TiO2 nanoparticles as an anode material for sodium ion batteries, J. Mater. Chem. A 2 (2014) 5182, doi: 10.1039/c4ta00041b.

[145] Y. Sun, L. Zhao, H. Pan, X. Lu, L. Gu, Y.-S. Hu, H. Li, M. Armand, Y. Ikuhara, L. Chen, X. Huang, Direct atomic-scale confirmation of three-phase storage mechanism in Li_4Ti O_{12} anodes for room-temperature sodium-ion batteries, Nat. Commun. 4 (2013) 1870, doi: 10.1038/ncomms2878.
[146] Q. Zhou, L. Liu, J. Tan, Z. Yan, Z. Huang, X. Wang, Synthesis of lithium titanate nanorods as anode materials for lithium and sodium ion batteries with superior electrochemical performance, J. Power Sources 283 (2015) 243–250, doi: 10.1016/j.jpowsour.2015.02.061.
[147] F. Zhao, P. Xue, H. Ge, L. Li, B. Wang, Na-doped Li 4 Ti 5 O 12 as an anode material for sodium-ion battery with superior rate and cycling performance, J. Electrochem. Soc. 163 (2016) A690–A695, doi: 10.1149/2.0781605jes.
[148] K.-T. Kim, C.-Y. Yu, C.S. Yoon, S.-J. Kim, Y.-K. Sun, S.-T. Myung, Carbon-coated Li4Ti5O12 nanowires showing high rate capability as an anode material for rechargeable sodium batteries, Nano Energy 12 (2015) 725–734, doi: 10.1016/j.nanoen.2015.01.034.
[149] X. Wang, Y. Li, Y. Gao, Z. Wang, L. Chen, Additive-free sodium titanate nanotube array as advanced electrode for sodium ion batteries, Nano Energy 13 (2015) 687–692, doi: 10.1016/j.nanoen.2015.03.029.
[150] J.-Y. Liao, A. Manthiram, High-performance Na2Ti2O5 nanowire arrays coated with VS2 nanosheets for sodium-ion storage, Nano Energy 18 (2015) 20–27, doi: 10.1016/j.nanoen.2015.09.014.
[151] S. Dong, L. Shen, H. Li, P. Nie, Y. Zhu, Q. Sheng, X. Zhang, Pseudocapacitive behaviours of Na 2 Ti 3 O 7 @CNT coaxial nanocables for high-performance sodium-ion capacitors, J. Mater. Chem. A 3 (2015) 21277–21283, doi: 10.1039/C5TA05714K.
[152] H. Li, H. Fei, X. Liu, J. Yang, M. Wei, In situ synthesis of Na2Ti7O15 nanotubes on a Ti net substrate as a high performance anode for Na-ion batteries, Chem. Commun. (Camb). 51 (2015) 9298–9300, doi: 10.1039/c5cc02612a.
[153] L. Li, L. Zhang, F. Chai, T. Wang, Z. Li, H. Xie, C. Wang, Z. Su, SnO 2 @N-doped carbon hollow nanoclusters for advanced lithium-ion battery anodes, Eur. J. Inorg. Chem. 2016 (2016) 812–817, doi: 10.1002/ejic.201501307.
[154] W.-D. Yang, H.-Y. Fang, R. Liu, Performance of SnO2/carbon nanotube composite electrode materials synthesised by the Pechini method, Micro Nano Lett. 11 (2016) 54–56, doi: 10.1049/mnl.2015.0296.
[155] C. Tan, S. Zhao, G. Yang, S. Hu, X. Qin, Facile and surfactant-free synthesis of SnO2-graphene hybrids as high performance anode for lithium-ion batteries, Ionics (Kiel). 21 (2014) 987–994, doi: 10.1007/s11581-014-1258-1.
[156] D. Zhou, W.-L. Song, L.-Z. Fan, Hollow core-shell SnO2/C fibers as highly stable anodes for lithium-ion batteries, ACS Appl. Mater. Interfaces 7 (2015) 21472–21478, doi: 10.1021/acsami.5b06512.
[157] P.-P. Dong, Y.-H. Sun, X. Lang, J.-M. Nan, H.-Y. Chen, Facile synthesis cuboid SnO2 nanoparticles and electrochemical properties as anode of lithium-ion battery, Russ. J. Electrochem. 51 (2015) 712–718, doi: 10.1134/S1023193515080042.
[158] S. Park, Novel SnO_2 nanowires as sodium ion battery anodes, in: 229th ECS Meet (May 29–June 2, 2016), Ecs, 2016.
[159] H. Bian, J. Zhang, M.-F. Yuen, W. Kang, Y. Zhan, D.Y.W. Yu, Z. Xu, Y.Y. Li, Anodic nanoporous SnO2 grown on Cu foils as superior binder-free Na-ion battery anodes, J. Power Sources 307 (2016) 634–640, doi: 10.1016/j.jpowsour.2015.12.118.
[160] X. Zhang, T. Wang, C. Jiang, F. Zhang, W. Li, Y. Tang, Manganese dioxide/carbon nanotubes composite with optimized microstructure via room temperature solution approach for high performance lithium-ion battery anodes, Electrochim. Acta 187 (2016) 465–472, doi: 10.1016/j.electacta.2015.11.084.
[161] S. Luo, Y. Yu, M. Li, H. Wu, F. Zhao, K. Jiang, J. Wang, F. Kang, S. Fan, Synergistic effect of manganese oxide nanoparticles and graphene nanosheets in composite anodes for lithium ion batteries, Mater. Res. Express. 2 (2015) 15503, doi: 10.1088/2053-1591/2/1/015503.
[162] Y. Zhang, Z. Luo, Q. Xiao, T. Sun, G. Lei, Z. Li, X. Li, Freestanding manganese dioxide nanosheet network grown on nickel/polyvinylidene fluoride coaxial fiber membrane as anode materials for high performance lithium ion batteries, J. Power Sources. 297 (2015) 442–448, doi: 10.1016/j.jpowsour.2015.08.012.
[163] S.-H. Park, W.-J. Lee, Hierarchically mesoporous carbon nanofiber/Mn3O4 coaxial nanocables as anodes in lithium ion batteries, J. Power Sources 281 (2015) 301–309, doi: 10.1016/j.jpowsour.2015.01.156.

[164] S.-K. Park, C.-Y. Seong, S. Yoo, Y. Piao, Porous Mn3O4 nanorod/reduced graphene oxide hybrid paper as a flexible and binder-free anode material for lithium ion battery, Energy 99 (2016) 266–273, doi: 10.1016/j.energy.2016.01.061.

[165] Z.-Y. Sui, C. Wang, K. Shu, Q.-S. Yang, Y. Ge, G.G. Wallace, B.-H. Han, Manganese dioxide-anchored three-dimensional nitrogen-doped graphene hybrid aerogels as excellent anode materials for lithium ion batteries, J. Mater. Chem. A 3 (2015) 10403–10412, doi: 10.1039/C5TA01508A.

[166] I.-H. Ko, S.-J. Kim, J. Lim, S.-H. Yu, J. Ahn, J.-K. Lee, Y.-E. Sung, Effect of PEDOT:PSS coating on manganese oxide nanowires for lithium ion battery anodes, Electrochim. Acta 187 (2016) 340–347, doi: 10.1016/j.electacta.2015.11.061.

[167] Y.-T. Weng, T.-Y. Huang, C.-H. Lim, P.-S. Shao, S. Hy, C.-Y. Kuo, J.-H. Cheng, B.-J. Hwang, J.-F. Lee, N.-L. Wu, An unexpected large capacity of ultrafine manganese oxide as a sodium-ion battery anode, Nanoscale 7 (2015) 20075–20081, doi: 10.1039/c5nr07100c.

[168] D. Larcher, C. Masquelier, D. Bonnin, Y. Chabre, V. Masson, J.-B. Leriche, J.-M. Tarascon, Effect of particle size on lithium intercalation into α-Fe[sub 2]O[sub 3], J. Electrochem. Soc. 150 (2003) A133, doi: 10.1149/1.1528941.

[169] X. Cui, Y. Zhu, F. Li, D. Liu, J. Chen, Y. Zhang, L.L. Zhang, J. Ji, Enhanced rate capability of a lithium ion battery anode based on liquid–solid-solution assembly of Fe 2 O 3 on crumpled graphene, RSC Adv. 6 (2016) 9007–9012, doi: 10.1039/C5RA22408J.

[170] Y. Huang, Z. Lin, M. Zheng, T. Wang, J. Yang, F. Yuan, X. Lu, L. Liu, D. Sun, Amorphous Fe2O3 nanoshells coated on carbonized bacterial cellulose nanofibers as a flexible anode for high- performance lithium ion batteries, J. Power Sources 307 (2016) 649–656, doi: 10.1016/j.jpowsour.2016.01.026.

[171] Y. Zuo, G. Wang, J. Peng, G. Li, Y. Ma, F. Yu, B. Dai, X. Guo, C.-P. Wong, Hybridization of graphene nanosheets and carbon-coated hollow Fe 3 O 4 nanoparticles as a high- performance anode material for lithium-ion batteries, J. Mater. Chem. A. 4 (2016) 2453–2460, doi: 10.1039/C5TA09742H.

[172] D. Cai, D. Li, L.-X. Ding, S. Wang, H. Wang, Interconnected α-Fe2O3 nanosheet arrays as high-performance anode materials for lithium-ion batteries, Electrochim. Acta 192 (2016) 407–413, doi: 10.1016/j.electacta.2016.02.010.

[173] M. Ren, M. Yang, W. Liu, M. Li, L. Su, C. Qiao, X. Wu, H. Ma, Ultra-small Fe3O4 nanocrystals decorated on 2D graphene nanosheets with excellent cycling stability as anode materials for lithium ion batteries, Electrochim. Acta 194 (2016) 219–227, doi: 10.1016/j.electacta.2016.02.091.

[174] C. Hu, J. Guo, H. Zhong, Z. Chen, J. Wen, The spherical α-Fe2O3 nanomaterials prepared by hydrolysis route and hydrothermal route for lithium-ion batteries, Mater. Res. Innov. 20 (2016) 76–80.

[175] F. Zheng, M. He, Y. Yang, Q. Chen, Nano electrochemical reactors of Fe 2 O 3 nanoparticles embedded in shells of nitrogen-doped hollow carbon spheres as high-performance anodes for lithium-ion batteries, Nanoscale 7 (2015) 3410–3417, doi: 10.1039/C4NR06321J.

[176] X. Cai, H. Lin, X. Zheng, X. Chen, P. Xia, X. Luo, X. Zhong, X. Li, W. Li, Facile synthesis of porous iron oxide rods coated with carbon as anode of high energy density lithium ion battery, Electrochim. Acta 191 (2016) 767–775, doi: 10.1016/j.electacta.2016.01.081.

[177] Y. Luo, L. Liu, W. Qiao, F. Liu, Y. Zhang, W. Tan, G. Qiu, Facile crystal-structure-controlled synthesis of iron oxides for adsorbents and anode materials of lithium batteries, Mater. Chem. Phys. 170 (2016) 239–245, doi: 10.1016/j.matchemphys.2015.12.044.

[178] M. Valvo, F. Lindgren, U. Lafont, F. Björefors, K. Edström, Towards more sustainable negative electrodes in Na-ion batteries via nanostructured iron oxide, J. Power Sources 245 (2014) 967–978, doi: 10.1016/j.jpowsour.2013.06.159.

[179] Y. Jiang, M. Hu, D. Zhang, T. Yuan, W. Sun, B. Xu, M. Yan, Transition metal oxides for high performance sodium ion battery anodes, Nano Energy 5 (2014) 60–66, doi: 10.1016/j.nanoen.2014.02.002.

[180] Z. Wu, Y. Zhong, J. Liu, J. Wu, X. Guo, B. Zhong, Z. Zhang, Subunits controlled synthesis of α- Fe 2 O 3 multi-shelled core–shell microspheres and their effects on lithium/sodium ion battery performances, J. Mater. Chem. A 3 (2015) 10092–10099, doi: 10.1039/C5TA01334H.

[181] B. Koo, S. Chattopadhyay, T. Shibata, V.B. Prakapenka, C.S. Johnson, T. Rajh, E.V. Shevchenko, Intercalation of sodium ions into hollow iron oxide nanoparticles, Chem. Mater. 25 (2013) 245–252, doi: 10.1021/cm303611z.

[182] L. Guo, Y. Ding, C. Qin, W. Li, J. Du, Z. Fu, W. Song, F. Wang, Nitrogen-doped porous carbon spheres anchored with Co3O4 nanoparticles as high-performance anode materials for lithium-ion batteries, Electrochim. Acta 187 (2016) 234–242, doi: 10.1016/j.electacta.2015.11.065.
[183] X. Xing, R. Liu, S. Liu, S. Xiao, Y. Xu, C. Wang, D. Wu, Surfactant-assisted hydrothermal synthesis of cobalt oxide/nitrogen-doped graphene framework for enhanced anodic performance in lithium ion batteries, Electrochim. Acta 194 (2016) 310–316, doi: 10.1016/j.electacta.2016.02.096.
[184] S. Harish, K. Silambarasan, G. Kalaiyarasan, A.V. Narendra Kumar, J. Joseph, Nanostructured porous cobalt oxide synthesis from Co3[Co(CN)6]2 and its possible applications in lithium battery, Mater. Lett. 165 (2016) 115–118, doi: 10.1016/j.matlet.2015.11.122.
[185] C. Shi, Q. Xia, X. Xue, Q. Liu, H.-J. Liu, Synthesis of cobalt-based layered coordination polymer nanosheets and their application in lithium-ion batteries as anode materials, RSC Adv. 6 (2016) 4442–4447, doi: 10.1039/C5RA22038F.
[186] A.K. Mondal, H. Liu, Z.-F. Li, G. Wang, Multiwall carbon nanotube-nickel cobalt oxide hybrid structure as high performance electrodes for supercapacitors and lithium ion batteries, Electrochim. Acta 190 (2016) 346–353, doi: 10.1016/j.electacta.2015.12.132.
[187] Q. Wang, B. Yu, X. Li, L. Xing, X. Xue, Core–shell Co 3 O 4 /ZnCo 2 O 4 coconut-like hollow spheres with extremely high performance as anode materials for lithium-ion batteries, J. Mater. Chem. A 4 (2016) 425–433, doi: 10.1039/C5TA06872J.
[188] Y. Zhang, Y. Wu, Y. Chu, L. Li, Q. Yu, Y. Zhu, G. Liu, Q. Hou, R. Zeng, L. Zhao, Self-assembled Co3O4 nanostructure with controllable morphology towards high performance anode for lithium ion batteries, Electrochim. Acta 188 (2016) 909–916, doi: 10.1016/j.electacta.2015.12.055.
[189] D. Darbar, M.V. Reddy, S. Sundarrajan, R. Pattabiraman, S. Ramakrishna, B.V.R. Chowdari, Anodic electrochemical performances of MgCo2O4 synthesized by oxalate decomposition method and electrospinning technique for Li-ion battery application, Mater. Res. Bull. 73 (2016) 369–376, doi: 10.1016/j.materresbull.2015.09.025.
[190] Q. Xie, D. Zeng, Y. Ma, L. Lin, L. Wang, D.-L. Peng, Synthesis of ZnO–ZnCo2O4 hybrid hollow microspheres with excellent lithium storage properties, Electrochim. Acta 169 (2015) 283–290, doi: 10.1016/j.electacta.2015.04.041.
[191] S.-H. Yu, S.H. Lee, D.J. Lee, Y.-E. Sung, T. Hyeon, Conversion reaction-based oxide nanomaterials for lithium ion battery anodes, Small 12 (2016) 2146–2172, doi: 10.1002/smll.201502299.
[192] R.M. Thankachan, M.M. Rahman, I. Sultana, A.M. Glushenkov, S. Thomas, N. Kalarikkal, Y. Chen, Enhanced lithium storage in ZnFe2O4–C nanocomposite produced by a low-energy ball milling, J. Power Sources 282 (2015) 462–470, doi: 10.1016/j.jpowsour.2015.02.039.
[193] S. Petnikota, S.K. Marka, V.V.S.S. Srikanth, M.V. Reddy, B.V.R. Chowdari, Elucidation of few layered graphene-complex metal oxide (A2Mo3O8, A = Co, Mn and Zn) composites as robust anode materials in Li ion batteries, Electrochim. Acta 178 (2015) 699–708, doi: 10.1016/j.electacta.2015.08.066.
[194] S. Sahoo, S.-H. Bae, Y.-S. Lee, J.-M. Lee, J.-M. Ahn, C.-G. Kim, I.-K. Oh, Defect-engineered mesoporous ternary nanoarchitecture of zinc-cobalt-oxide/nitrogen-doped graphene as anode material in lithium ion batteries, Carbon N.Y. 94 (2015) 455–463, doi: 10.1016/j.carbon.2015.07.024.
[195] W.-J. Li, Q.-R. Yang, S.-L. Chou, J.-Z. Wang, H.-K. Liu, Cobalt phosphide as a new anode material for sodium storage, J. Power Sources 294 (2015) 627–632, doi: 10.1016/j.jpowsour.2015.06.097.
[196] J. Jiang, W. Wang, C. Wang, L. Zhang, K. Tang, J. Zuo, Q. Yang, Electrochemical performance of iron diphosphide/carbon tube nanohybrids in lithium-ion batteries, Electrochim. Acta 170 (2015) 140–145, doi: 10.1016/j.electacta.2015.04.115.
[197] Y. Zhang, H. Zhang, Y. Feng, L. Liu, Y. Wang, Unique Fe2P nanoparticles enveloped in sandwichlike graphited carbon sheets as excellent hydrogen evolution reaction catalyst and lithium-ion battery anode, ACS Appl. Mater. Interfaces 7 (2015) 26684–26690, doi: 10.1021/acsami.5b08620.
[198] W. Li, H. Li, Z. Lu, L. Gan, L. Ke, T. Zhai, H. Zhou, Layered phosphorus-like GeP 5: a promising anode candidate with high initial coulombic efficiency and large capacity for lithium ion batteries, Energy Environ. Sci. 8 (2015) 3629–3636, doi: 10.1039/C5EE02524A.
[199] G. Park, S. Sim, J. Lee, S.-M. Lee, Effect of silicon doping on the electrochemical properties of MoP2 nano-cluster anode for lithium ion batteries, J. Alloys Compd. 639 (2015) 296–300, doi: 10.1016/j.jallcom.2015.03.146.

[200] S. Liu, H. Zhang, L. Xu, L. Ma, Synthesis of hollow spherical tin phosphides (Sn4P3) and their high adsorptive and electrochemical performance, J. Cryst. Growth 438 (2016) 31–37, doi: 10.1016/j.jcrysgro.2015.12.018.

[201] C. Choi, S.-D. Seo, H.-W. Shim, M.A. Dar, I.S. Cho, D.-W. Kim, Facile synthesis and electroactivity of 3-D hierarchically superstructured cobalt orthophosphate for lithium-ion batteries, J. Alloys Compd. 652 (2015) 100–105, doi: 10.1016/j.jallcom.2015.08.203.

[202] J. Yang, Y. Zhang, C. Sun, H. Liu, L. Li, W. Si, W. Huang, Q. Yan, X. Dong, Graphene and cobalt phosphide nanowire composite as an anode material for high performance lithium-ion batteries, Nano Res. 9 (2016) 612–621, doi: 10.1007/s12274-015-0941-5.

[203] Y. Bai, H. Zhang, L. Liu, H. Xu, Y. Wang, Tunable and specific formation of C@NiCoP peapods with enhanced HER activity and lithium storage performance, Chemistry 22 (2016) 1021–1029, doi: 10.1002/chem.201504154.

[204] S. Liu, J. Feng, X. Bian, J. Liu, H. Xu, Electroless deposition of Ni 3 P–Ni arrays on 3-D nickel foam as a high performance anode for lithium-ion batteries, RSC Adv. 5 (2015) 60870–60875, doi: 10.1039/C5RA08926C.

[205] M. Walter, M.I. Bodnarchuk, K.V. Kravchyk, M.V. Kovalenko, Evaluation of metal phosphide nanocrystals as anode materials for Na-ion batteries, Chim. Int. J. Chem. 69 (2015) 724–728, doi: 10.2533/chimia.2015.724.

[206] H. Usui, T. Sakata, M. Shimizu, H. Sakaguchi, Electrochemical Na-insertion/Extraction Properties of Sn–P Anodes, Electrochemistry 83 (2015) 810–812, doi: 10.5796/electrochemistry.83.810.

[207] J. Song, Z. Yu, M.L. Gordin, X. Li, H. Peng, D. Wang, Advanced sodium ion battery anode constructed via chemical bonding between phosphorus, carbon nanotube, and cross-linked polymer binder, ACS Nano 9 (2015) 11933–11941, doi: 10.1021/acsnano.5b04474.

[208] J. Liu, P. Kopold, C. Wu, P.A. van Aken, J. Maier, Y. Yu, Uniform yolk–shell Sn 4 3 @C nanospheres as high-capacity and cycle-stable anode materials for sodium-ion batteries, Energy Environ. Sci. 8 (2015) 3531–3538, doi: 10.1039/C5EE02074C.

[209] S.-O. Kim, A. Manthiram, The facile synthesis and enhanced sodium-storage performance of a chemically bonded CuP 2 /C hybrid anode, Chem. Commun. 52 (2016) 4337–4340, doi: 10.1039/C5CC10585D.

[210] Y. Xia, B. Wang, X. Zhao, G. Wang, H. Wang, Core-shell composite of hierarchical MoS2 nanosheets supported on graphitized hollow carbon microspheres for high performance lithium-ion batteries, Electrochim. Acta 187 (2016) 55–64, doi: 10.1016/j.electacta.2015.11.047.

[211] X. Xu, D. Yu, H. Zhou, L. Zhang, C. Xiao, C. Guo, S. Guo, S. Ding, MoS2 nanosheets grown on amorphous carbon nanotubes as advanced anode materials for sodium-ion batteries, J. Mater. Chem. A 4 (2016) 4375–4379, doi: 10.1039/C6TA00068A.

[212] H. Jiang, D. Ren, H. Wang, Y. Hu, S. Guo, H. Yuan, P. Hu, L. Zhang, C. Li, 2D monolayer MoS_2-carbon interoverlapped superstructure: engineering ideal atomic interface for lithium ion storage, Adv. Mater. 27 (2015) 3687–3695, doi: 10.1002/adma.201501059.

[213] J. Kong, C. Zhao, Y. Wei, X. Lu, MoS2 nanosheets hosted in polydopamine-derived mesoporous carbon nanofibers as lithium-ion battery anodes: enhanced MoS2 capacity utilization and underlying mechanism, ACS Appl. Mater. Interfaces 7 (2015) 24279–24287, doi: 10.1021/acsami.5b07950.

[214] X.-Y. Yu, H. Hu, Y. Wang, H. Chen, X.W.D. Lou, Ultrathin MoS_2 nanosheets supported on N-doped carbon nanoboxes with enhanced lithium storage and electrocatalytic properties, Angew. Chem. Int. Ed. Engl. 54 (2015) 7395–7398, doi: 10.1002/anie.201502117.

[215] J. Zhou, J. Qin, X. Zhang, C. Shi, E. Liu, J. Li, N. Zhao, C. He, 2D space-confined synthesis of few-layer MoS2 anchored on carbon nanosheet for lithium-ion battery anode, ACS Nano 9 (2015) 3837–3848, doi: 10.1021/nn506850e.

[216] Z. Hu, Q. Liu, W. Sun, W. Li, Z. Tao, S.-L. Chou, J. Chen, S.-X. Dou, MoS 2 with an intercalation reaction as a long-life anode material for lithium ion batteries, Inorg. Chem. Front. 3 (2016) 532–535, doi: 10.1039/C5QI00237K.

[217] S. Zhou, J. Chen, L. Gan, Q. Zhang, Z. Zheng, H. Li, T. Zhai, Scalable production of self- supported WS2/CNFs by electrospinning as the anode for high-performance lithium-ion batteries, Sci. Bull. 61 (2016) 227–235, doi: 10.1007/s11434-015-0992-8.

[218] L. Zhou, S. Yan, L. Pan, X. Wang, Y. Wang, Y. Shi, A scalable sulfuration of WS2 to improve cyclability and capability of lithium-ion batteries, Nano Res. 9 (2016) 857–865, doi: 10.1007/s12274-015-0966-9.
[219] S. Yu, J.-W. Jung, I.-D. Kim, Single layers of WS2 nanoplates embedded in nitrogen-doped carbon nanofibers as anode materials for lithium-ion batteries, Nanoscale 7 (2015) 11945–11950, doi: 10.1039/c5nr02425k.
[220] Y. Du, X. Zhu, L. Si, Y. Li, X. Zhou, J. Bao, Improving the anode performance of WS 2 through a self-assembled double carbon coating, J. Phys. Chem. C 119 (2015) 15874–15881, doi: 10.1021/acs.jpcc.5b03540.
[221] Y. Zhu, Y. Chu, J. Liang, Y. Li, Z. Yuan, W. Li, Y. Zhang, X. Pan, S.-L. Chou, L. Zhao, R. Zeng, Tucked flower-like SnS2/Co3O4 composite for high-performance anode material in lithium-ion batteries, Electrochim. Acta 190 (2016) 843–851, doi: 10.1016/j.electacta.2015.12.127.
[222] M.-S. Balogun, W. Qiu, J. Jian, Y. Huang, Y. Luo, H. Yang, C. Liang, X. Lu, Y. Tong, Vanadium nitride nanowire supported SnS2 nanosheets with high reversible capacity as anode material for lithium ion batteries, ACS Appl. Mater. Interfaces 7 (2015) 23205–23215, doi: 10.1021/acsami.5b07044.
[223] H. Tang, X. Qi, W. Han, L. Ren, Y. Liu, X. Wang, J. Zhong, SnS2 nanoplates embedded in 3D interconnected graphene network as anode material with superior lithium storage performance, Appl. Surf. Sci. 355 (2015) 7–13, doi: 10.1016/j.apsusc.2015.07.091.
[224] Y.N. Ko, Y.C. Kang, Co9S8–carbon composite as anode materials with improved Na-storage performance, Carbon N.Y. 94 (2015) 85–90, doi: 10.1016/j.carbon.2015.06.064.
[225] H. Fei, Z. Feng, X. Liu, Novel sodium bismuth sulfide nanostructures: a promising anode materials for sodium-ion batteries with high capacity, Ionics (Kiel). 21 (2014) 1967–1972, doi: 10.1007/s11581-014-1356-0.
[226] W. Sun, X. Rui, D. Yang, Z. Sun, B. Li, W. Zhang, Y. Zong, S. Madhavi, S. Dou, Q. Yan, Two-dimensional tin disulfide nanosheets for enhanced sodium storage, ACS Nano 9 (2015) 11371–11381, doi: 10.1021/acsnano.5b05229.
[227] W. Qin, T. Chen, T. Lu, D.H.C. Chua, L. Pan, Layered nickel sulfide-reduced graphene oxide composites synthesized via microwave-assisted method as high performance anode materials of sodium-ion batteries, J. Power Sources 302 (2016) 202–209, doi: 10.1016/j.jpowsour.2015.10.064.
[228] W. Sun, X. Rui, D. Zhang, Y. Jiang, Z. Sun, H. Liu, S. Dou, Bismuth sulfide: a high-capacity anode for sodium-ion batteries, J. Power Sources 309 (2016) 135–140, doi: 10.1016/j.jpowsour.2016.01.092.
[229] L. Wu, H. Lu, L. Xiao, X. Ai, H. Yang, Y. Cao, Improved sodium-storage performance of stannous sulfide@reduced graphene oxide composite as high capacity anodes for sodium- ion batteries, J. Power Sources 293 (2015) 784–789, doi: 10.1016/j.jpowsour.2015.06.015.
[230] S.H. Choi, Y.C. Kang, Aerosol-assisted rapid synthesis of SnS-C composite microspheres as anode material for Na-ion batteries, Nano Res 8 (2015) 1595–1603, doi: 10.1007/s12274-014-0648-z.
[231] S. Peng, X. Han, L. Li, Z. Zhu, F. Cheng, M. Srinivansan, S. Adams, S. Ramakrishna, Unique cobalt sulfide/reduced graphene oxide composite as an anode for sodium-ion batteries with superior rate capability and long cycling stability, Small 12 (2016) 1359–1368, doi: 10.1002/smll.201502788.
[232] S.D. Lacey, J. Wan, A. von Wald Cresce, S.M. Russell, J. Dai, W. Bao, K. Xu, L. Hu, Atomic force microscopy studies on molybdenum disulfide flakes as sodium-ion anodes, Nano Lett. 15 (2015) 1018–1024, doi: 10.1021/nl503871s.

CHAPTER THREE

Review of Nanotechnology for Cathode Materials in Batteries

S.-J. Cho, M.-J. Uddin, P. Alaboina
North Carolina A&T State University, Greensboro, NC, United States

Contents

Nanotechnology is an emerging and highly promising technology to develop efficient and enhanced batteries. This chapter discusses the nanotechnology-driven synthesis and surface modification techniques for cathode materials to improve lithium-ion batteries (LIBs). Various approaches for synthesis of nanostructured cathode materials with enhanced properties and nanoscale surface modification on cathodes with improved electrochemical performances are described. The chapter concludes with a comparison of the different techniques discussed and addresses the importance of nanotechnology in the race to build improved batteries.

1 INTRODUCTION

Batteries have evolved rapidly over recent years. Nickel cadmium batteries, nickel–metal hydride batteries, and lithium-ion batteries are some of the different types of popular batteries that are now commercially available. However, the true potential of these batteries is not fully utilized yet, their purposes can be enhanced with a little help from nanotechnology. Out of all the different types of batteries,

Emerging Nanotechnologies in Rechargeable Energy Storage Systems

LIBs have gained a significant position in the market and are the most promising and the fastest growing battery technology. They offer very high energy densities, have low self-discharge, and can operate at higher voltages compared to other battery candidates. They are widely used and commercially accepted in applications ranging from portable devices, electric automobiles, and aerospace applications. Thus, in this chapter LIBs will be placed in the spotlight and given the most focus along with nanotechnology.

The power of nanotechnology should not be considered to be limited to just LIBs, however, its scope and potential includes several other battery technologies. Although batteries, especially LIBs, are present in almost all portable electronics around the globe, which suggests that battery technology may have reached its peak, there are still several technical limitations to battery performance including energy, power, and long-lasting life and also safety, which are mostly attributed to cathode material in LIB technologies. In order to improve battery cathode materials, nanotechnology-driven design, synthesis, and engineering modification play a significant role. In this chapter, we summarize and highlight the emerging nanotechnologies applied to cathode materials for LIBs. In the next section, we focus on nanostructural design and synthesis of cathode material and discuss how the nanostructural design and synthesis effectively influence their performance in LIB. Section 3 presents an overview of popular nanoscale surface deposition methods, which are used to modify cathodes with a protective surface layer to overcome the capacity limitations related to side reactions with electrolytes, reduce self-discharge reactions, improve thermal and structural stability, and further enhance the battery's performance.

2 NANOSTRUCTURAL DESIGN AND SYNTHESIS OF CATHODE MATERIALS FOR LITHIUM-ION BATTERIES

Since the design and capacity of most LIBs are cathode limited, the key to improving LIBs is the controlled design of cathode materials to enhance performance. Nanotechnology can offer fundamentally new and different ways to design and engineer cathode materials for such a need. In this section, we discuss the different synthesis methods utilized for LIB cathode materials powered by nanotechnology. From the nanoprocessing point of view, most of the techniques utilized for nanosynthesis of cathode materials are bottom-up techniques with zero-dimensional or one-dimensional cathode nanostructures as the final product. Liquid phase methods are most common, although solid-state methods are also tried. Among the different routes to synthesize nanostructures, template-based methods, solvothermal/hydrothermal methods, coprecipitation methods, and solid-state methods have proved to be useful for cathode material development. Thus, we have discussed these methods in terms of discharge capacity, capacity retention, and rate performance of the resulting nanostructured cathode materials.

2.1 Nanotemplate methods

Nanotemplate method is a straightforward template-directed synthesis method to generate highly controllable nanostructure with intended morphology [1–6]. Generally, a template is simply used to provide the necessary scaffold within or around which the desired nanostructure is formed with a morphology corresponding to that of the template. Include nanoporous channels in a porous material, surfactant, or block copolymer assisted self-assembled nanostructure, or simply any existing nanostructure. In order to obtain a pure nanostructure it is often necessary to remove the template at the end of the synthesis by chemical etching or calcination [1–5]. However, it is also possible to design the synthesis method such that the template will be eliminated as the reaction proceeds, and a pure product will be obtained.

Nanotemplating is preferable method when a controlled nanostructure with porous and/or hierarchical structure is required. Vu et al. have synthesized a 3D, hierarchically porous multiconstituent $LiFePO_4$–carbon (LFP/C) composite cathode material using colloidal crystal method [4]. They have used a dual templating method to obtain mesopores and macropores in the structure. For mesopores, they have used nonionic surfactant Pluronic F127 from BASF, poly(ethylene oxide)–poly(propylene oxide)–poly(ethylene oxide) triblock copolymer ($PEO_{106}PPO_{70}PEO_{106}$). For macropores, they have used poly(methyl methacrylate) (PMMA), colloidal crystals. They have featured a multiconstituent synthesis method where all the required precursor materials for $LiFePO_4$ (LFP) and carbon, as well as the phenol-formaldehyde sol and a nonionic surfactant were infiltrated at the same time into the colloidal crystal template. After multiple heat treatments at slow ramp rates, a final pyrolysis during 10 h at high temperature (600, 700, or 800°C) yields a three-dimensionally ordered macroporous and meso-/microporous (3DOM/m) structure of LFP/C monolithic composite. Scanning electron microscope (SEM) images of the 3DOM/m LFP/C composite pyrolyzed at 600°C showed a well-ordered interconnected macroporous network structure resulting from the colloidal crystal template, where the wall of the composite contains meso-/micropores resulted from the triblock copolymer nonionic surfactant (Fig. 3.1A-1–A-2). The average diameters of the macro- and micropores were 285 and 90 nm, respectively, and the Brunauer–Emmett–Teller (BET) surface area was 200 m^2/g for the composite. Transmission electron microscope (TEM) images suggested that the composite is well ordered with an inverse replica of face-centered cubic (FCC) colloidal crystal template (Fig. 3.1B-1–B-2). The contrast difference in TEM images suggested the presence of LFP in the octahedral sites and mesoporous carbon links in the tetrahedral sites.

The interconnected ordered macroporous structure with walls containing meso-/micropores, provide a 3D interconnected network that facilitate the penetration of electrolyte, arrange for large accessible surface area for Li^+ exchange between electrolyte and electrode, offer shorter Li^+ diffusion path (less than 180 nm), and form an effectively wired network of conductive carbon of hundreds of micrometer in size [4]. The

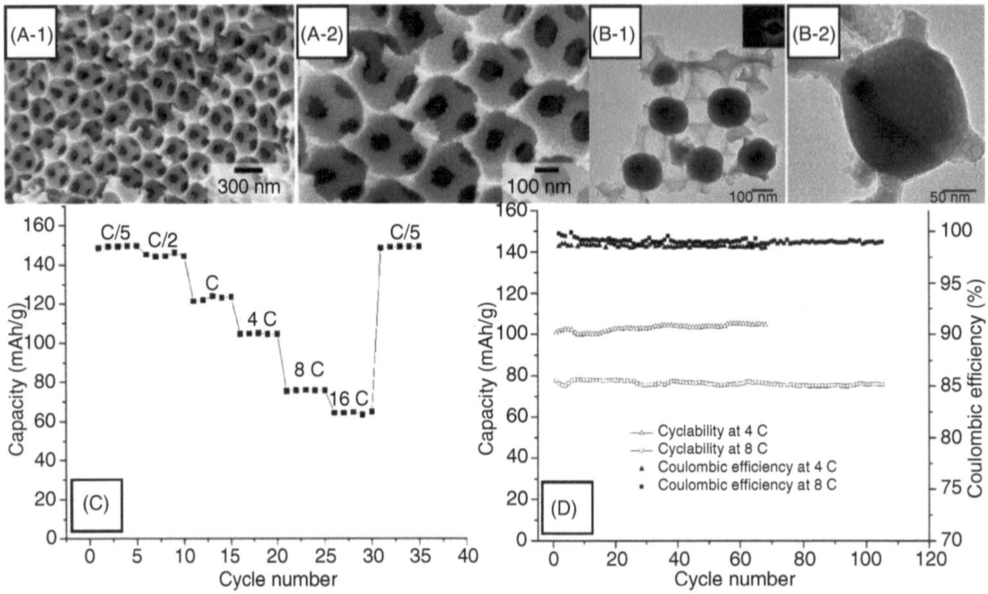

Figure 3.1 (A-1) Low- and (A-2) high-magnification SEM images of $LiFePO_4$–carbon (LFP/C) pyrolyzed sample at 600°C; (B-1) low- and (B-2) high-resolution TEM images of the same sample; (C) rate capability; and (D) cycling behavior of LFP/C composite cathode material. *(Reprinted with permission from A. Vu, A. Stein, Multiconstituent synthesis of $LiFePO_4$/C composites with hierarchical porosity as cathode materials for lithium ion batteries, Chem. Mater. 23 (2011) 3237–3245 [4], Copyright © 2011 American Chemical Society).*

combination of these properties resulted in a high performance, high capacity, and high rate capability of the 3DOM/m LFP/C composite cathodes. The composite showed discharge capacity of 150 mAh/g at C/5, 123 mAh/g at 1 C, 78 mAh/g at 8 C, and 64 mAh/g at 16 C rate (Fig. 3.1C). Without adding any carbon black additive or binder, which was possible for the monolithic nature of the composite, the composite showed high coulombic efficiency (near 99%) and excellent capacity retention (no fading over 100 cycles, as shown in Fig. 3.1D). The composite was able to support current density as high as 2720 mA/g (16 C rate) [4].

In similar work, triblock copolymer F127 was utilized as a layered template to synthesize flake-stacked monoclinic $Li[Li_{0.2}Mn_{0.54}Ni_{0.13}Co_{0.13}]O_2$ (LMR-NMC) nanoparticles for high capacity and high power Li-ion cathode [3]. A triconstituent coassembly of surfactant F127, resols, and metal nitrates was used as precursors to form the layered structure. The hydrophilic poly(ethylene oxide) (PEO) layer of F127 gets attached with the phenolic resol and the metal nitrates, and the hydrophobic poly(propylene oxide) (PPO) layer remains outside forming a hydrophobic layer by self-assembly. The PEO layer was reported to further grow by attaching more and more metals due to its hydrophilic nature. Eventually a calcination at high temperature (850°C) yields in the flake-stacked LMR-NMC nanoparticles (Fig. 3.2A–D). The SEM image showed that the cathode material is composed of uniformly distributed homogenous flakes of 100–200 nm size. The nanoflake morphology of LMR-NMC facilitates short Li^+ diffusion path and fast Li^+ insertion/extraction reactions.

The TEM images along with the X-ray diffraction (XRD) results revealed that the synthesized LMR-NMC cathode material is mostly composed of monoclinic Li_2MnO_3 structure (space group C2/m) instead of the trigonal $LiMO_2$ (M = Mn, Ni, Co) structure (*R3m* symmetry). The dominance of the monoclinic structures in the synthesized cathode material can help to stabilize the structure of the cathode, and thus to improve the cycling stability. The flake-stacked structure of LMR-NMC nanoparticles, in combination with a monoclinic structure, help resolve most of the issue of LMR-NMC cathode materials, such as low initial coulombic efficiency (60–70%), rapid capacity fading, poor rate capability, and so forth, and take benefit of their high theoretical capacity (>250 mAh/g). The synthesized material showed better coulombic efficiency, reversible capacity, cycling performance, and rate capability. The material showed high initial discharge capacity of 277.4 mAh/g at C/15 rate, and discharge capacity of 188.6 mAh/g at 1 C, and 131.2 mAh/g at 4 C. Also, as the current density is put back to C/6 rate, the capacity returns to 239 mAh/g. The coulombic efficiency of the material was found to be as high as 87.3%, and capacity fade was very small (0.046% per cycle over 100 cycles).

Zhao et al. have also tried to improve the electrochemical performance of LMR-NMC using a template-based synthesis method of the cathode material [1]. In their work, they have used β-MnO_2 nanorods as a sacrificial template to synthesize LMR-NMC cathode material by Li and other metal ion insertion into the MnO_2 nanorods during preparation (Fig. 3.2E). The metal ions inserted rodlike structure is thermodynamically unstable and it goes to crystal growth during calcination. Different structures, nanorods, and polyhedrons were synthesized at different calcination temperature of 750, 800, 850, or 900°C. Higher calcination temperature turns the rodlike structure gradually to polyhedron-like structure (rod, rod, and polyhedron structure at 700, 800, and 850°C, respectively). Also, it was reported that the average crystallite size increases with calcination temperature. XRD results suggest improvement of a layered structure with calcination temperature.

With improved and orderly layered structure of the cathode material, a solid framework of LMR-NMC can readily improve the discharge capacity by preventing the initial irreversible capacity loss. However, increasing the particle size may block the effect Li^+ diffusion and decrease the capacity. Hence, it is critical to determine the optimum calcination temperature that provides the best achievable properties. Analyzing the electrochemical performance data, it seemed that the calcination at 850°C provided the best LMR-NMC cathode material. The initial discharge capacity at 20 mA/g is 207.0, 227.9, 239.2, and 205.7 mAh/g with a coulombic efficiency of 61.7, 69.0, 71.6, and 71.7% for the cathode material calcined at 700, 800, 850, and 900°C, respectively. The capacity

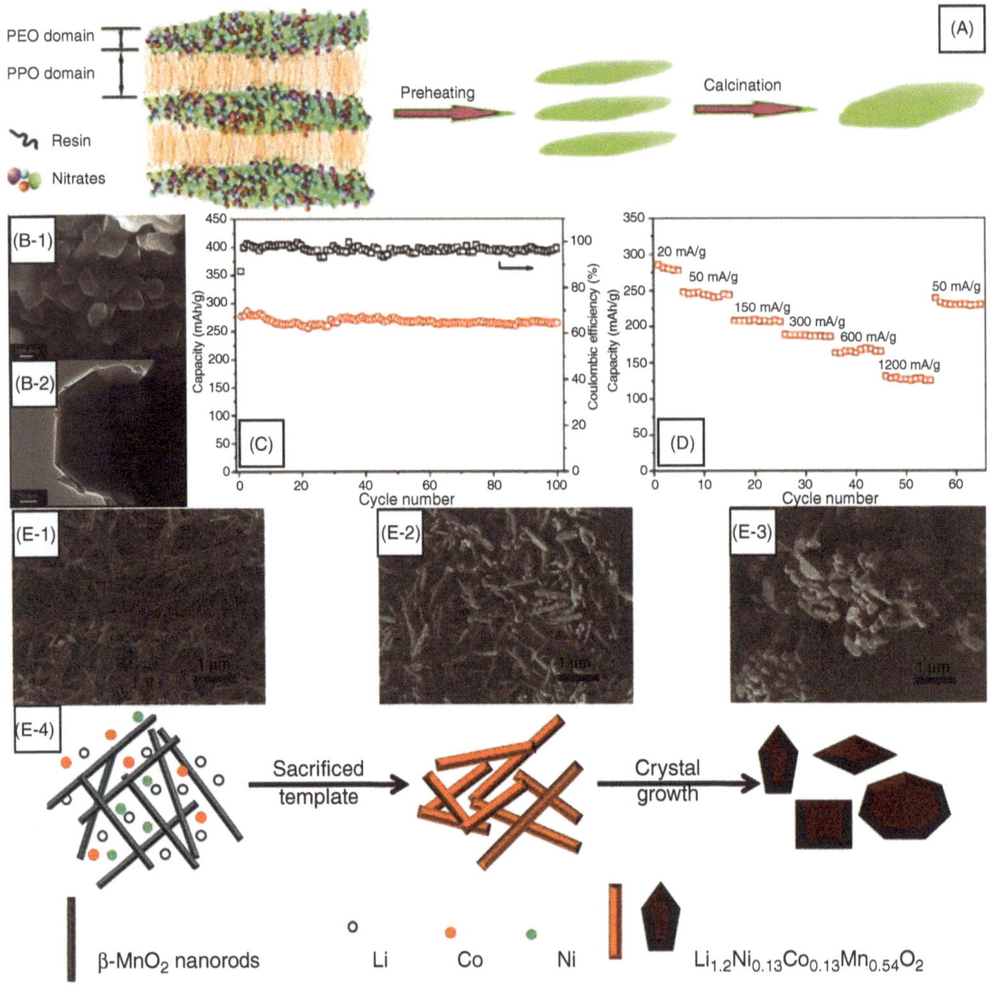

Figure 3.2 (A) Schematic diagram of preparation process of the flake-stacked LMR-NMC; (B-1) SEM and (B-2) TEM images; (C) cycling and (D) rate performance of LMR-NMC; SEM images of (E-1) MnO_2 nanorod, (E-2) rod-like LMR-NMC, (E-3) polyhedral structure; and (E-4) the schematic illustration of the preparation of LMR-NMC. *(Part A–D, reproduced from S. Qiu, Z. Chen, F. Pei, F. Wu, Y. Wu, X. Ai, et al., Synthesis of monoclinic $Li[Li_{0.2}Mn_{0.54}Ni_{0.13}Co_{0.13}]O_2$ nanoparticles by a layered-template route for high-performance Li-ion batteries, Eur. J. Inorg. Chem. 2013 (2013) 2887–2892 [3] with permission, Copyright © 2013 Wiley-VCH Verlag GmbH & Co. KGaA; part E, reproduced from C. Zhao, X. Wang, R. Liu, F. Xu, Q. Shen, β-MnO_2 sacrificial template synthesis of $Li_{1.2}Ni_{0.13}Co_{0.13}Mn_{0.54}O_2$ for lithium ion battery cathodes, RSC Adv. 4 (2014) 7154 [1] with permission from The Royal Society of Chemistry, Copyright © 2014).*

retention is 70.7, 86.7, 90.0, and 95.9% for over 70 cycles for cathode material calcined at 700, 800, 850, and 900°C, respectively. The material calcined at 850°C also showed the best rate capability; discharge capacity of 128.4 and 92.8 mAh/g were shown at 400 and 1000 mA/g, respectively. The exceptional electrochemical performance of polyhedron-like nanoparticles produced by the calcination at 850°C was attributed to their improved layered structure and more appropriate particle size.

The sacrificial templates have been utilized to create nanostructures of other cathode materials as well, such as $LiMn_2O_4$ [5] and $LiCoO_2$ [2]. Xie at al. has synthesized single-crystalline spinel $LiMn_2O_4$ cathode material from self-sacrificial Mn_3O_4 nanorod [5]; the preparation schematic is shown in Fig. 3.3A. Porous Mn_3O_4 nanorods were used, and LiOH was impregnated into the nanorod templates. Finally, a calcination resulted in single crystal spinel $LiMn_2O_4$ structure. TEM result showed that the nanorods have a width of ~200 nm, and they are single crystalline (Fig. 3.3B-1–B-2). Also, the

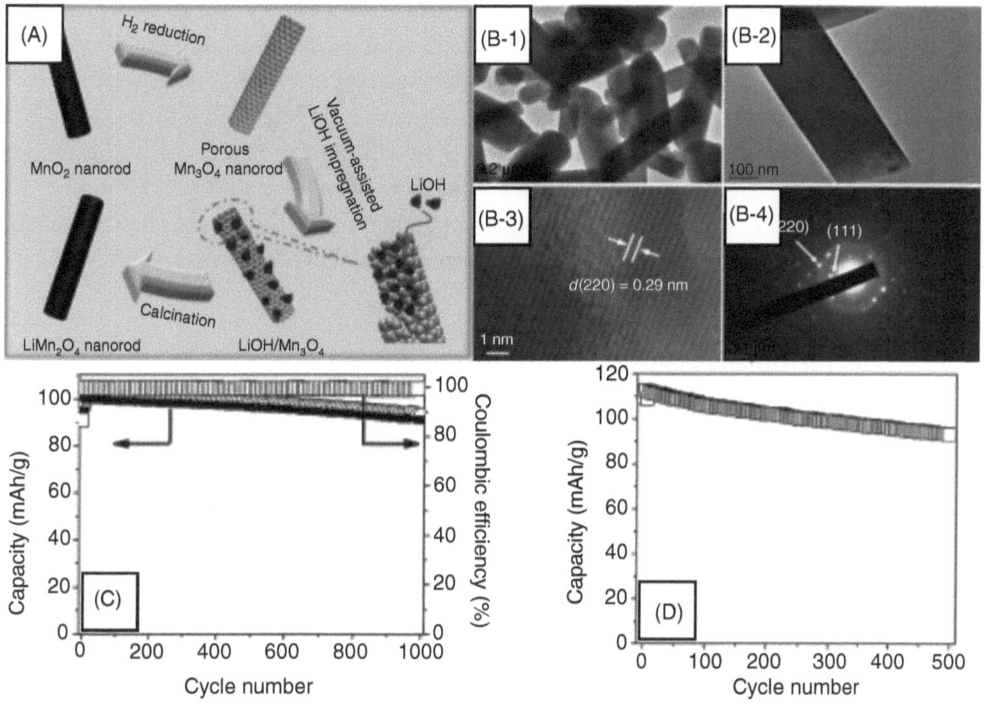

Figure 3.3 (A) Preparation procedure of $LiMn_2O_4$ nanorods; (B-1) low- and (B-2) high-magnification TEM images; (B-3) lattice-resolved high-resolution TEM image; (B-4) SAED pattern of the nanorods; (C) cycle performance and coulombic efficiency of $LiMn_2O_4$ nanorods from Li:Mn = 1.2:2 precursor ratio; and (D) cycle performance of as prepared $LiMn_2O_4$ at 60°C. *(Reproduced from X. Xie, D. Su, B. Sun, J. Zhang, C. Wang, G. Wang, Synthesis of single-crystalline spinel $LiMn_2O_4$ nanorods for lithium-ion batteries with high rate capability and long cycle life, Chem. A Eur. J. 20 (2014) 17125–17131 [5] with permission, Copyright © 2014 Wiley-VCH Verlag GmbH & Co. KGaA).*

lattice-resolved high-resolution TEM image (Fig. 3.3B-3) and the selected area electron diffraction (SAED) pattern (Fig. 3.3B-4) confirmed the single crystalline spinel structure of $LiMn_2O_4$. The prepared $LiMn_2O_4$ nanorods showed good rate performance; discharge capacities of 114, 114, 111, 108, 103, 101, and 91 mAh/g were found at 1 C, 2 C, 3 C, 5 C, 8 C, 10 C, and 20 C, respectively. The initial discharge capacity and columbic efficiency of the $LiMn_2O_4$ nanorods prepared from Li:Mn = 1.2:2 precursor ratio was 101 mAh/g and 86.3%, respectively, at 3 C rate, and excellent retention of 95.6% was reported for 1000 cycles (Fig. 3.3C). Also, the single-crystalline nanorods showed an initial capacity of 109 mAh/g at 3 C rate at 60°C (Fig. 3.3D).

Nanotemplating method provides a simple and straightforward way to synthesize highly controllable nanostructures of cathode materials with enhanced properties. It allows researchers to produce nanostructures with complex morphology and hierarchical design from template directly in a single step. The throughput is usually very high, and it is a cost effective method to produce nanostructure with high purity. On the other hand, the drawback of this method is that the amount of nanostructures that can be produced in a templating method is usually very low, and the synthesized structures are often polycrystalline [6]. Nevertheless, nanotemplating offers us to synthesize nanostructure materials with complex morphology, which would otherwise be impossible to produce. Xia et al. in their review article provide an overview of different template-assisted methods for the synthesis of different nanostructures that offers more discussion about nanotemplate methods. The enhanced properties of the synthesized cathode nanostructures can be attributed to their improved structure and reduced Li^+ diffusion path lengths facilitating fast Li^+ insertion/extraction reactions [1–5].

2.2 Solvothermal/hydrothermal methods

The solvothermal method utilizes the solubility of the precursors in a solvent at high temperatures and pressures, following the crystallization of nanostructures from the solvent. When water is used as the solvent, the method is called the hydrothermal method [7–18]. At elevated temperature, the solubility and reactivity of the precursors in the solvent and their transformation into nanostructure depend on various synthesis parameters, such as reaction temperature, pressure, time, and the nature of the solvent. Thus, it is possible to tune the parameters such that a high nucleation rate and good size distribution can be realized providing good control over size, crystallinity, and shape distribution of cathode metal oxide nanostructures.

Zhu et al. have attempted to understand the mechanism of the solvothermal reaction of LFP nanoparticles formation [7]. They have used a mixed solvent of water and tri(ethylene glycol) and tried to observe the crystal nucleation, morphological evolution, and electrochemical characteristics with increasing reaction duration.

They observed that the short reaction time results in smaller size polycrystalline material, whereas longer reaction time results in the single crystal due to Ostwald ripening

(Fig. 3.4A). As the reaction time increases, the particle morphology turns from diamond-like shape to polygonal-prism-shaped particles. Also, the size of the primary particles increases, whereas the overall secondary particle size decreases. This happens because the nucleation of LFP particles results in a decrease of pH and surface charges on the particles, which enable the primary LFP particles to approach each other easily and form secondary particles. Further reduction of pH results in increasing the LFP solubility and

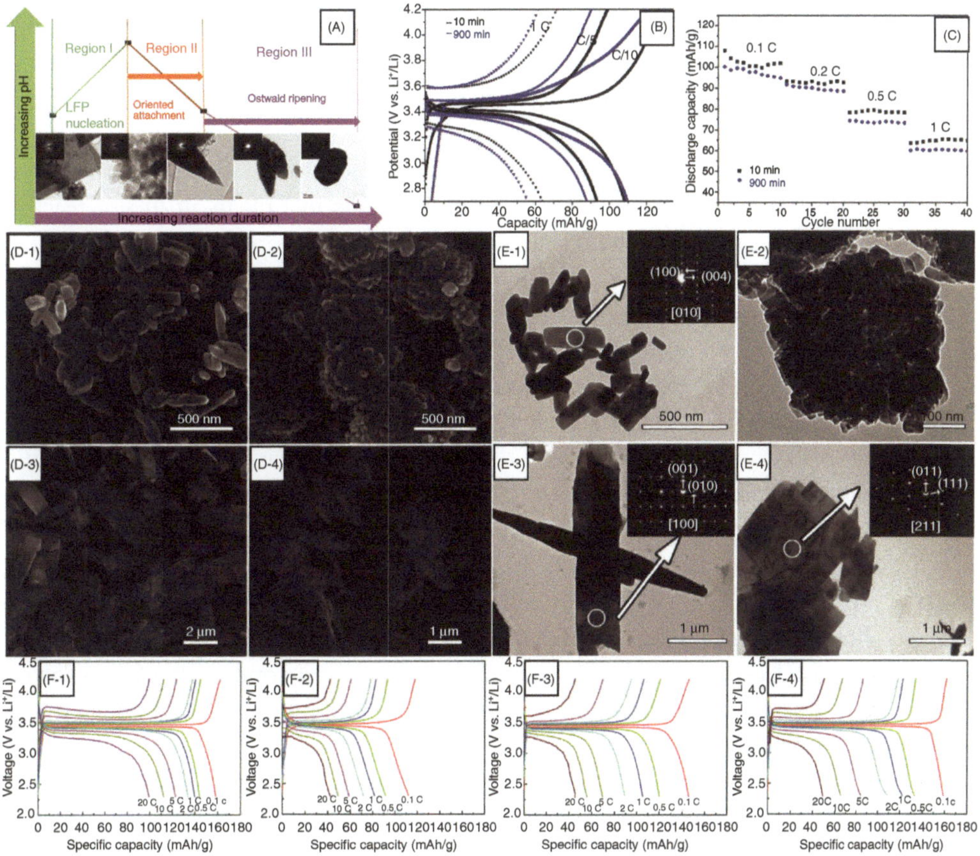

Figure 3.4 (A) Crystal growth evolution in solvothermal method for LFP nanostructures with reaction time and pH; (B) charge–discharge profiles, and (C) cycle performance at various rates from 2.7 to 4.5 V; (D-1–D-4) SEM images, (E-1–E-4) TEM images, and (F-1–F-4) charge/discharge curves at different rates for different synthesis conditions, H-UR, S-UR, H-EP, and S-EP. *(Part A–C, reprinted with permission from J. Zhu, J. Fiore, D. Li, N.M. Kinsinger, Q. Wang, E. DiMasi, et al., Solvothermal synthesis, development, and performance of $LiFePO_4$ nanostructures, Cryst. Growth Des. 13 (2013) 4659–4666 [7], Copyright © 2013 American Chemical Society; part D–E, reproduced from Z. Li, K. Zhu, J. Li, X. Wang, Morphological and orientational diversity of $LiFePO_4$ crystallites: remarkable reaction path dependence in hydrothermal/solvothermal syntheses, CrystEngComm 16 (2014) 10112–10122 [8] with permission from The Royal Society of Chemistry, Copyright © 2014).*

activates a dissolution-crystallization (Ostwald ripening) mechanism. Thus, the crystals grow in a polygonal drumlike shape of single crystals. The as-synthesized LFP particles showed good electrochemical properties: more than 100 mAh/g discharge capacity at C/10 rate. Polycrystalline diamond-like particles with smaller primary crystallite size showed better electrochemical performance than polygonal drumlike single crystal particles. This happened because the small size primary particles in larger secondary particles resulted in short diffusion path of Li-ion and better kinetics while maintaining a high tap density.

Li et al. have further studied the effect of reaction path (hydrothermal/solvothermal, excess H_3PO_4) on the morphological and orientational diversity, and their electrochemical properties [8]. They have used an equal volume of water and ethylene glycol, or only ethylene glycol as a solvent for hydrothermal, or solvothermal route, respectively, additionally with or without excess H_3PO_4 in the precursors. The solvothermal synthesis with usual precursor ratio (Li:Fe:P = 3:1:1) results in LFP in an in situ reaction mechanism, whereas the other three routes follow an aggregation growth mechanism by generating different intermediates, resulting in the diverse morphology of LFP nanoparticles (Fig. 3.4D–F). Hydrothermal synthesis with usual precursor ratio (H-UR) resulted in rodlike shaped crystallites oriented in [010] direction; solvothermal synthesis of the same ratio (S-UR) resulted in aggregated ultrafine flakelets of 20-nm size. With the presence of excess H_3PO_4, hydrothermal route (H-EP) resulted in big slab-like particles made up of small crystallites oriented at [100] direction, whereas the solvothermal route (S-EP) resulted in interlaced flake morphologies with [211] as the predominant orientation. The electrochemical properties of the particles that followed aggregation growth mechanism (H-UR, H-EP, and S-EP) showed better properties, with H-UR showing the best properties in terms of initial discharge capacity and rate performance.

Morphological control of the LFP nanoparticles with solvothermal/hydrothermal synthesis has also been reported by varying different reaction conditions, such as solvent type, volume ratio and concentration, precursor source and mixing sequence, and drying temperature [9–13,19].

Yang et al. have followed a solvothermal route to synthesize LFP particles with water-polyethylene glycol (PEG) solvent. Three different variants of LFP particles were formed—nanoparticles (~50 nm) (Fig. 3.5A-1), nanoplates (100 nm thick) (Fig. 3.5A-2), and microplates (0.3 μm thick) (Fig. 3.5A-3) [9]. The variation of morphology and size of LFP particles found to be strongly dependent on different synthesis parameters, such as PEG to water ratio, concentration, reaction temperature, and feeding sequence. LFP nanoparticles with the smallest size [Fig. 3.5B(a) and C(a)] showed better electrochemical properties then compared to nanoplates [Fig. 3.5B(b) and C(b)] and microplates [Fig. 3.5B(c) and C(c)], especially for high rate capacity.

Saravanan et al. have synthesized LFP/C nanoplates by a solvothermal method using ethylene glycol as solvent [19]. They have controlled the morphology of the LFP/C

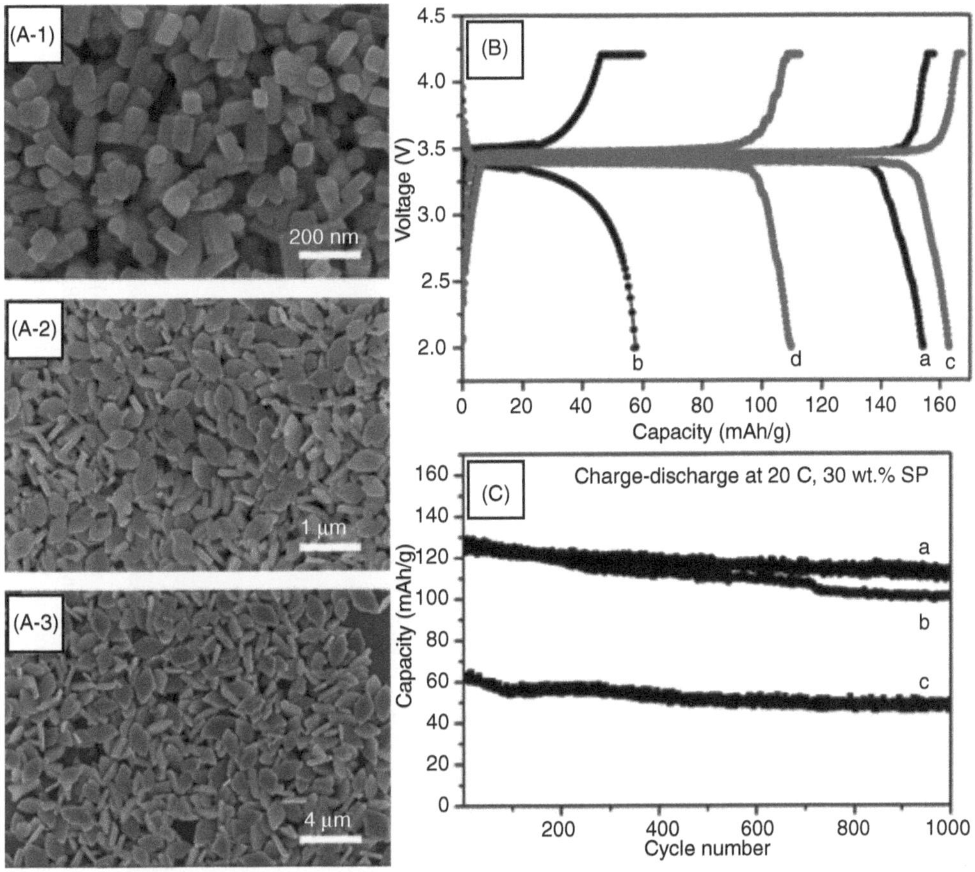

Figure 3.5 (A) SEM images at different synthesis conditions: (A-1) nanoparticles prepared at 180°C for 9 h, (A-2) nanoplates prepared at 140°C for 24 h, (A-3) microplates prepared at 180°C for 9 h with switched feeding sequence of $FeSO_4$ and LiOH; (B) charge/discharge curves for (a) carbon-coated nanoparticles, (b) nanoparticle without coating, (c) carbon-coated nanoplates, and (d) carbon-coated microplates; (C) cycle performance of LFP: (a) nanoparticles, (b) nanoplates, and (c) microplates. *(Reproduced from S. Yang, X. Zhou, J. Zhang, Z. Liu, Morphology-controlled solvothermal synthesis of $LiFePO_4$ as a cathode material for lithium-ion batteries, J. Mater. Chem. 20 (2010) 8086 [9] with permission from The Royal Society of Chemistry, Copyright © 2010).*

nanoplates by using different iron precursors (Fig. 3.6A-1–A-8). Iron oxalate resulted in hierarchical-LFP nanoplates, whereas iron acetate resulted in spindlelike shape. 1D coordination polymer, Fe(PMIDA), resulted in thick plate morphology, whereas iron gluconate resulted in diamond-shaped microcrystals. Hierarchical LFP/C nanoplates showed the best electrochemical performance, which was attributed to their smallest plate thickness Fig. 3.6B.

Park et al. have also varied the mixing sequence of precursors to change the morphology of LFP particles in a hydrothermal synthesis method with phosphate ester

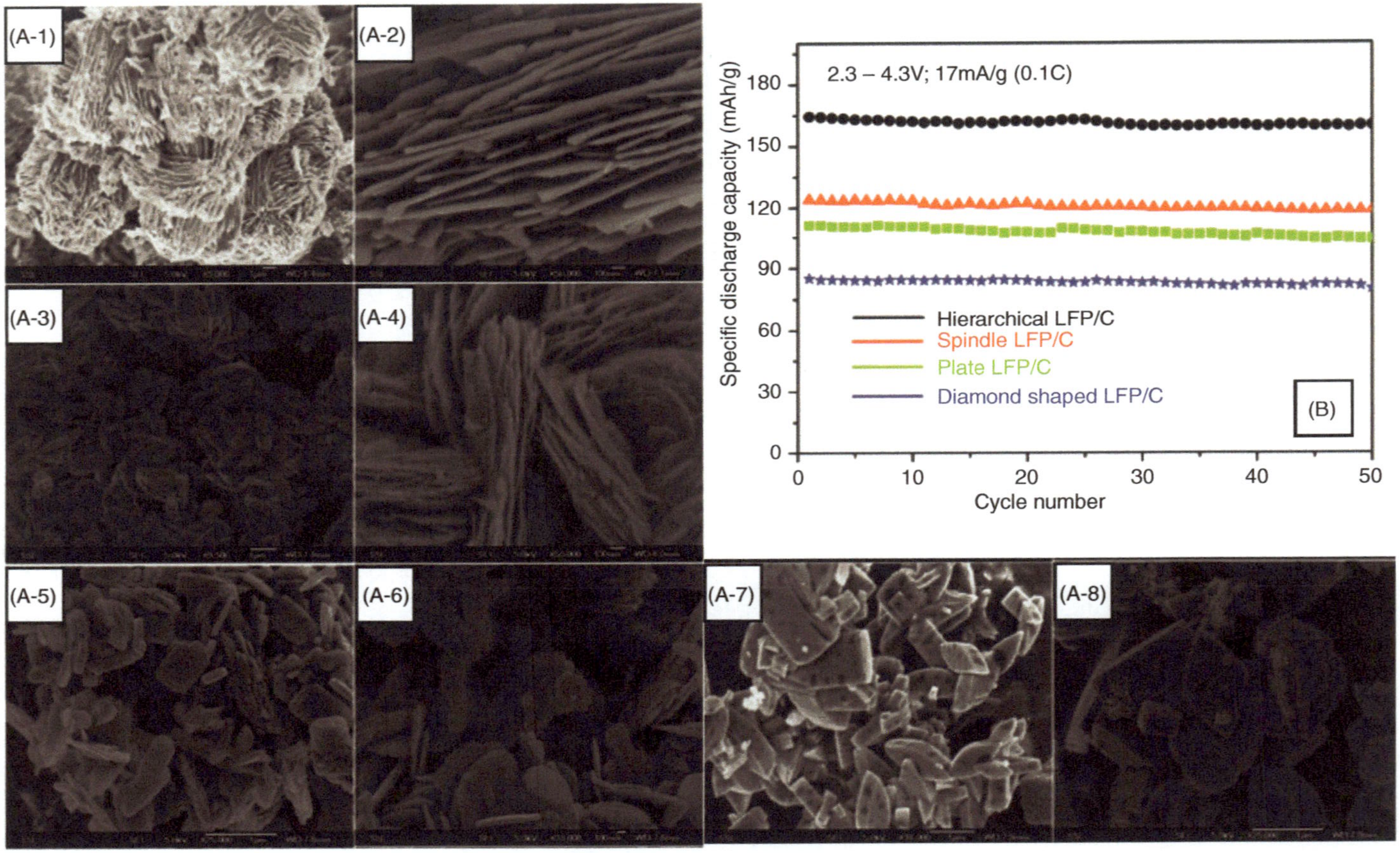

Figure 3.6 (A) SEM images of LFP/C nanoplates from different precursors at low and high magnifications (A-1–A-2) Fe(oxalate); (A-3–A-4) $Fe(acac)_3$; (A-5–A-6) Fe-PMIDA; and (A-7–A-8) $Fe(gluconate)_2$. *(Reproduced from K. Saravanan, P. Balaya, M. V. Reddy, B.V.R. Chowdari, J.J. Vittal, Morphology controlled synthesis of $LiFePO_4$/C nanoplates for Li-ion batteries, Energy Environ. Sci. 3 (2010) 457 [19] with permission from The Royal Society of Chemistry, Copyright © 2010).*

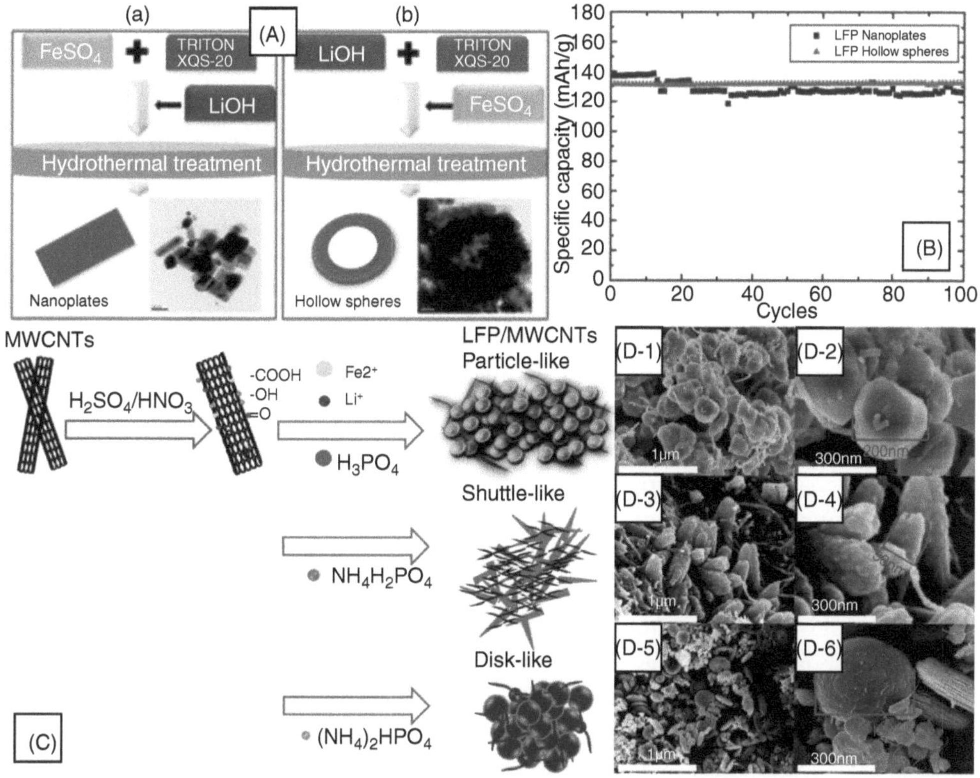

Figure 3.7 (A) Schematic illustration of synthesis of LFP nanoplates and hollow spheres, (a) nanoplates, (b) hollow spheres; (B) discharge capacities of the carbon-coated LFP nanoplates and hollow spheres upon cycling between 2.0 and 4.3 V at 0.5 C; (C) A schematic illustration of the synthesis procedure of the LFP/MWCNT nanoparticles; (D) SEM images of the LFP/MWCNT nanoparticles (D-1–D-2) particle-like, (D-3–D-4) shuttle-like, and (D-5–D-6) disk-like. *(Part A–B, reproduced from Y. Park, K.C. Roh, W. Shin, J. Lee, Novel morphology-controlled synthesis of homogeneous $LiFePO_4$ for Li-ion batteries using an organic phosphate source, RSC Adv. 3 (2013) 14263 [10] with permission from The Royal Society of Chemistry, Copyright © 2013; part C–D, reproduced from G. Qin, S. Xue, Q. Ma, C. Wang, The morphology controlled synthesis of 3D networking $LiFePO_4$ with multiwalled-carbon nanotubes for Li-ion batteries, CrystEngComm 16 (2014) 260–269 [11] with permission from The Royal Society of Chemistry, Copyright © 2014).*

source [10]. They have obtained nanoplates by mixing the iron source first, and hollow spheres by mixing the lithium source first (Fig. 3.7A). Both types of particles provided with good electrochemical properties (Fig. 3.7A). Qin et al. have synthesized 3D hierarchical LFP structures using multiwalled carbon nanotube in hydrothermal route [11] and the schematic illustration of the process is shown in Fig. 3.7C. They have reported different morphology of the networked structure by changing the phosphorous source (Fig. 3.7C). Among the different morphologies [particle-like (Fig. 3.7D-1–D-2), shuttle-like (Fig. 3.7D-3–D-4), and disk-like (Fig. 3.7D-5–D-6)], disk-like crystals offer the shortest Li^+ transfer distance and thus has shown the best properties (168.3 mAh/g at a rate of 0.1 C and 121.5 mAh/g at 30 C).

Ma et al. have studied the morphological and electrochemical effect of the ratio of solvents on the synthesis of LFP particles [12]. The schematic of the method employed is illustrated in Fig. 3.8A. They have used glycerol, water, and a mixture of glycerol and water as a solvent to synthesize nanoplate, hexagonal prism nanorod, and rectangular prism nanorod, respectively (Fig. 3.8A–D). The rectangular prism nanorods showed initial discharge capacity of 163.8 mAh/g at C/5 and over 75 mAh/g at 20 C; their performance has been attributed to shorter Li^+ diffusion length along [010] direction and moderate particle size. Song et al. have also studied the effect of solvent ratio on final LFP particle morphology and electrochemical property [13]. By changing the ratio of ethylene glycol and oleic acid, they have shown the morphological evolution from nanoplates to nanorods (Fig. 3.8E–F). Electrochemical properties of nanorods were excellent; they showed a discharge capacity of 155 mAh/g at 0.5 C and capacity retention of 80% at 5 C. The improved performance was attributed to the short Li^+ diffusion pathway and electron transportation due to their smaller size.

Zhou et al. have obtained a 3D hierarchical flowerlike LFP/C mesocrystalline structure using LiH_2PO_4 and ferrous sulfate heptahydrate, $FeSO_4{\cdot}7H_2O$, as starting material, and dimethylacetamide as a cosolvent with water and ethylene glycol [14]. First a nucleation-growth of LFP nanoplates occurred via Li_3PO_4, then the LFP nanoplates combined together due to interface energy or ion diffusivity to form flowerlike LFP mesocrystals as illustrated in Fig. 3.9A. Fig. 3.9B shows the SEM images of the pure (Fig. 3.9B-1–B-2) and carbon-coated (Fig. 3.9B-3–B-4) flowerlike LFP mesocrystals. The flowerlike LFP mesocrystals showed superior electrochemical performance. The carbon-coated LFP mesocrystals showed the high initial capacity of 161 mAh/g at 0.1 C, they also showed high-rate performance and cycle stability.

Lou et al. have synthesized microspheres of LFP/C cathode materials composed of densely packed nanoplates from the precursor $FePO_4{\cdot}2H_2O$ of same morphology [15]. Both the precursor and the final LFP/C cathode material have quasi-microspheres diameter of 1 μm that is composed of 30-nm thick nanoplates—the LFP/C cathode materials has an amorphous carbon coating layer of ~2 nm at the surface (Fig. 3.9C-1 and C-2, respectively). The $FePO_4{\cdot}2H_2O$ precursor has been synthesized by sodium dodecyl sulfate assisted a hydrothermal method. The final LFP/C composite was synthesized by a rheological phase reaction method (extended reading [20]) followed by calcination at 650°C. The synthesized LFP/C composite showed excellent electrochemical properties and exhibited high-rate capabilities with a capacity of 116, 96, and 75 mAh/g at 10 C, 20 C, and 30 C, respectively (Fig. 3.9D-1–D-3). Also, the LFP/C cathode materials showed high tap density. Wang et al. have followed a very similar method to obtain the micro-nano hierarchical structure of LFP/C composite for high-rate performance and high tap density [16]. A monoclinic phase $FePO_4{\cdot}2H_2O$ precursor (SEM images of the

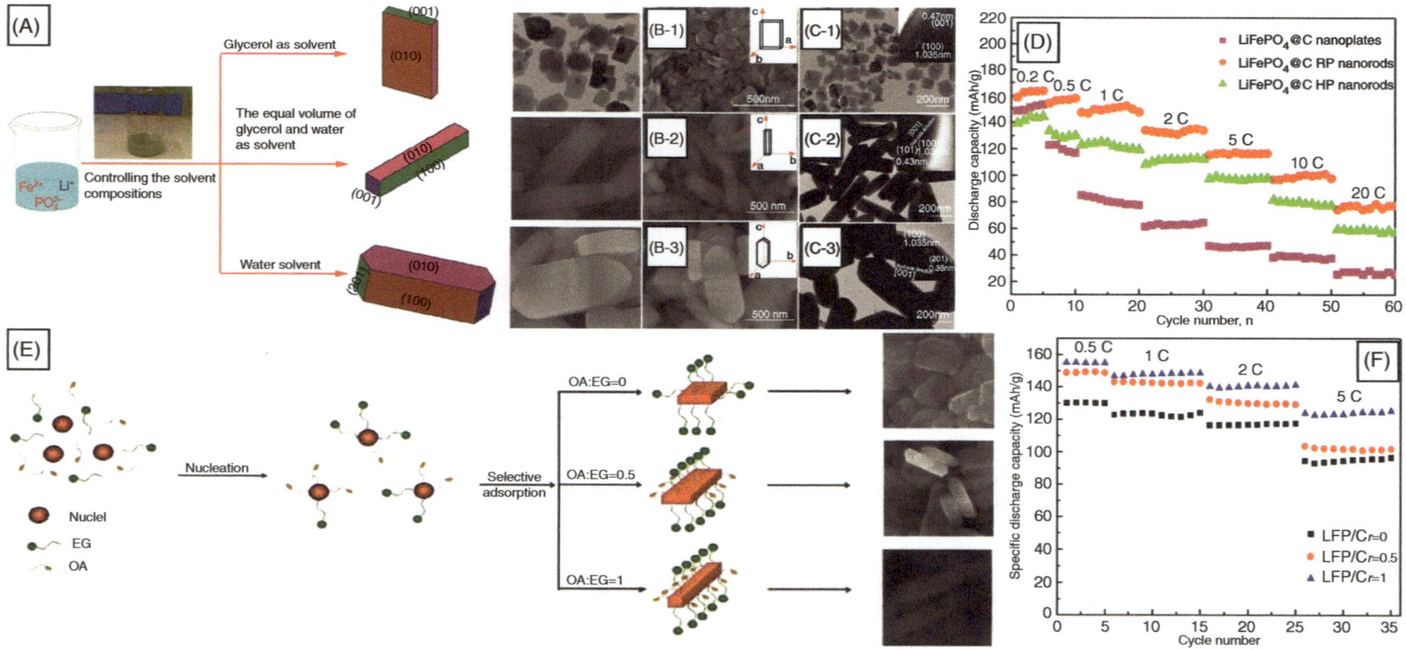

Figure 3.8 (A) Effects of the solvent compositions on the morphological transformations of LFP nanoparticles; SEM and TEM images of samples prepared from different ratio of glycerol to water (B-1, C-1) 2:0, (B-2, C-2) 1:1, (B-3, C-3) 0:2; (D) rate performance of LFP/C nanoplates, rectangular prism nanorods, and hexagonal prism nanorods; (E) morphology evolution mechanism with different volume ratios of OA/EG and (F) their cycle performance at different rate. *(Part A–D, reprinted with permission from Z. Ma, G. Shao, Y. Fan, G. Wang, J. Song, T. Liu, Tunable morphology synthesis of $LiFePO_4$ nanoparticles as cathode materials for lithium ion batteries, ACS Appl. Mater. Interfaces 6 (2014) 9236–9244 [12], Copyright © 2011 American Chemical Society; part E, reproduced from J. Song, L. Wang, G. Shao, M. Shi, Z. Ma, G. Wang, et al., Controllable synthesis, morphology evolution and electrochemical properties of $LiFePO_4$ cathode materials for Li-ion batteries, Phys. Chem. Chem. Phys. 16 (2014) 7728 [13] with permission from The Royal Society of Chemistry, Copyright © 2014).*

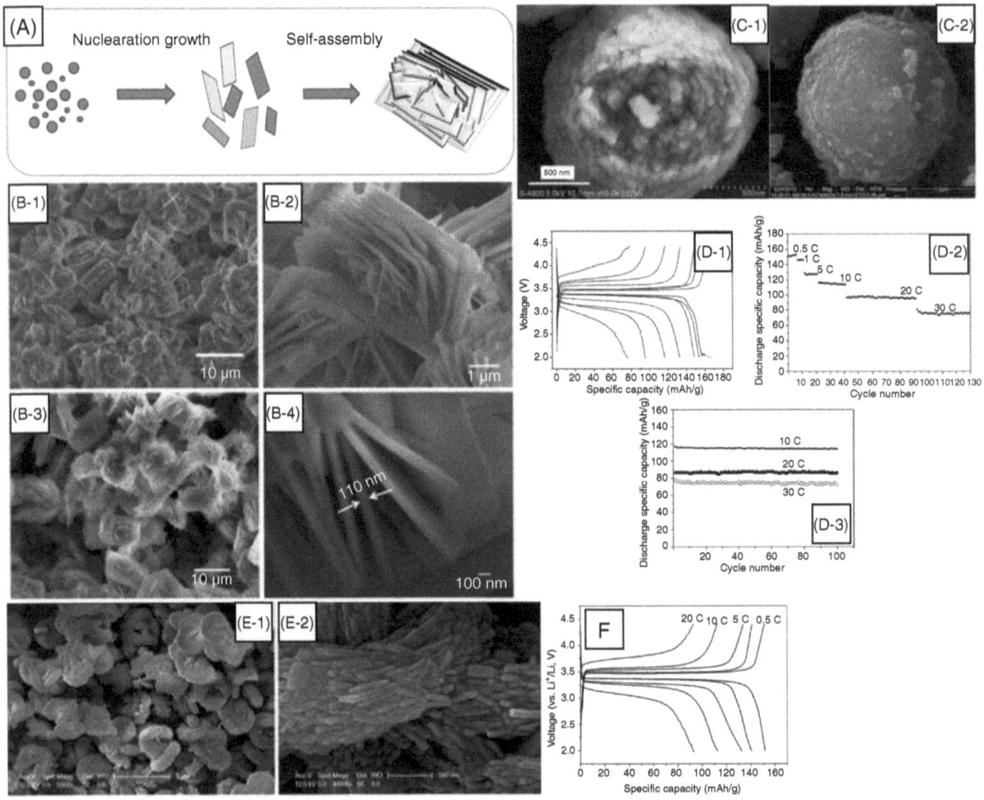

Figure 3.9 (A) Schematic illustration of the formation of hierarchical flowerlike LFP mesocrystals; (B) SEM images of pure (B-1–B-2) and carbon-coated (B-3–B-4) hierarchical flowerlike LFP mesocrystals; (C) SEM images of (C-1) precursor and (C-2) LFP microsphere; (D-1) charge–discharge curves at various current rates; (D-2) the specific capacities at different rates; (D-3) the high rates capacity retention for the LFP/C sample at 2.0–4.4 V; (E) SEM image of the precursor at (E-1) low and (E-2) high magnification; (F) charge/discharge profile at various rate. *(Part A–B, reproduced from N. Zhou, E. Uchaker, H.-Y. Wang, M. Zhang, S.-Q. Liu, Y.-N. Liu, et al., Additive-free solvothermal synthesis of hierarchical flower-like $LiFePO_4$/C mesocrystal and its electrochemical performance, RSC Adv. 3 (2013) 19366 [14] with permission from The Royal Society of Chemistry, Copyright © 2013; part C–D, reproduced from X. Lou, Y. Zhang, Synthesis of $LiFePO_4$/C cathode materials with both high-rate capability and high tap density for lithium-ion batteries, J. Mater. Chem. 21 (2011) 4156 [15] with permission from The Royal Society of Chemistry, Copyright © 2011; part E–F, reproduced from M. Wang, Y. Yang, Y. Zhang, Synthesis of micro-nano hierarchical structured $LiFePO_4$/C composite with both superior high-rate performance and high tap density, Nanoscale 3 (2011) 4434 [16] with permission from The Royal Society of Chemistry, Copyright © 2011).*

precursor at low and high resolution is shown in Fig. 3.9E-1 and E-2, respectively) with micro-nano hierarchical structure is first synthesized by hydrothermal method. Then, the final LFP/C composite of the same morphology is obtained by a simple rheological phase method. The synthesized LFP/C composite showed the high capacity of 120, 110, and 90 mAh/g at 5 C, 10 C, and 20 C, respectively (Fig. 3.9F). Also, the composite showed high tap density. With improved rate capabilities and high tap density, LFP/C composites are even closer to large-scale applications.

Hydrothermal/solvothermal methods are also employed for synthesizing other cathode materials. For $LiNi_{0.5}Mn_{1.5}O_4$ (LNMO) cathode materials, the hydrothermal method can result in better dispersity of the reagents and smaller particle size [17,18]. These effectively reduce the impurity contents and shortens the Li^+ diffusion path length. Thus, they result in higher capacity and better cycling stability.

Xue et al. have synthesized LNMO nanoparticles for long-term cyclability at a high rate with an ethanol-assisted hydrothermal method [17]. For comparison, they have also synthesized LNMO cathode material with precipitation and hydrothermal method. The ethanol-assisted hydrothermal method resulted in smaller particle size (Fig. 3.10A) than compared to precipitation (Fig. 3.10B) and hydrothermal synthesized particles (Fig. 3.10C), which had bigger particles of around 2 μm. The addition of ethanol reduces the solvent dielectric constant, which results in the smaller particle sizes as reported in [21,22]. Also, ethanol resulted in LNMO with lower secondary phase or Mn^{3+} impurity content by improving the dispersity of the reagents and distributing them homogenously. Smaller particle size led the LNMO to shorter Li^+ diffusion path, and the lower impurity resulted in improved Li^+ mobility. Thus, the synthesized cathode material showed excellent long-term cyclability with high rate performance. The LNMO cathode material showed discharge capacity as high as 81.7 mAh/g at 20 C. At 5 C rate, the LNMO cathode showed a capacity of 102.1 mAh/g after 1000 cycles with a capacity retention of 88.1%.

Similarly, Wang et al. have synthesized peanutlike LNMO cathode material in an ethylene glycol (EG) assisted hydrothermal method [18]. Fig. 3.10D shows the SEM images if the synthesized LNMO nanoparticles. Ethylene glycol also improves the dispersity of the solution and mixes the reagents more homogenously. This resulted in smaller particle size, as well as higher crystallinity, low secondary phase, and lower Mn^{3+} impurity contents. The synthesized LNMO cathode materials showed better capacity retention and higher rate performance compared to the LNMO, made a hydrothermal method in the absence of EG as shown in Fig. 3.10E. The material synthesized by EG-assisted hydrothermal method showed 96.4% capacity retention after 100 cycles at 1 C (Fig. 3.10F). Also, they showed discharge capacity of 121.4 mAh/g at 10 C.

The hierarchical structure of LMR-NMC synthesized by the solvothermal method is also beneficial for their electrochemical properties since they can effectively shorten the Li^+ diffusion lengths as well as provides good structural stability [23]. Zhang et al. have synthesized a peanutlike hierarchical micro/nano LMR-NMC cathode material by solvothermal method followed by a calcination process. Fig. 3.10G shows the SEM image of the synthesized cathode material. The precursor oxide materials were first prepared by a solvothermal method with 2:1 ratio of ethylene glycol and ethanol, which showed peanutlike morphology. After calcination with $LiOH.H_2O$, the morphology is still the same with larger primary particles (100–300 nm) and pores (>50 nm). For

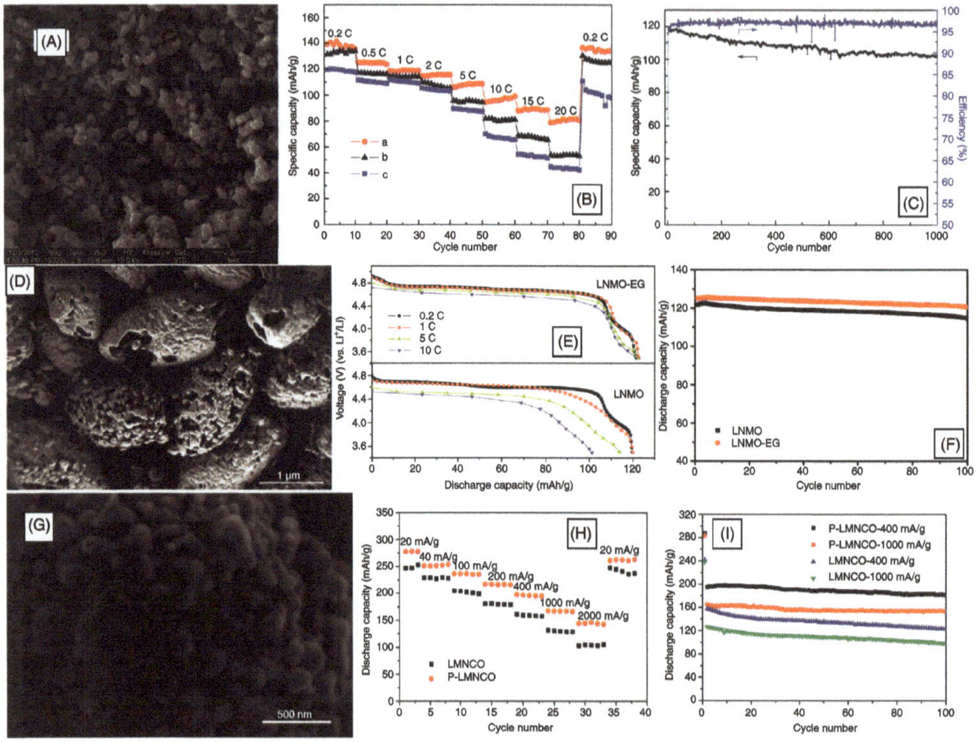

Figure 3.10 (A) SEM image of synthesized LNMO nanoparticles; (B) rate performance at different rates, and (C) cycle performance at 5 C; (D) SEM image of synthesized LNMO-EG; (E) charge/discharge curve at different rate; and (F) cycle performance curve; (G) SEM image of the synthesized cathode material; (H) rate performance; and (I) cycle performance at different rates. *(Part A–C, reproduced from Y. Xue, Z. Wang, F. Yu, Y. Zhang, G. Yin, Ethanol-assisted hydrothermal synthesis of $LiNi_{0.5}Mn_{1.5}O_4$ with excellent long-term cyclability at high rate for lithium-ion batteries, J. Mater. Chem. A 2 (2014) 4185 [17] with permission from The Royal Society of Chemistry, Copyright © 2014; part D–F, reproduced from L. Wang, G. Liu, W. Wu, D. Chen, G. Liang, Synthesis of porous peanut-like $LiNi_{0.5}Mn_{1.5}O_4$ cathode materials through an ethylene glycol-assisted hydrothermal method using urea as a precipitant, J. Mater. Chem. A. 3 (2015) 19497–19506, Y. Xue, Z. Wang, F. Yu, Y. Zhang, G. Yin; Ethanol-assisted hydrothermal synthesis of $LiNi_{0.5}Mn_{1.5}O_4$ with excellent long-term cyclability at high rate for lithium-ion batteries, J. Mater. Chem. A 2 (2014) 4185 [17,18] with permission from The Royal Society of Chemistry, Copyright © 2015; part G–I, reproduced from Y. Zhang, Y. Li, X. Niu, D. Wang, D. Zhou, X. Wang, et al., A peanut-like hierarchical micro/nano-$Li_{1.2}Mn_{0.54}Ni_{0.18}Co_{0.08}O_2$ cathode material for lithium-ion batteries with enhanced electrochemical performance, J. Mater. Chem. A 3 (2015) 14291–14297 [23] with permission from The Royal Society of Chemistry, Copyright © 2015).*

comparison, LMR-NMC cathode materials were also synthesized with the sol–gel process. The peanutlike LMR-NMC cathode materials were found to be highly pure and with better hexagonal crystalline ordering. The prepared LMR-NMC cathode material showed the high initial capacity of 229.9 mAh/g at 200 mA/g compared to the sol–gel process cathode material that showed a capacity of only 174.1 mAh/g (Fig. 3.10H). Also, the capacity retention for peanutlike LMR NMC was 94.2% (Fig. 3.10I). Even at a high rate, such as 2000 mA/g, the peanutlike LMR-NMC showed high discharge capacity of 145 mAh/g as displayed in Fig. 3.10H.

Solvothermal/hydrothermal synthesis is a simple and low-cost route to synthesize cathode nanostructures with high yield and particle size as low as 20 nm [8]. The enhanced performance of the cathode materials is due to the reduced Li^+ diffusion path and better electron transport within the synthesized nanostructures. Hydrothermal synthesis method with ethanol or ethylene glycol additive can help to further reduce the particle size [17]. However, additive-free methods have also been reported to synthesize finer and desired nanostructures [14]. The major drawback is that the size and shape control of the final nanostructure is somewhat difficult to obtain because of multiple process variables (type, volume ratio, and concentration; precursor source and mixing sequence; process time, pressure, and drying temperature) that are involved (for extended reading, see Refs. [9–13,19]). Nonetheless, the solvothermal/hydrothermal method has been considered in recent time as one of the most popular methods to synthesize nanostructured cathode materials, especially for $LiFePO_4$ (LFP) based cathodes [7–15,19,20].

2.3 Solid-state reaction methods

Solid-state reaction is a common synthesis method to obtain polycrystalline material from solid reagents. For the reaction to occur usually a very high temperature is employed. Factors that affect solid-state reaction are chemical and morphological properties of the reagents including the reactivity, surface area, and free energy change with the solid-state reaction, and other reaction conditions, such as the temperature, pressure, and the environment of the reaction. The advantage of solid-state reaction method includes the simplicity and large-scale production [24].

Li et al. have synthesized nanoparticles of LFP/C composites from different surfactants and their combinations [25]. They found that the amount of graphitic carbon and the particle sizes of the LFP/C composites are affected by the structures of the surfactants. Briefly, the surfactant with longer chain length (Tween 80) effectively prevented particle growth and reduced the particle size, whereas the shorter surfactant (Tween 20) formed more carbon during pyrolysis as illustrated in Fig. 3.11A. The TEM images of the LFP/C composite synthesized with different surfactants Tween 80, Tween 40, and Tween 20 are shown in Fig. 3.11B-1, B-2, and B-3, respectively. Fig. 3.11C shows that the combination of Tween 80 to Tween 20 in a ratio of 1.5:1 resulted in smaller size of particles (Fig. 3.11C-2) as well as with graphitelike carbon layer, and showed high electrochemical performance (discharge capacity of 167.3 mAh/g at 0.1 C, 144.4 mAh/g at 1 C, and 129.4 mAh/g at 5 C with good retention up to 100 cycles).

The successful synthesis of LNMO cathode materials with hollow structures has also been reported by solid-state reaction [26–28]. Zhou et al. have prepared LNMO hollow microspheres/microcubes by impregnation of LiOH and $Ni(NO_3)_2$ in MnO_2 microspheres/microcubes followed by the solid-state reaction [26]. Fig. 3.12A shows

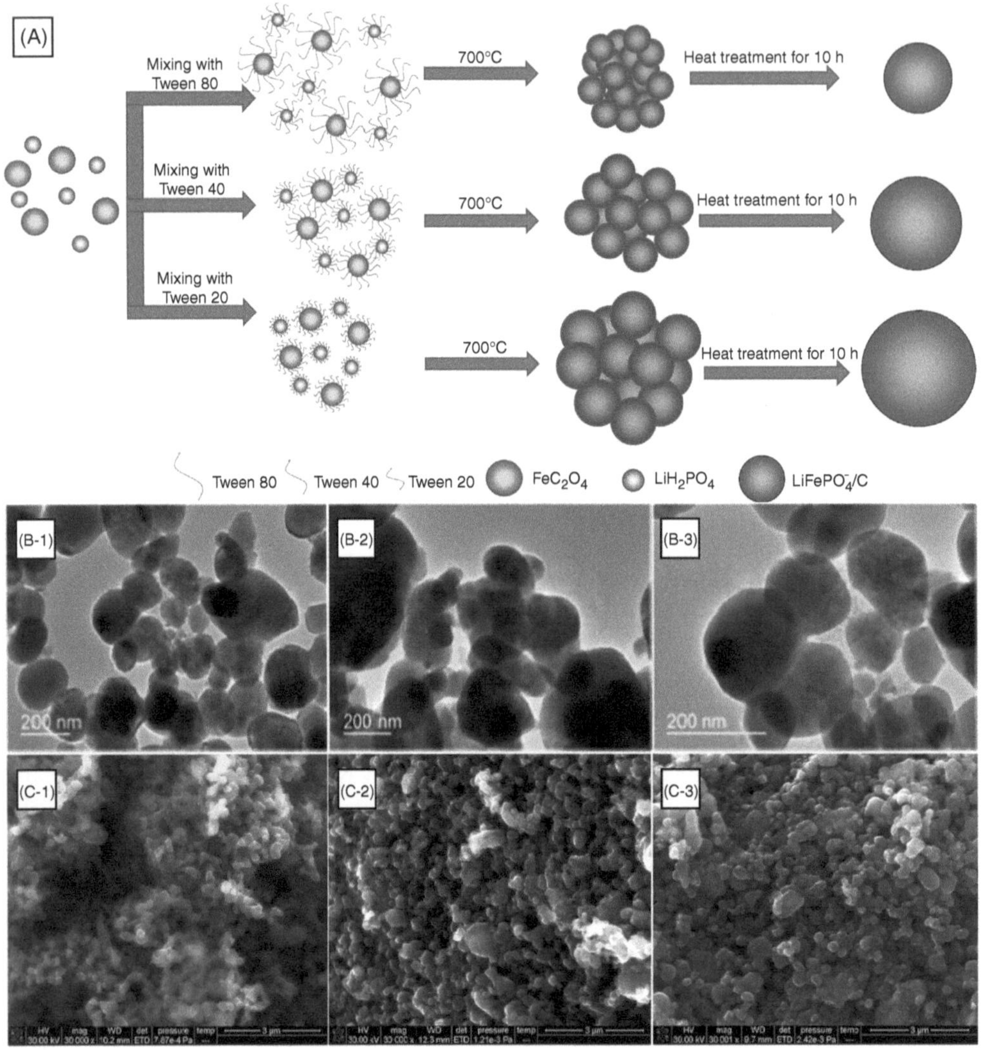

Figure 3.11 (A) Preparation process of the nano-LFP/C composite with different surfactants; (B) TEM images of the LFP/C composite synthesized with different surfactants (B-1) Tween 80, (B-2) Tween 40, and (B-3) Tween 20; (C) SEM images of LFP/C composites synthesized with different ratios of Tween 80 to Tween 20 (C-1) 2.5, (C-2) 1.5, and (C-3) 0.5. *(Reproduced from Q. Li, F. Zheng, Y. Huang, X. Zhang, Q. Wu, D. Fu, et al., Surfactants assisted synthesis of nano-$LiFePO_4$/C composite as cathode materials for lithium-ion batteries, J. Mater. Chem. A 3 (2015) 2025–2035 [25] with permission from The Royal Society of Chemistry, Copyright © 2015).*

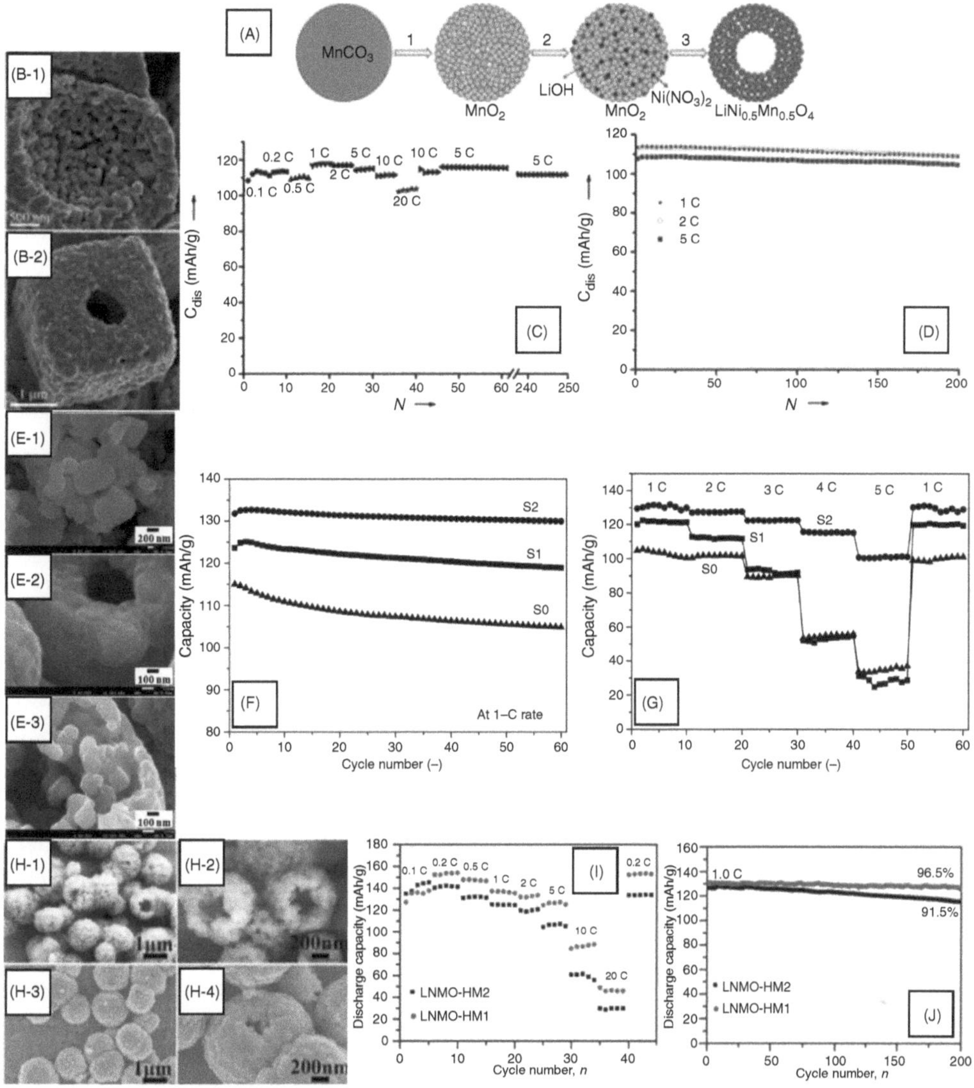

Figure 3.12 (A) Illustration of the fabrication of LNMO hollow microstructures; (B) SEM images of (B-1) hollow microspheres, and (B-2) hollow microcubes; (C) cycling performance at various rates (0.1 C–20 C) and (D) cycling performance at 1 C, 2 C, and 5 C for 200 cycles; (E) SEM images of LNMO made in three different methods (E-1) traditional method, (E-2) dense Mn_2O_3 microspheres, and (E-3) hollow Mn_2O_3 microspheres; (F) cycle performance, and (G) rate performance of different LNMO samples; (H) SEM images of LNMO made from (H-1–H-2) Mn_2O_3 microspheres and (H-3–H-4) MnO_2 microspheres; (I) rate capability and (J) cycle performance of both LNMO samples. *(Part A–D, reproduced from L. Zhou, D. Zhao, X. Lou, $LiNi_{0.5}Mn_{1.5}O_4$ hollow structures as high-performance cathodes for lithium-ion batteries, Angew. Chem. Int. Ed. 51 (2012) 239–241 [26] with permission, Copyright © 2012 Wiley–VCH Verlag GmbH & Co. KGaA; part E–G, reproduced from C. Zhu, T. Akiyama, Designed synthesis of $LiNi_{0.5}Mn_{1.5}O_4$ hollow microspheres with superior electrochemical properties as high-voltage cathode materials for lithium-ion batteries, RSC Adv. 4 (2014) 10151 [27] with permission from The Royal Society of Chemistry, Copyright © 2014; part H–J, reproduced from H. Luo, P. Nie, L. Shen, H. Li, H. Deng, Y. Zhu, et al., Synthesis of $LiNi_{0.5}Mn_{1.5}O_4$ hollow microspheres and their lithium-storage properties, ChemElectroChem 2 (2015) 127–133 [28] with permission, Copyright © 2015 Wiley–VCH Verlag GmbH & Co. KGaA).*

the schematic of the process employed to prepare hollow nanostructures. A mechanism analogous to Kirkendall effect is suggested to explain the formation of hollow structure. The fast outward diffusion of Mn and Ni atoms and the slow inward diffusion of O atoms are proposed to be responsible for the formation of the hollow cavity in the LNMO microspheres/microcubes. The hollowness was confirmed by the SEM images shown in Fig. 3.12B-1–B-2. The LNMO microspheres showed excellent cycling and rate performance. At 1 C, 2 C, and 5 C, the discharge capacity of 118, 117, and 115 mAh/g was reported (Fig. 3.12C). At 2 C rate, capacity retention after 200 cycles was as high as 96.6%. The microcubes also resulted in high rate and cycling performance. At 2 C rate, the LNMO microcubes showed the initial capacity of 124 mAh/g and 97.6% capacity retention up to 200 cycles as shown in Fig. 3.12D.

Zhu et al. have also synthesized LNMO hollow microspheres in a similar route [27]. They have used Mn_2O_3 dense and hollow microspheres as starting material where the impregnation of Li and Ni precursors were done to synthesize LNMO hollow microspheres in solid-state reaction (Fig. 3.12E-1–E-3). The LNMO hollow microspheres synthesized from Mn_2O_3 hollow microspheres are composed of porous walls with subparticles of 50–200 nm size. Also, the undesired grain growth was subsidized. The porous structure effectively facilitates the rapid transfer of Li^+ insertion/extraction, and the smaller size reduces the Li^+ diffusion length. Also, the porous framework can accommodate any strain or volume change during Li^+ transfer. These resulted in high rate capability and long-term cyclability of LNMO cathode materials made from them. The initial discharge capacity of 131.77 mAh/g at 1 C rate were reached with 98.6% capacity retention after 60 cycles (Fig. 3.12F). The material also showed discharge capacity as high as 100.5 mAh/g at a 5 C rate, as illustrated in Fig. 3.12G.

Recently, Luo et al. have synthesized LNMO hollow microspheres from $MnCO_3$ dense microspheres or MnO_2 hollow microspheres as starting materials [28]. Impregnation of Li and Ni precursors followed by the solid-state reaction is employed to synthesize porous hollow microspheres of LNMO cathode material. Fig. 3.12H-1–H-2 shows the SEM images of the synthesized LNMO from $MnCO_3$, and the LNMO synthesized from MnO_2 are shown in SEM images of Fig. 3.12H-3–H-4. LNMO prepared from $MnCO_3$ showed high initial discharge capacity of 137.3 mAh/g at 0.1 C and 96.5% capacity retention at 1 C up to 200 cycles as shown in Fig. 3.12I and J, respectively. MnO_2 precursor route resulted in LNMO material that showed initial discharge capacity of 140.4 mAh/g at 0.1 C and 91.5% capacity retention at 1 C up to 200 cycles.

In short, solid-state reactions method can be widely used to produce polycrystalline cathode materials with high-throughput that can be obtained in a continuous manner. Hollow and porous cathode nanostructures can be easily synthesized with short Li^+ diffusion path lengths for high rate and long-term cycling stability. However, in terms of nanostructure synthesis, this method lacks of good control over the final size, and shape of the material. Also, it is difficult to obtain nanostructured materials with well-controlled morphology since the starting reagent materials are solids and do not always mix well. The readers are recommended to reference [24] for further learning about solid-state reactions.

2.4 Coprecipitation methods

Chemical coprecipitation method can be realized by the kinetics of nucleation and growth in a homogeneous solution. The understanding of supersaturation, nucleation, and growth and postgrowth (Ostwald ripening) is critical to synthesize ordered nanostructures. Multiple well-studied factors, such as temperature, pressure, concentration, pH, stirring speed, reaction time, as well as the nature of the precursors and the presence of surfactants or other growth inhibitor play an important role in controlling the size, shape and the size distribution of the final nanostructures. It is one of the most common wet chemical methods to synthesize cathode materials, yet it is difficult to synthesize controllable nanostructure using coprecipitation method.

Chemical coprecipitation is a well-studied method for cathode material synthesis, as it is also simple, straightforward, and low cost [29]. Hua et al. have utilized a modified carbonate coprecipitation method with a vacuum distillation-crystallization process to synthesize NMC cathode material and found that the modified method resulted in a better hexagonal layered structure with smaller particle size [30]. For comparison, they also synthesized NMC cathode material by regular coprecipitation route. They found that the modified method resulted in a hexagonal layered structure with the spherical morphology of porous structure of the NMC cathode materials of 110.5 nm size (Fig. 3.13A-2), whereas the regular coprecipitation method resulted in a denser and larger particle morphology (Fig. 3.13A-1). The modified method resulted in superior NMC cathode material in terms of their rate capability and cyclability. At the different rate of 5 C, 10 C, and 20 C, the material showed discharge capacity of 134.96, 123.02, and 112.64 mAh/g (Fig. 3.13B, labeled as LNCM-2). Also, at 1 C, the NMC cathode material synthesized by modified coprecipitation method showed high capacity retention of 94.02% after 100 cycles (Fig. 3.13C labeled as LNCM-2), comparing to 86.26% in the cathode made by the regular method (Fig. 3.13C labeled as LNCM-1). Smaller particle size and bigger Li^+ transfer area have been attributed to the improved performance of the nanoarchitectured NMC cathode material.

Using coprecipitation method, Yang et al. has synthesized core–shell structure of Li-rich cathode material with $Li_{1.15}[Ni_{1/3}Co_{1/3}Mn_{1/3}]_{0.85}O_2$ in core and $Li_{1.15}[Ni_{1/4}Mn_{3/4}]_{0.85}O_2$ in the shell [31]. The approach is illustrated in the Fig. 3.13D. Due to the small size, spherical morphology of both the core and shell material and the synergetic effect of the two materials achieved by the formation of the core–shell architecture, the core–shell cathode showed higher energy density, longer cycle life, and better rate performance. The initial discharge capacity of 242 mAh/g at a rate of 0.1 C and a capacity retention of 95.6% at 0.5 C rate up to 50 cycles have been reported (Fig. 3.13E).

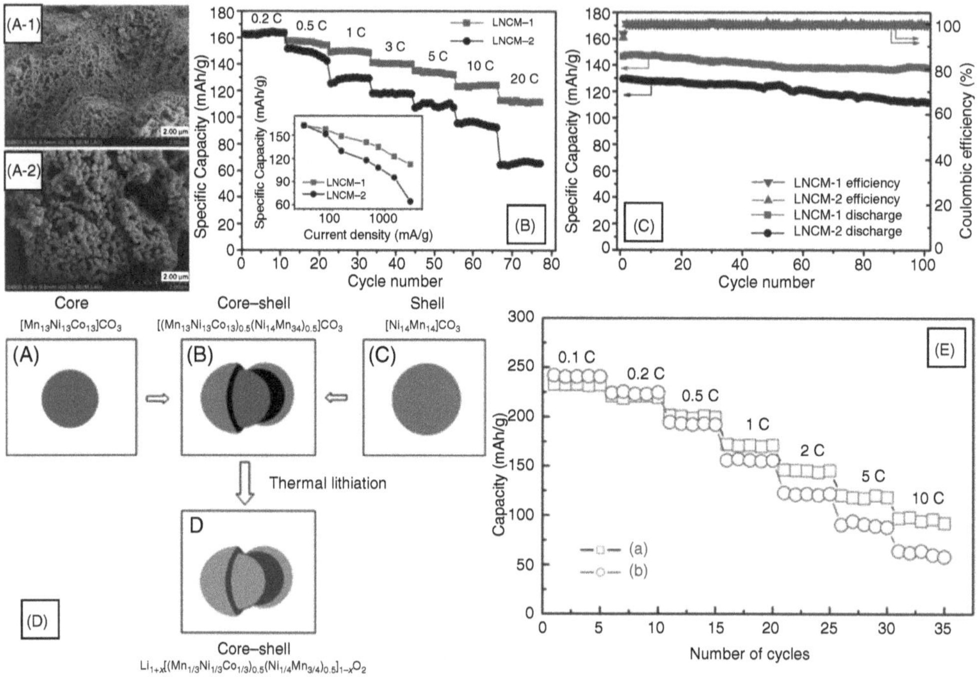

Figure 3.13 (A) SEM images of (A-1) NMC precursor and (A-2) cathode material made by modified coprecipitation method; (B) rate capability and (C) cycle performance of NMC made by modified and traditional method; (D) schematic diagram of the formation process of the spherical core–shell structure; (E) capacity retention at different rates for (a) core–shell structure and (b) regular structure (core only). *(Part A–C, reproduced from W. Hua, Y. Wang, Y. Zhong, G. Wang, B. Zhong, B. Fang, et al., An approach towards synthesis of nanoarchitectured $LiNi_{1/3}Co_{1/3}Mn_{1/3}O_2$ cathode material for lithium ion batteries, Chin. J. Chem. 33 (2015) 261–267 [30] with permission, Copyright © 2015 Wiley-VCH Verlag GmbH & Co. KGaA; part D–E, reproduced from X. Yang, X. Wang, Q. Wei, H. Shu, L. Liu, S. Yang, et al., Synthesis and characterization of a Li-rich layered cathode material $Li_{1.15}[(Mn_{1/3}Ni_{1/3}Co_{1/3})_{0.5}(Ni_{1/4}Mn_{3/4})_{0.5}]_{0.85}O_2$ with spherical core–shell structure, J. Mater. Chem. 22 (2012) 19666 [31] with permission from The Royal Society of Chemistry, Copyright © 2012).*

Using a combination of coprecipitation and solid-state method, Remith et al. synthesized hierarchically arranged nanoparticles constructed Li-rich cathode microspheres consisting of both the rhombohedral and monoclinic phase [32]. MnO_2 microspheres are formed as an intermediate, which produces the porous and hollow microspheres of $Li_{1.2}Mn_{0.6}Ni_{0.1}Co_{0.1}O_2$ upon calcination. The SEM images of the synthesized material and the schematic of the synthesis process is shown in Fig. 3.14A and B, respectively. The growth mechanism is again explained by Kirkendall effect and Ostwald ripening. The hierarchically arranged and interconnected nanostructure showed a higher steady-state capacity of 180 mAh/g up to 35 cycles and reaching 97% coulombic efficiency, which is attributed to the synergistic effect of the two phases and the unique morphology

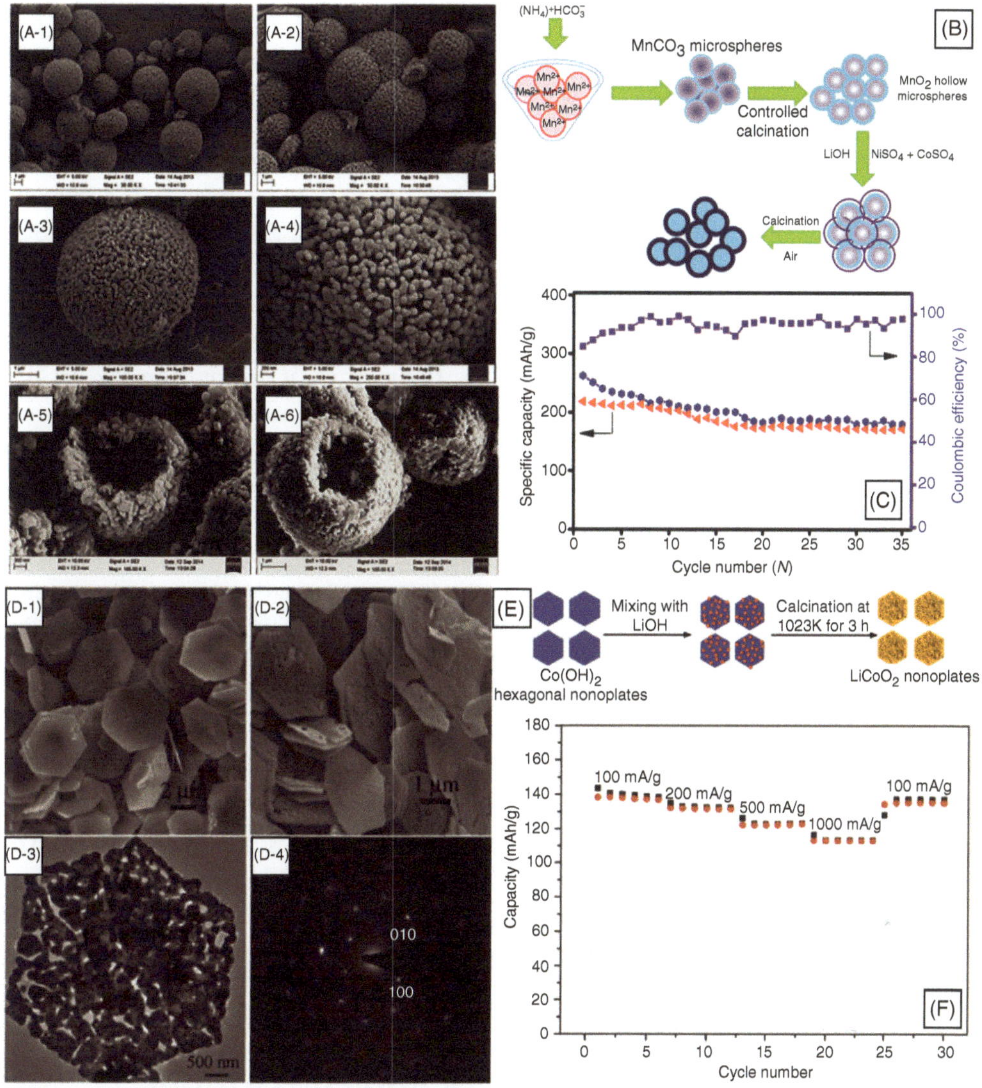

Figure 3.14 (A) SEM images of synthesized hollow microspheres at different resolutions; (B) synthesis process diagram of the synthesized material; (C) cycle performance and coulombic efficiency of the synthesized cathode; (D) SEM image of (D-1) β-$Co(OH)_2$, and (D-2) LCO nanoplates; (D-3) HRTEM image of LCO nanoplates; (D-4) SAED image of LCO nanoplates; (E) two-step synthesis route of LCO nanoplates; and (F) rate capability of LCO nanoplates. *(Part A–C, reproduced from P. Remith, N. Kalaiselvi, $Li_{1.2}Mn_{0.6}Ni_{0.1}Co_{0.1}O_2$ microspheres constructed by hierarchically arranged nanoparticles as lithium battery cathode with enhanced electrochemical performance, Nanoscale 6 (2014) 14724–14732 [32] with permission from The Royal Society of Chemistry, Copyright © 2014; part D–F, reprinted from X. Xiao, X. Liu, L. Wang, H. Zhao, Z. Hu, X. He, et al., $LiCoO_2$ nanoplates with exposed (001) planes and high rate capability for lithium-ion batteries, Nano Res. 5 (2012) 395–401 [33], Copyright © 2012, with permission from Springer).*

of the structure. The as-synthesized cathode showed discharge capacity of 242 mAh/g at 50 mA/g as illustrated in Fig. 3.14C.

Xiao et al. have also used a combination of coprecipitation and solid-state method to synthesize nanostructured cathode material [33]. Fig. 3.14D-1 shows the SEM image of precursor β-$Co(OH)_2$ nanoplates. SEM, HRTEM, and SAED image of the synthesized LCO nanoplates are shown in Fig. 3.14D-2, D-3, and D-4, respectively. The schematic of the synthesis steps is shown in Fig. 3.14E. They reported the synthesis of $LiCoO_2$ nanoplates with exposed (001) planes, which facilitate better Li^+ diffusion to its perpendicular planes, and thus resulted in better electrochemical properties. The as-synthesized material showed discharge capacity of 140 mAh/g at 100 mA/g rate; at a high rate of 1000 mA/g, the discharge capacity is as high as 113 mAh/g, as displayed in Fig. 3.14F.

Coprecipitation is a simple and low cost method for synthesizing cathode nanostructures with enhanced properties but is influenced by multiple process factors to determine the final morphology (Refs. [29,30] for extended reading). It is possible to yield cathode nanostructures with good control and it utilizes the homogeneous mixing of the reactants that can avoid high-temperature sintering. On the other side, since multiple metal ions are involved in coprecipitation, which may possess different solubility and precipitation rate, it is difficult to control the precipitation. In general, with coprecipitation method, it is difficult to obtain nanostructure of well-controlled size, shape, and purity.

In summary, nanosynthesis has resulted in the exceptional performance of cathode materials especially by introducing large Li^+ ion exchange area, and proving short Li^+ diffusion paths. Advancements in the nanosynthesis methods have gained precise control over size, shape, and reproducibility with low or medium processing costs. The big challenge for nanotechnology in LIBs still remains in the development of transferable mass production technology. Other challenges include the lack of mechanical strength and lower throughput. However, nanotechnology is a fast-growing field, and the gaps are minimizing rapidly.

3 NANOSCALE SURFACE MODIFICATION ON CATHODE MATERIALS FOR LITHIUM-ION BATTERIES

Surface modification on cathodes is the simplest way to overcome the cathode capacity limitations related to side reactions with electrolytes and surface structural changes; and will make better cathodes with enhanced electrochemical performance. Surface modification in simple words is surface coating or growing a thin protective layer either on the cathode powder or on the surface of the cathode electrode. Surface coatings can be performed with chemically inert materials [34–38] or conductive materials [39–41] as a protective layer on the cathode. The coating needs to be continuous and conformal. The protective and enhancing ability of the surface coating also majorly depends on the type, amount, and thickness of the coating material. The trend is moving toward growing nanoscale conformal coatings for improved cathode performance. In the later sections of this chapter, some of the promising nanoscale thin film deposition techniques are introduced, which includes atomic layer deposition (ALD), chemical vapor deposition (CVD), sputtering/physical vapor deposition (PVD), and wet-coating/sol–gel methods.

3.1 Atomic layer deposition

ALD has the potential to offer true nanoscale-thick surface depositions, which are highly conformal and pinhole free [34–37]. ALD is an ultrathin film deposition technique that is controlled by gas phase and sequential self-limiting chemical reactions of the precursors at the material surface. Most ALD processes typically require two precursors, which are supplied in sequence one at a time to contribute to surface coating. Fig. 3.15A shows the schematic of Al_2O_3 ALD on $LiCoO_2$ (LCO) material using trimethylaluminum [$Al(CH_3)_3$] and water (H_2O) as the precursors in sequence. The precursors sequence cycle can be repeated to deposit thicker coatings if needed.

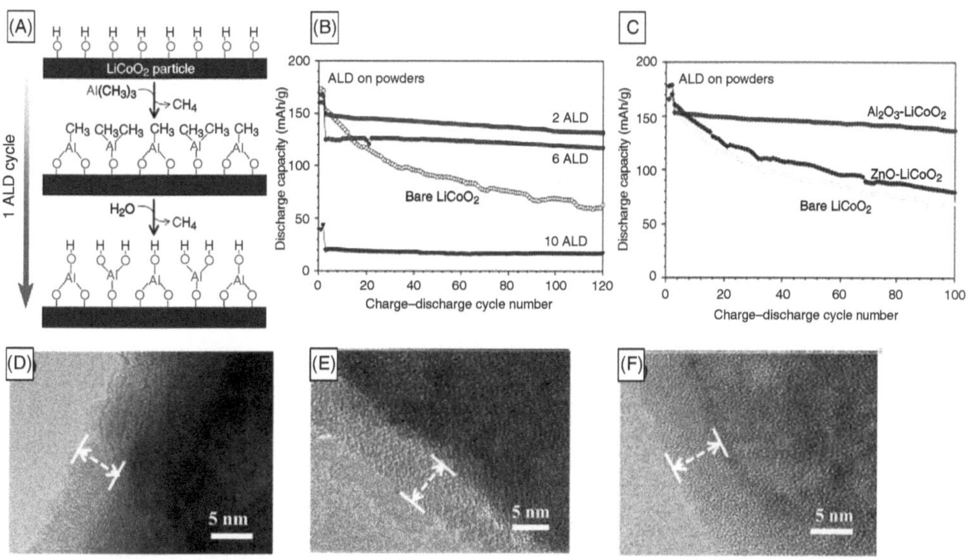

Figure 3.15 (A) Schematic representation of ALD–Al_2O_3 coating on $LiCoO_2$; (B) cycling performance of the bare $LiCoO_2$ powders and the Al_2O_3 ALD-coated $LiCoO_2$ powders using 2, 6, and 10 ALD cycles; (C) the bare, Al_2O_3 ALD, and ZnO ALD-coated $LiCoO_2$ powders using 4 ALD cycles; high-resolution TEM images of the $LiCoO_2$ cathodes by 50-ALD-cycles coating layers: (D) TiO_2; (E) ZrO_2; and (F) Al_2O_3. *(Part A–C, reprinted with permission from Y.-S.S. Jung, A.S. Cavanagh, A.C. Dillon, M.D. Groner, S.M. George, S.-H. Lee, Enhanced stability of $LiCoO_2$ cathodes in lithium-ion batteries using surface modification by atomic layer deposition, J. Electrochem. Soc. 157 (2010) A75 [34], Copyright © 2010, The Electrochemical Society; part D–F, reprinted with permission from X. Li, J. Liu, X. Meng, Y. Tang, M.N. Banis, J. Yang, et al., Significant impact on cathode performance of lithium-ion batteries by precisely controlled metal oxide nanocoatings via atomic layer deposition, J. Power Sources 247 (2014) 57–69 [35], Copyright © 2014, Elsevier).*

ALD offers highly conformal, pinhole free, and angstrom (Å) level surface deposition. It provides a true nanoscale level coating layer, and the thickness is self-controlled for it is based on self-limiting surface reactions. The thickness achieved with each ALD cycle is process specific but it is usually in subnanometers. Thicker coatings are easily obtained by simply repeating the ALD cycles. ALD is cost effective considering its precise control and high throughput, making it the most competitive surface film deposition technique. It has the ability to easily produce coatings on powder samples as well as on large area surfaces. In addition, all of these are achieved at low operation temperatures. However, it is a slow deposition process and demands ultraclean surfaces, which increases the cost. It also requires elimination of the precursor while switching to the next precursor in sequence.

ALD is an emerging technique used to modify surface of cathode electrodes with ultrathin conformal protection layers to mainly prevent electrode/electrolyte side reactions and improve electrochemical cycling performance. Se-Hee Lee et al. reported a capacity retention of 89% after coating Li_2CoO_2 powders with 2 ALD cycles of Al_2O_3, while the bare LCO powders showed only 45% of capacity retention, after 120 cycles in the 3.3–4.5 V (vs. Li/Li^+) voltage window, as shown in Fig. 3.15B. However, coating LCO powders with ZnO ALD layers did not show significant improvement in performance as depicted in Fig. 3.15C [34].

High-resolution TEM (HRTEM) images in Fig. 3.15D–F give a closer look at the conformal and nanoscale ALD coating layers on LCO cathode with different materials, TiO_2 (Fig. 3.15D), ZrO_2 (Fig. 3.15E), and Al_2O_3 (Fig. 3.15F). The improvement in the electrochemical performance is dependent upon the type of coating material. LCO powders with Al_2O_3 ALD films displayed better cycling ability while with ZrO_2 ALD films resulted in better rate performance. It was also disclosed that compared to ALD-coated LCO powders, the ALD coating on the surface of prefabricated LCO electrodes resulted in much better electrochemical performance [35].

Coating $LiMnO_4$ (LMO) particles with cerium dioxide (CeO_2) ALD film compared to bare, Al_2O_3-coated LMO, and ZrO_2-coated LMO, displayed improved capacity and cycling performance. TEM images of the bare (Fig. 3.16A) and CeO_2 ALD-coated (Fig. 3.16B–D) $LiMnO_4$ particles. The coating with CeO_2, 50 ALD cycles, increased the capacity by 24% compared to the bare. Also, it exhibited 96 and 95% capacity retention with 1 C rate even after 1000 battery cycles at room temperature (Fig. 3.16E) and 55°C (Fig. 3.16F), respectively [37].

Similarly, in other work, amorphous $FePO_4$ ALD coating on nitrogen-doped carbon nanotubes (NCNTs) was performed. High-resolution TEM images in Fig. 3.17A–D shows different coating thickness of $FePO_4$ obtained at different temperatures of 200, 250, 300, and 300°C, suggesting the dependence of ALD growth with temperature. The cathode material with 500 ALD cycles of $FePO_4$ on NCNTs showed improved battery activity and was able to deliver a discharge capacity of 141 mAh/g at a 1 C rate even after 100 battery cycles, as displayed in Fig. 3.17E [36].

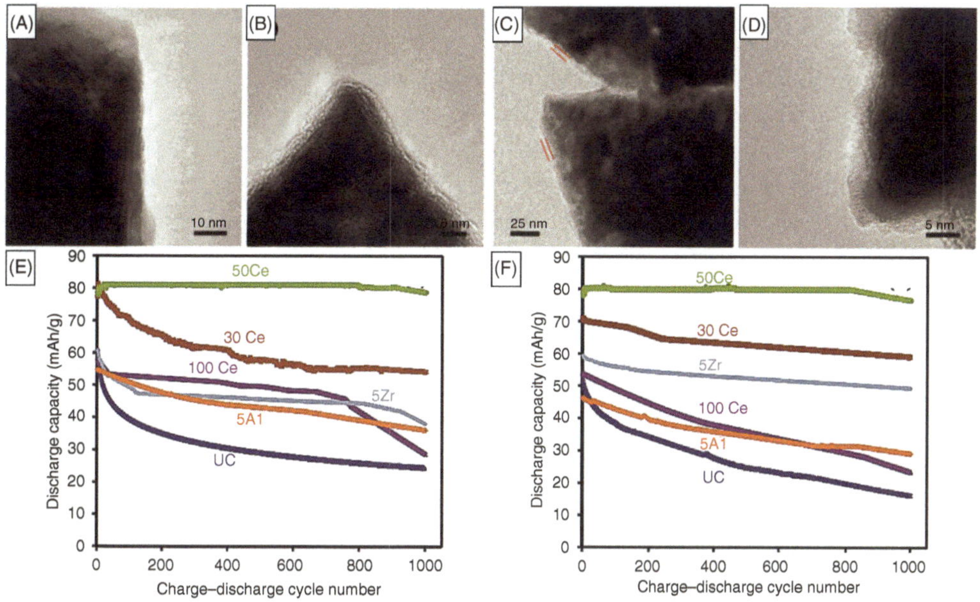

Figure 3.16 TEM images of (A) the uncoated (UC); (B) 50 cycles; (C–D) 100 cycles of CeO_2 ALD-coated $LiMn_2O_4$ particles; charge–discharge cycle curves of cells coated with various oxides at a 1 C rate in a voltage range between 3.4 and 4.5 V (E) at room temperature; and (F) at 55°C. *(Reprinted with permission from R.L. Patel, H. Xie, J. Park, H.Y. Asl, A. Choudhury, X. Liang, Significant capacity and cycle-life improvement of lithium-ion batteries through ultrathin conductive film stabilized cathode particles, Adv. Mater. Interfaces. 2 (2015) n/a [37], Copyright © 2015, Wiley).*

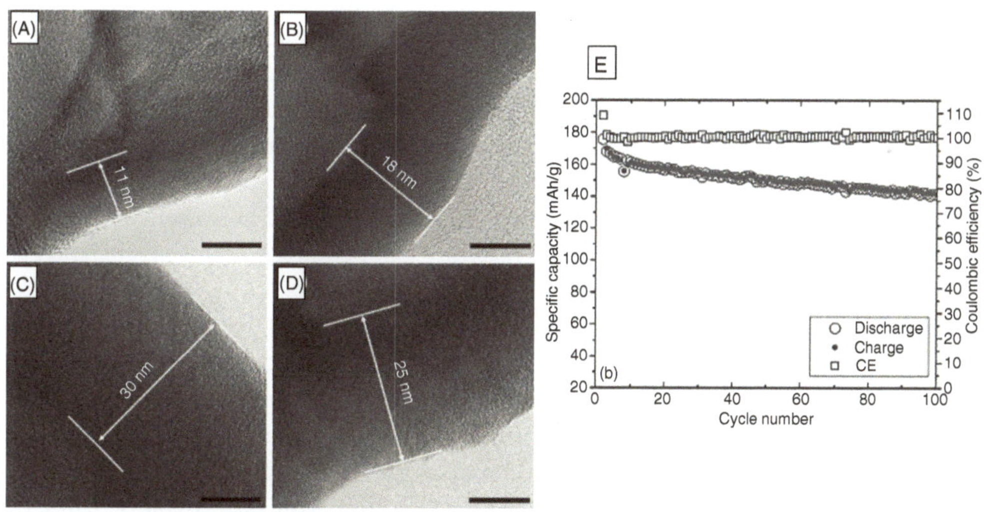

Figure 3.17 HRTEM images of 500 ALD cycles of $FePO_4$ on nitrogen-doped carbon nanotubes (NCNTs) at (A) 200; (B) 250; (C) 300; and (D) 350°C, respectively; and (E) cycling stability of 500-ALD cycles $FePO_4$/NCNTs at a current density of 178 mA/g (1 C). *(Reprinted with permission from J. Liu, B. Xiao, M.N. Banis, R. Li, T.-K. Sham, X. Sun, Atomic layer deposition of amorphous iron phosphates on carbon nanotubes as cathode materials for lithium-ion batteries, Electrochim. Acta 162 (2015) 275–281 [36], Copyright © 2015, Elsevier).*

Adapting ALD for the surface treatment with chemically inert or conductive material as a thin protective layer on the cathode material would be the best bet. However, it should be noted that the improvement in electrochemical performance is mainly dependent upon the thickness [34] and the material [35] of the coating. Each self-limiting ALD cycle has the ability to deposit coatings of thickness from angstrom level to subnanolevel. More dense ALD coatings can be easily obtained by merely increasing the number of ALD cycles, but the disadvantage of having thicker ALD coatings on cathodes would be reflected in the rise of surface impedance and the fall in lithium diffusion. The readers are suggested to refer [42,43] as extended reading to learn more about the fundamentals and application of ALD.

3.2 Chemical vapor deposition

CVD is very popular among thin film deposition techniques in semiconductor industries, and it has been gaining importance in the battery domain to surface-modify electrodes [38,40,44]. CVD is a thin film deposition technique that is enabled by the surface chemical reactions between the volatile precursors and the material surfaces. The volatile precursors when exposed to the material surface, undergo a chemical reaction and decompose to produce thin films on the surface. CVD can be practiced in a variety of forms, and can be categorized based on the triggering or initiation of the surface deposition chemical reactions. In battery applications, for surface treatment of cathodes, plasma-enhanced CVD is widely accepted, which utilizes plasma to initiate the chemical deposition processes. In addition, this technique has gained its importance for it can be performed at lower temperatures.

CVD is the most common tool used to grow thin films and has a number of advantages. The primary advantage of using CVD is that good conformal coatings can be easily obtained at fast deposition rates. Moreover, this technique can be easily applied to deposit a wide variety of materials with high purity. Regarding the disadvantages, CVD typically requires elevated operation temperatures and requires volatile precursors, which are mostly toxic, corrosive, or explosive gases [45]. However, with plasma enhanced CVD [38,40,44], the chemical reactions can be initiated with plasma and can be carried out at much lower temperatures.

Joong Kee Lee et al. surface modified LCO particles by ZnO coating using plasma-enhanced CVD. The plasma was generated by a 200 W radio frequency power and performed at 25°C. The resultant cathode showed an improvement in the cycling and thermal stability. The coatings were performed with different amounts of ZnO, and the SEM images of the bare and coated particles are displayed in Fig. 3.18A–E [38].

In the electrochemical performance test, the coating with 0.21 wt.% of ZnO on LCO showed comparatively better cycling performance indicating the cathode's performance dependence on the type and amount of the protective coating material. However, with 0.49 wt.% ZnO coating, there was a decrease in the battery performance, as shown in Fig. 3.18F. The reason for the decrease was reported to be the agglomeration of ZnO

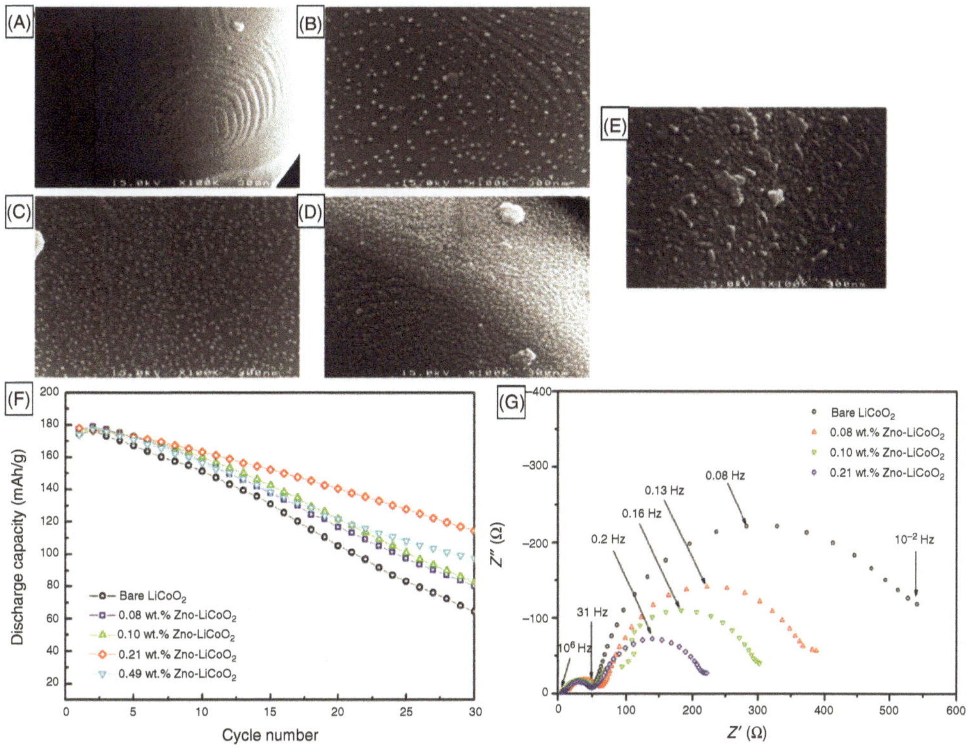

Figure 3.18 SEM images of bare and different amounts of ZnO-coated $LiCoO_2$ particles: (A) bare, (B) 0.08 wt.%, (C) 0.10 wt.%, (D) 0.21 wt.%, and (E) 0.49 wt.% ZnO-coated $LiCoO_2$; (F) cycling performance of bare and ZnO-coated $LiCoO_2$ with different coating amounts between 3 and 4.5 V at 1 C rate; and (G) impedance spectra of bare $LiCoO_2$ and ZnO-coated $LiCoO_2$ electrodes with different coating amounts obtained after 30th cycles. *(Reprinted with permission from W. Chang, J.-W. Choi, J.-C. Im, J.K. Lee, Effects of ZnO coating on electrochemical performance and thermal stability of $LiCoO_2$ as cathode material for lithium-ion batteries, J. Power Sources 195 (2010) 320–326 [38], Copyright © 2010, Elsevier).*

particles with an increase in ZnO amount, which can be observed in Fig. 3.18E. It was also reported that ZnO coating effectively reduced the cell impedance, as illustrated in the impedance spectra in Fig. 3.18G [38].

Robert M. Kostecki et al. successfully coated graphitic carbon on $LiNi_{1/3}Co_{1/3}Mn_{1/3}O_2$ (NMC) using one-step microwave plasma chemical vapor deposition (MPCVD). Fig. 3.19A–B shows the MPCVD system and the schematic of the deposition of carbon on NMC powder, respectively. The process was initiated by a 2.54 GHz, 1200 W microwave magnetron, to generate the plasma and applied for 2 s, which produced an average coating thickness of approximately 10–20 nm. Anthracene was used as the precursor, and the HRTEM images of the carbon coating on LMNO using MPCVD are displayed in Fig. 3.19C. The carbon coating improved the electrical contact between the cathode primary particles, improved the Li+ diffusion, and enhanced the

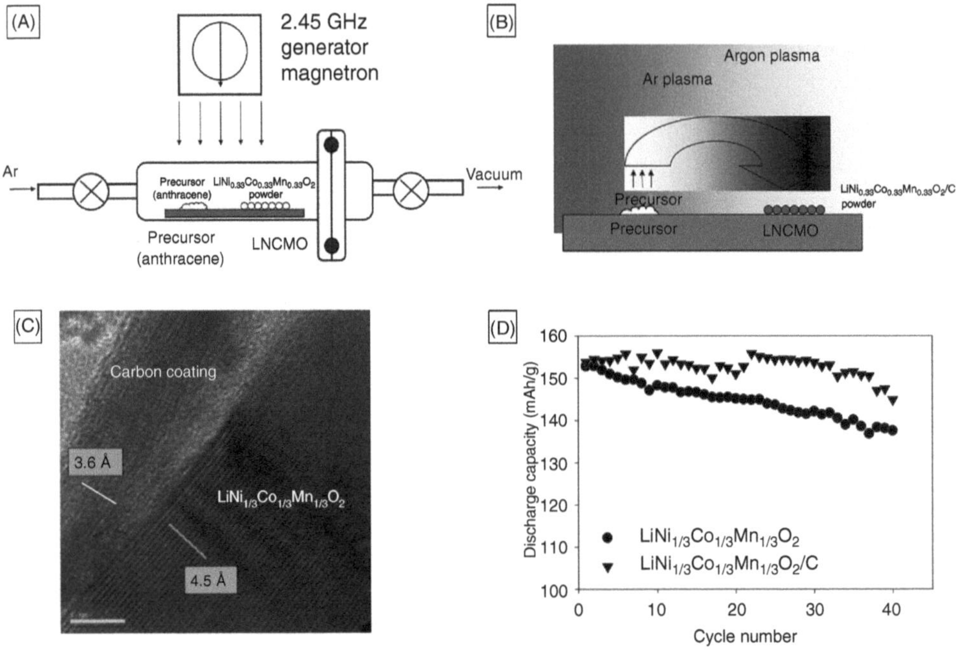

Figure 3.19 (A) Microwave plasma CVD (MPCVD) system; (B) schematic representation of deposition from the precursor; (C) high-resolution TEM image of MPCVD carbon on $LiNi_{1/3}Co_{1/3}Mn_{1/3}O_2$ (NMC); and (D) cycling of carbon-coated and bare NMC during cycling at room temperature at C/5 rate between 4.3 and 2.0 V. *(Reprinted with permission from M.L. Marcinek, J.W. Wilcox, M.M. Doeff, R.M. Kostecki, Microwave plasma chemical vapor deposition of carbon coatings on $LiNi_{1/3}Co_{1/3}Mn_{1/3}O_2$ for Li-ion battery composite cathodes, J. Electrochem. Soc. 156 (2009) A48 [40], Copyright © 2009, The Electrochemical Society).*

electrochemical performance of the battery. The improvement in the cycling ability is shown in Fig. 3.19D. Carbon-coated NMC showed 4% capacity fading compared to 10% capacity fading for bare LMNO material, after 40 cycles at 0.2 C rate [40].

At the electrode level, Wonchang Choi et al. employed SnO_2 deposition on tape-cast $LiNi_{0.5}Mn_{1.5}O_4$ (LNMO) electrodes using a similar version of plasma-CVD method, which was named as electron cyclotron resonance–metal organic chemical vapor deposition (ECR-MOCVD). The schematic of the deposition is shown in Fig. 3.20A [44].

The objective of the work was to reduce side reactions and self-discharge reactions to make the cathode material suitable for high voltage lithium-ion battery applications such as electric vehicles and grids. In the electrochemical performance tests at room temperature, SnO_2-coated electrodes showed better rate properties than the bare LNMO electrode. To evaluate the self-discharge property, Storage test for 2 weeks was performed at 60°C in the fully charged state which revealed that coating SnO_2 on LNMO reduced the self-discharge to significant amount compared to the pristine LNMO. Fig. 3.20B–C

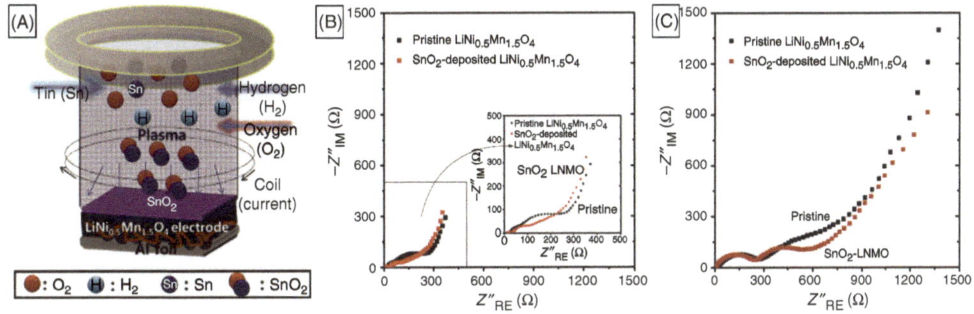

Figure 3.20 (A) Schematic of SnO_2 deposition process on the 5 V Spinel $LiNi_{0.5}Mn_{1.5}O_4$ (LNMO) cathode electrodes by electron cyclotron resonance–metal organic CVD (ECR-MOCVD); electrochemical impedance spectra of pristine and SnO_2-coated LNMO, (B) after 3 cycles at room temperature; and (C) after storage test for 2 weeks at 60°C. *(Reprinted with permission from Y. Lee, T.Y. Kim, D.-W. Kim, J.K. Lee, W. Choi, Coating of spinel $LiNi_{0.5}Mn_{1.5}O_4$ cathodes with SnO_2 by an electron cyclotron resonance metal–organic chemical vapor deposition method for high-voltage applications in lithium ion batteries, J. Electroanal. Chem. 736 (2015) 16–21 [44], Copyright © 2015, Elsevier).*

shows that the charge transfer resistance of the pristine electrode increased tremendously after the storage test while that of the coated electrode did not change significantly. The charge transfer resistance for pristine LNMO was 214.5 Ω in fully charged state after 3 cycles and enormously increased to 716.5 Ω after the storage test. Whereas, in the case of SnO_2-coated LNMO, the charge transfer resistance was 254.9 Ω, initially, and after the storage test it increased to 301.9 Ω only. This suggested very low sufferings of self-discharge in coated electrodes than compared to the pristine electrodes.

CVD can be widely used to form good conformal coatings on cathode material to reduce the electrolyte/electrode side reactions, reduce self-discharge reactions, improve thermal and structural stability, increase the conductivity of cathodes, and further enhance the battery performance. Low temperature deposition is possible also using plasma enhanced CVD. However, similar to ALD, the improvement in electrochemical performance is again dependent on type, amount and thickness of the coating material (for extended reading about CVD, Refs. [43,45] are suggested).

3.3 Sputtering

Sputtering is a physical vapor deposition (PVD) method to grow thin films. Similar to CVD, sputtering also finds potential application for thin film growth in the semiconductor field, and also has been gaining a significant position in energy and battery applications [39,46,47]. Sputtering is a process in which the material to be deposited (target) is bombarded with high energy ions in a plasma, and the atoms ejected deposit onto the substrate surface forming thin films. The ejected target atoms travel through the plasma in the line of sight and physically deposit on the material. Unlike CVD, there is no chemical reaction between the deposited target atoms and the materials surface.

Sputtering can be employed on a wide variety of materials, including those with high melting points. It does not demand the use of toxic or specialized precursor as required for CVD and offers good coating adhesion even though there are no surface chemical reactions involved. The disadvantages of using sputtering include the line-of-sight travel of the ejected atoms, making it difficult for selective paths to form coating layers. However, the lift-off path can be made selective by using a magnetic field, which would be then called magnetron-sputtering method [39,46].

Graphite layer was deposited by Kaili Zhang et al. to directly cover the surface of $LiMn_2O_4$ (LMO) based electrodes using DC magnetron sputtering. Fig. 3.21A shows the procedure followed for graphite coating on LMO electrodes. The blended slurry obtained by mixing LMO, polyvinylidene fluoride (PVDF), and Super P carbon black (SP) in 90, 10, and 10 wt.%, respectively, in *N*-2-methyl pyrrolidone (NMP) solvent was coated on aluminum foil. It was followed by a heat treatment at 120°C for 3 h for drying to obtain bare LMO electrodes (BLMO). The as obtained BLMO electrodes were placed in magnetron plasma excited by 0.4 A, 200 V DC power. The rate of deposition of the graphite layer was around 100 nm/h [39].

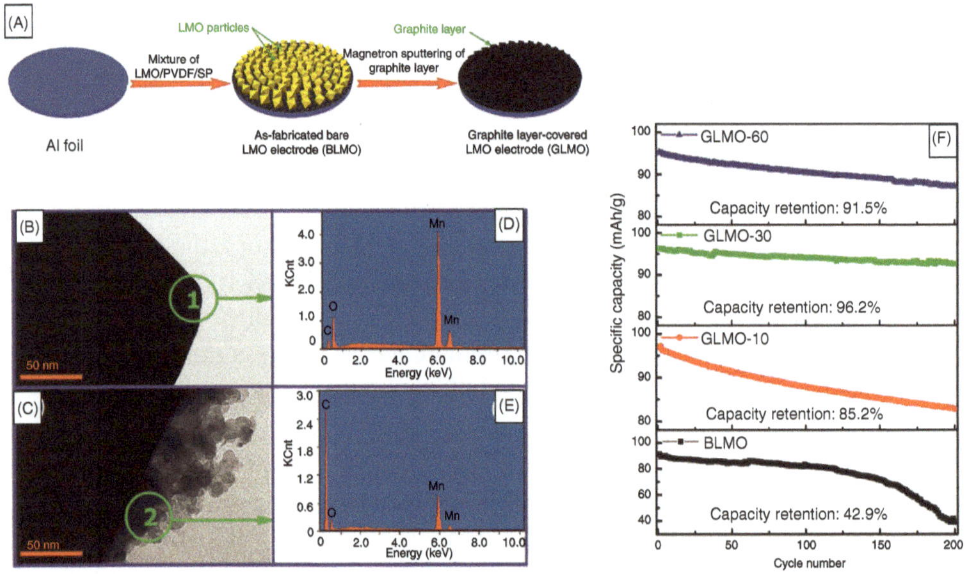

Figure 3.21 (A) Schematic diagram of graphite coating on $LiMn_2O_4$ (LMO) using DC magnetron sputtering; TEM images of (B) bare LMO electrodes (BLMO); (C) graphite-coated LMO-30 min growth time (GLMO-30); EDS analysis of (D) BLMO; (E) GLMO-30; and (F) cycling evaluation test at 1 C rate under 55°C. *(Adapted from J. Wang, Q. Zhang, X. Li, Z. Wang, H. Guo, D. Xu, et al., Sputtering graphite coating to improve the elevated-temperature cycling ability of the $LiMn_2O_4$ electrode, Phys. Chem. Chem. Phys. 16 (2014) 16021–16029 [39] with permission from the PCCP Owner Societies).*

Different samples of graphite-coated LMO electrodes (GLMO) were made by sputtering for different growth times, 10 min (GLMO-10), 20 min (GLMO-20), 30 min (GLMO-30), and 60 min (GLMO-60). The coating was found to act as a barrier layer suppressing the side reactions including corrosion and metallic dissolution. Fig. 3.21B–E shows the TEM images of the BLMO and GLMO-30 electrodes, and the energy dispersive spectroscopy (EDS) results confirming the graphitic carbon coating on GLMO-30. The electrochemical performance test evaluation was cycled at 1 C rate between the potential window 3.3–4.35 V, at 55°C. It was revealed that all the GLMO electrodes showed better cycling performance than the BLMO electrodes and the GLMO-30 showed the best capacity retention of 96.2% after 200 cycles, as shown in Fig. 3.21F [39].

Similarly, radio frequency (RF) magnetron sputtering was used to make coralline glassy lithium phosphate-coated $LiFePO_4$ (GLP-coated LFP) cathodes by Renjie Chen et al. The schematic of the GLP-coated LFP electrode preparation is shown in Fig. 3.22A. The target and substrate used were Li_3PO_4 pellet and $LiFePO_4$ (bare LFP electrode), respectively. Sputtering was carried out in Argon plasma environment at an RF power of 100 W, and the deposition time was used to control the coating thickness. For example, a thickness of 200 nm was obtained for a deposition time of 20 min. The GLP coating on LFP enhanced the ionic and electronic transport properties of the cathode and facilitated fast kinetics suitable for high power applications. Moreover, GLP coating on cathode hindered any side reactions and provided extra structural stability. Furthermore, the GLP-coated LFP displayed high capacity, high efficiency, stable cycle performance, and improved rate performance compared to the bare LFP, as shown in Fig. 3.22B–C [47].

Recently, in 2014, Zhaoping Liu et al. successfully surface modified lithium rich-layered, $0.3Li_2MnO_3-{}_{0.7}LiNi_{5/21}Co_{5/21}Mn_{11/21}O_2$ (LMO-NCM) with ZnO film coatings using the reactive magnetron sputtering (RMS) for the first time. The schematic of the deposition process using RMS is shown in Fig. 3.22D. The system used metallic Zn as the target. In contrast to typical magnetron sputtering, which uses only one process gas, in RMS the deposition was conducted in a mixture of 40% argon and 60% oxygen. Argon was used as the process gas to eject Zn atoms from the target. Oxygen was used as the reactive gas, which reacts with the ejected Zn atoms to form ZnO and finally deposit on the substrate as ZnO thin films. Fig. 3.22E–F shows the HRTEM images of the pristine layered electrode and ultrathin ZnO-coated electrode using RMS-4 min deposition time, respectively [46].

Electrochemical impedance spectra (EIS) of the pristine and RMS ZnO-coated electrodes with deposition times of 1, 2, and 4 min are shown in Fig. 3.22G. It can be observed from the graphs that ZnO coating reduced the charge transfer resistance compared to the bare, and can be optimized to an RMS deposition time of 2 min. ZnO-coated LMO-NCM using RMS-2 min deposition time electrode exhibited high capacity, and improved rate and cycle performance compared to the bare LMO-NCM [46].

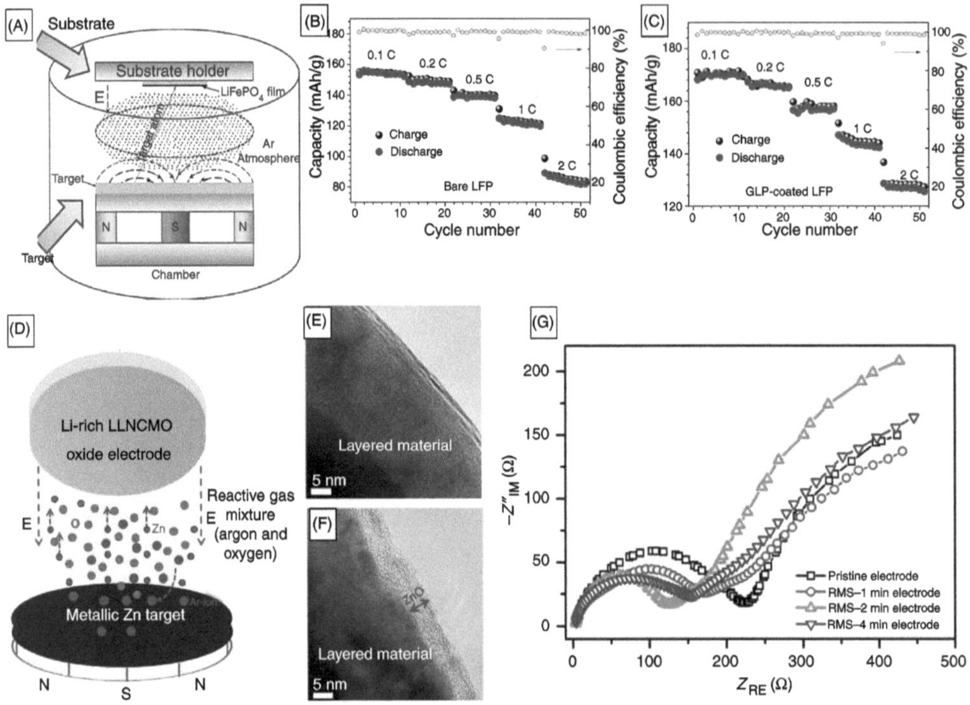

Figure 3.22 (A) Schematic of Li_3PO_4 (GLP) coating on $LiFePO_4$ (LFP) substrate using RF magnetron sputtering; rate performance test of (B) bare LFP; and (C) GLP-coated LFP (20 min) electrodes; (D) schematic of ZnO coating on $0.3Li_2MnO_{3-0.7}LiNi_{5/21}Co_{5/21}Mn_{11/21}O_2$ (LMO-NCM) using RMS; HR-TEM images of (E) pristine LMO-NCM electrode; and (F) ZnO-coated LMO-NCM electrode using (reactive magnetron sputtering) RMS-4 min deposition time; and (G) EIS of the pristine and ZnO-coated LMO-NCM electrodes after initially charged to 4.5 V at the current density of 25 mA/g. *(Part A–C, reprinted with permission from G. Tan, F. Wu, L. Li, R. Chen, S. Chen, Coralline glassy lithium phosphate-coated $LiFePO_4$ cathodes with improved power capability for lithium ion batteries, J. Phys. Chem. C 117 (2013) 6013–6021 [47], Copyright © 2013, American Chemical Society; part D–G, reprinted with permission from B. Qiu, J. Wang, Y. Xia, Z. Wei, S. Han, Z. Liu, Enhanced electrochemical performance with surface coating by reactive magnetron sputtering on lithium-rich layered oxide electrodes, ACS Appl. Mater. Interfaces 6 (2014) 9185–9193 [46], Copyright © 2014, American Chemical Society).*

Sputtering can be used for the surface treatment of cathode materials when achieving ultrathin protective coatings is not one of the major priority. If so, sputtering can be easily employed, and is simple and fast deposition method. It can be performed in a variety of forms depending on the gas used in plasma and the method used to accelerate the plasma ions. The most common method used to accelerate the plasma ions is to apply an electric field. Further, the atoms ejected after getting a strike from plasma ions can be driven in selective coating paths instead of line-of-sight by employing magnetic field as in magnetron sputtering. Over the recent years, magnetron sputtering or its variations have been extensively employed for surface coating the electrodes for battery applications [39,46,47].

3.4 Wet-coating/sol–gel method

Wet coating is a simple yet powerful technique to grow thin films. It has been very commonly used since very long time and had been employed in a variety of forms: sol–gel, dip coating, spray coating, and so forth. There has been significant progress and advancement in wet-coating methods to grow uniform surface coatings and precisely control the thickness [41,48–52].

Wet coating is a common thin film deposition technique in which the material is dispersed in a liquid precursor solution, followed by heat treatment, which ends up as a thin layer coating on the materials surface. It is named wet coating because most of the deposition process are carried out in the liquid phase. Sol–gel is a promising wet-coating method, which is well known for its ability to grow high-quality thin films. In this method, the liquid precursor solution polymerizes to gel phase forming a continuous layer on the materials surface. The sol–gel technique can be performed at low temperatures, and can produce thin to thick coatings (nm to μm) with good adhesion and high purity.

Sol–gel and other wet-coating methods are widely accepted in lithium-ion battery industry for making surface modified cathode electrodes. S.B. Majumder et al. demonstrated zirconia (ZrO_2) surface coating on $0.25Li_2MnO_3$–$0.75Li(Mn_{0.375}Ni_{0.375}Co_{0.25})O_2$ particles using the wet-coating method. The particles are dispersed in a liquid precursor solution of zirconium (IV) butoxide dissolved in 2-methoxyethanol. The prepared solution was ultrasonicated and calcined to polymerize zirconia as gel forming a coating on the active material. Zirconia coating thickness of around ~5 nm was reported. The coating was optimized to 2.5 wt.% of zirconia, and the modified cathodes exhibited excellent cycling ability and rate performance compared to the bare electrodes, as shown in Fig. 3.23. The cells were cycled at 10 mA/g rate in the potential window, 2.5–4.8 V. The inset in Fig. 3.23A shows high columbic efficiency, close to 99% achieved with 2.5 wt.% zirconia coating compared to the bare. It was also disclosed that zirconia coating reduced impedance growth at the cathode–electrolyte interface and reduced capacity fading. The rate performance test is shown in Fig. 3.23B, which was cycled in the voltage window 2.5–4.5 V, at cycling rates from C/10 to 5 C and back to C/10. The nominal discharge rate was considered as 230 mAh/g to determine the C rate value. It can be noted that 2.5 wt.% zirconia-coated samples displayed better rate performance compared to the other sample candidates [51].

Jaephil Cho et al. worked on surface treatment of $LiNi_{0.6}Co_{0.2}Mn_{0.2}O_2$ (NCM) particles using the wet-coating method [48]. It was reported that NCM particles (secondary particles of NCM–Ni-rich) suffer from crack generation and structural changes on the surface at high temperatures and high voltages. The cracks cause the connection loss between the primary particles leading to oxygen generation and structural degradation. The focus of the work reported was to coat the primary particles of NCM to suppress structural degradation. Lithium and cobalt precursors dissolved in ethanol solution were the liquid phase precursor, which was mixed with NCM particles. The obtained solution

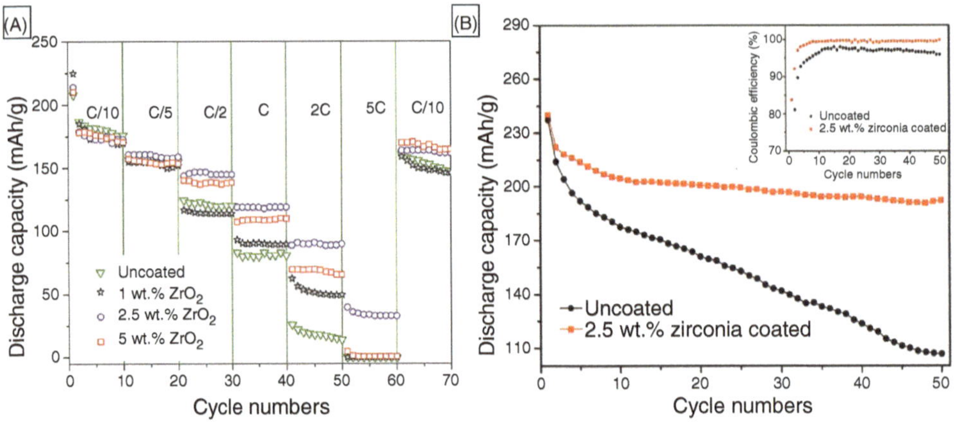

Figure 3.23 (A) Cycling performance of bare $0.25Li_2MnO_3$-$0.75Li(Mn_{0.375}Ni_{0.375}Co_{0.25})O_2$ and 2.5 wt.% zirconia-coated sample in the voltage window 2.5–4.8 V using 10 mA/g rate; and (B) rate performance test of different samples in the voltage window 2.5 to 4.5 V at 25°C. *(Reprinted with permission from C. Ghanty, P.P. Dahiya, R.N. Basu, J.-K. Chang, S.B. Majumder, Improvement of the electrochemical characteristics of lithium and manganese rich layered cathode materials: effect of surface coating, J. Electrochem. Soc. 162 (2015) A1957–A1965 [51], Copyright © 2015, The Electrochemical Society).*

was dried and subjected to calcination at 800°C to obtain surface-treated NCM (ST-NCM). The protective surface coating layer-cation mixing phase (*Fm3m*) was formed on the layered structure phase (*R3m*) of the pristine NCM particles. The schematic of the coating process is shown in Fig. 3.24A. The ST-NCM secondary particle SEM image is shown in Fig. 3.24B. Scanning transmission electron microscope (STEM) of primary particle indicated by a small rectangle in SEM image is shown in Fig. 3.24C. The energy-dispersive spectroscopy (EDS) analysis of the ST-NCM corresponding to the arrow region indicated in STEM images confirms higher content of Co than Mn and Ni in the interior, up to depth of ~60 nm, as shown in Fig. 3.24D. The ST-NCM was found to reduce cracks and structural changes, as displayed in Fig. 3.24E–F. In addition, it also reduced oxygen evolution at high temperatures due to presence of higher oxidation state of Mn^{+4} on the surface of ST-NCM. It was reported that surface treatment of primary particles can be realized to suppress structural changes, improve the electrochemical performance and increase the thermal stability [48].

In other work by Jieun Lee and Wonchang Choi, overlithiated layered oxide (OLO)-$Li_{1.17}Mn_{0.56}Co_{0.095}Ni_{0.175}O_2$ powders were coated with a conductive polymer, poly(3,4-ethylenedioxythiophene) polystyrene sulfonate (PEDOT: PSS) using a water based wet-coating method. The thin surface coating was obtained at low-temperature heat treatment at 110°C. Samples with 1 and 2.5 wt.% of PEDOT:PSS produced surface coatings of thickness around ~5.3 and ~7.27 nm, respectively, which is shown in the TEM images of Fig. 3.25A–C. It was found that PEDOT:PSS surface-coated OLO

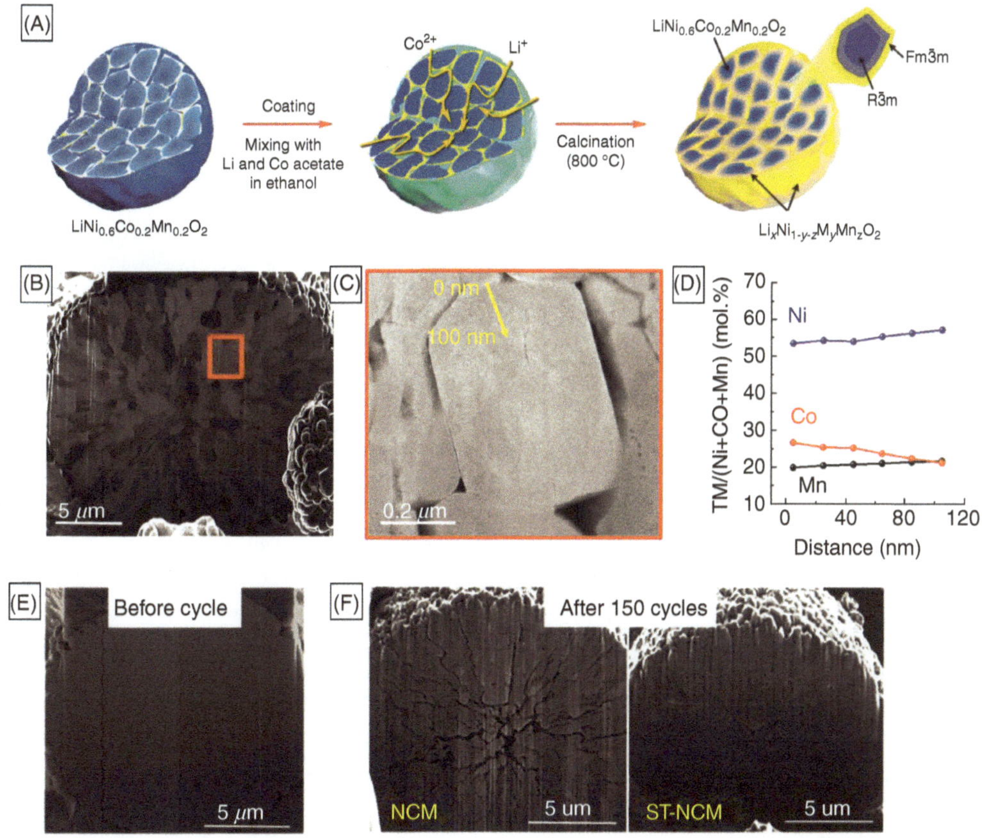

Figure 3.24 (A) Schematic of surface treatment of primary particles of $LiNi_{0.6}Co_{0.2}Mn_{0.2}O_2$ (NCM); (B) SEM image of secondary particle of surface-treated NCM (ST-NCM); (C) STEM of the primary particle of ST-NCM indicated as *rectangle box* in (B); (D) EDS analysis corresponding to the *arrow* indicated in (C); (E) before cycling SEM image of bare; and (F) after cycling SEM image comparison between the bare NCM and surface-treated NCM. *(Reprinted with permission from H. Kim, M.G. Kim, H.Y. Jeong, H. Nam, J. Cho, A New Coating Method for Alleviating Surface Degradation of* $LiNi_{0.6}Co_{0.2}Mn_{0.2}O_2$ *Cathode Material: Nanoscale Surface Treatment of primary particles, Nano Lett. 15 (2015) 2111–2119 [48], Copyright © 2014, American Chemical Society).*

samples exhibited high capacity and better rate performance compared to the pristine OLO. For instance, in rate performance curves of Fig. 3.25D, at 3 C rate evaluation, surface-coated samples delivered a discharge capacity of 83 mAh/g while the pristine samples delivered just 20 mAh/g. It can also be observed that 1 and 2 wt.% PEDOT:PSS samples exhibited similar performance up to 6 C rate. However, at 9 C rate 1 wt.% PEDOT:PSS showed comparatively better rate performance [41].

George Ting-Kuo Fey et al. surface modified $LiCoO_2$ (LCO) particles with a coating of lanthanum aluminum garnets ($3LaAlO_3$:Al_2O_3 or LAG) using an in situ sol–gel method. The LCO particles were dispersed in liquid precursor solution containing

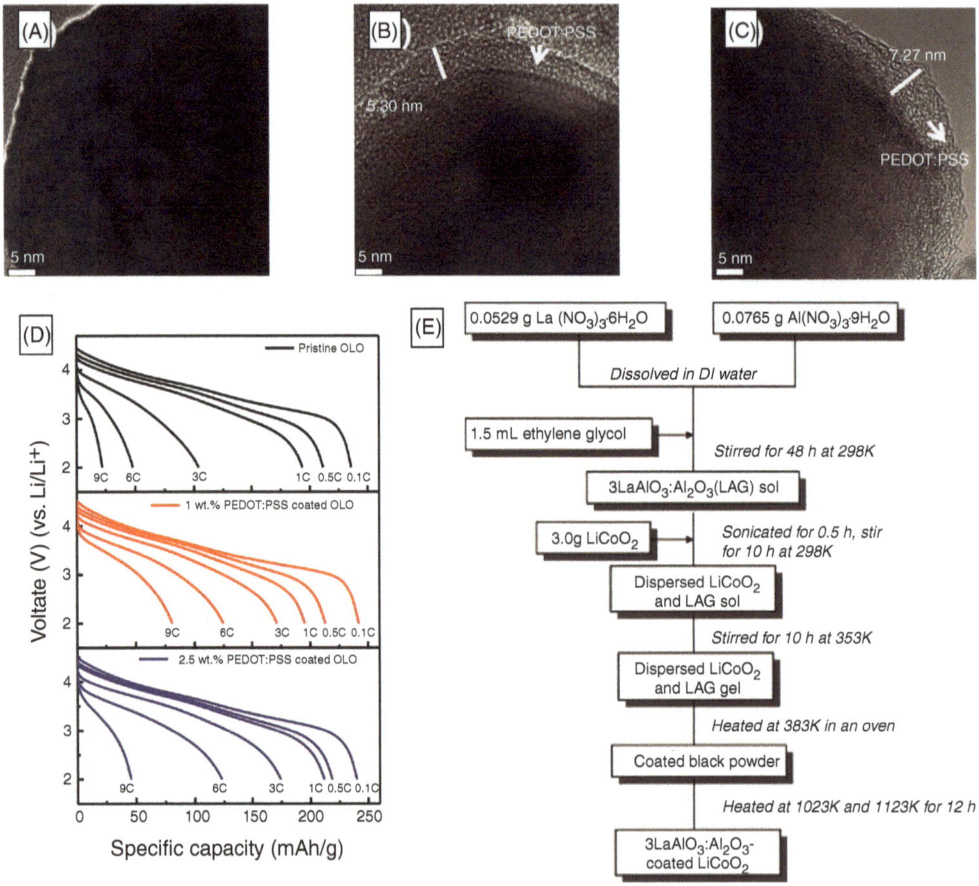

Figure 3.25 TEM images of (A) pristine over lithiated layered oxide—$Li_{1.17}Mn_{0.56}Co_{0.095}Ni_{0.175}O_2$ (OLO), (B) 1 wt.% PEDOT:PSS-coated OLO, (C) 2.5 wt.% PEDOT:PSS-coated OLO; and (D) rate capability evaluation for the pristine OLO, 1 wt.% PEDOT:PSS-coated OLO, and 2.5 wt.% PEDOT:PSS-coated OLO electrode; and (E) schematic for the surface deposition of $3LaAlO_3$:Al_2O_3 on $LiCoO_2$ cathode materials by in situ sol-gel process. *(Part A–D, reprinted with permission from J. Lee, W. Choi, Surface modification of over-lithiated layered oxides with PEDOT:PSS conducting polymer in lithium-ion batteries, J. Electrochem. Soc. 162 (2015) A743–A748 [41], Copyright © 2015, The Electrochemical Society; part E, reprinted with permission from C.Z. Lu, J.M. Chen, Y. Da Cho, W.H. Hsu, P. Muralidharan, G.T.K. Fey, Electrochemical performance of $LiCoO_2$ cathodes by surface modification using lanthanum aluminum garnet, J. Power Sources 184 (2008) 392–401 [50], Copyright © 2008, Elsevier).*

lanthanum nitrate hexahydrate [$La(NO_3)_3 \cdot 6H_2O$] and aluminum nitrate nonahydrate [$Al(NO)_3 \cdot 9H_2O$] dissolved in ethylene glycol. The prepared solution was heat-treated at around 850°C or 1123 K in air to produce a surface coating of lanthanum aluminum garnet (LAG, $3LaAlO_3$:Al_2O_3). The deposition process steps are shown as a flow chart in Fig. 3.25E. TEM images show the bare LCO and the thin coating of LAG on LCO particles in Fig. 3.26A and B, respectively. It was reported that the LAG coatings obtained on LCO particles had an average thickness of 20–50 nm. EDS characterization in Fig. 3.26C confirms the composition of the LAG coating on LCO particles. The samples were prepared for various wt.% of LAG, and 1 wt.% LAG-coated LCO particles were found to be better. Surface-coated LCO showed stable cycling compared to the bare LCO, as shown in Fig. 3.26D. Bare LCO electrodes showed gradual fading during cycling while surface-coated LCO maintained stable cycling suggesting the suppression of impedance growth in surface-coated samples [50].

The sol–gel method was used in similar works; to grow surface protective layers on $LiFePO_4$/carbon composite electrodes (LFP/C) and coatings on $LiNi_{0.5}Mn_{1.5}O_{4-\delta}$ (LNMO)–based electrodes. Bart M. Bartlett, and other coworkers, surface-coated LNMO with lithium titanate (LTO) and/or anatase (TiO_2) layer. The coating thickness was controlled in the range from 5 to 30 nm depending on the amount of precursor used. HRTEM images in Fig. 3.26E–I shows the different coating thickness obtained. In the electrochemical performance evaluation at high temperatures, it was reported that surface coating on sample reduced capacity fading and improved structural integrity [52]. Similarly, Guiling Wang et al. surface modified LFP/C with $Li_3V_2(PO_4)_3$ (LVP) surface coating using sol–gel combustion method. The liquid precursor was NH_4VO_3 and LiH_2PO_4 in citric acid and ethylene glycol solution. Heat treatment was performed at 700°C in nitrogen gas to produce LVP-coated LFP/C. The surface coating was reported to suppress the charge transfer resistance of the electrodes, resulting in excellent electrochemical performance compared to the bare. Fig. 3.26J displays the rate comparison of bare composite and LVP-coated LFP/C composite, which shows the higher rate performance of surface-coated samples compared to its bare LFP/C counterpart [49].

Wet coating/sol–gel is a low cost and a low-temperature method to deposit surface protected coatings on the cathode materials. It is also the most commercially employed surface deposition technique and offers high purity and high adhesion coatings. However, it requires high-temperature posttreatment to grow crystalline coatings. This may lead to phase composition change, more defects, and an increase in cost. In addition, sol–gel suffers from large shrinkages and cracking during drying or heat treatment processing. For more information about sol–gel science [53] is suggested.

In summary, nanoscale surface modification of cathodes with surface coatings improves the overall electrochemical performance. Surface protected cathodes become resistant to electrolyte/electrode side reactions, show reduced self-discharge behavior, exhibit thermal and structural stability. In addition, the protective layer could be used to increase the conductivity of cathodes, suppress the impedance growth and further enhance the battery performance. It should be noted that these properties are dependent on the physicochemical properties, its amount, and thickness of the coating material. Sometimes, processing temperature also requires optimization. Nanoscale surface modification on cathodes can be easily employed in a number of ways as discussed in the chapter. ALD is a promising method if looking for angstrom (Å) level conformal surface

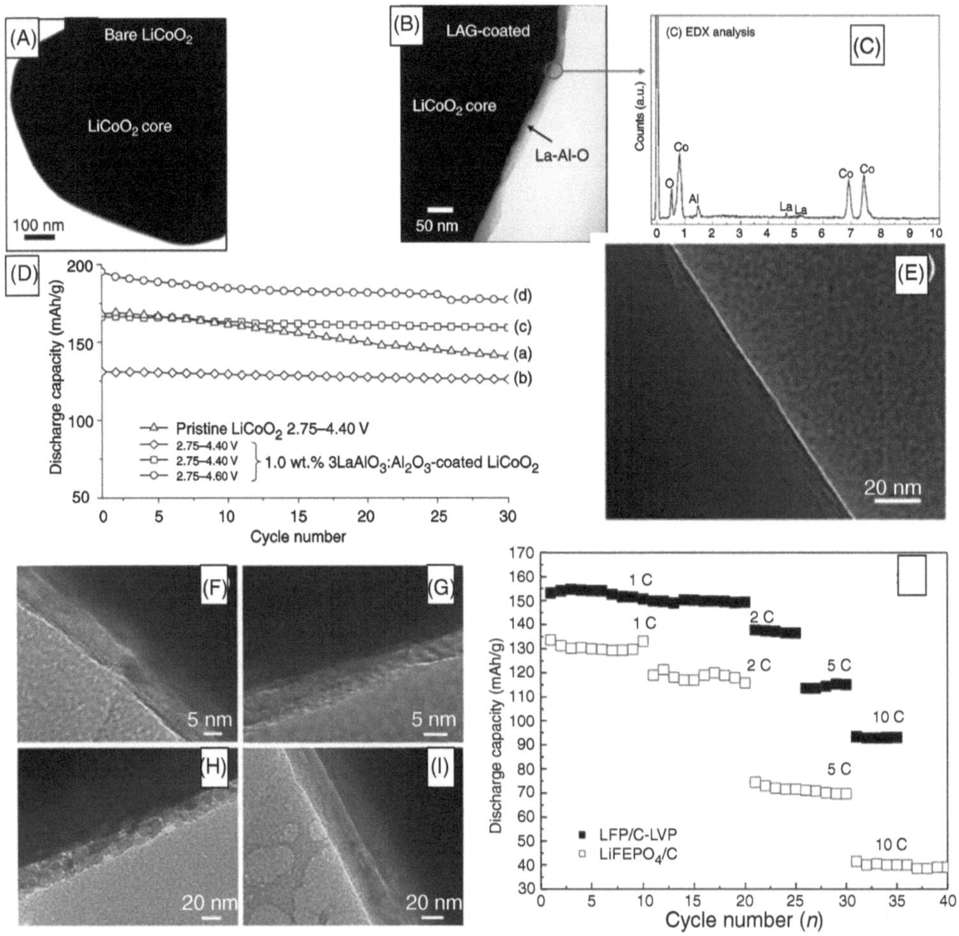

Figure 3.26 TEM image of (A) bare $LiCoO_2$ (LCO), (B) lanthanum–aluminum–garnet-coated LCO (LAG-coated LCO); (C) EDS of LAG-coated LCO corresponding to *circle* indicated in (B); and (D) cycling performance of 1 wt.% LAG-coated LCO for different voltage windows at 0.2 C rate; HR-TEM images of (E) bare $LiNi_{0.5}Mn_{1.5}O_{4-\delta}$ (LNMO); (F) 5–10 nm (lithium titanate) LTO-coated LNMO; (G) 5–10 nm TiO_2-coated LNMO; (H) 20–30 nm LTO-coated LNMO; and (I) 20–30 nm TiO_2-coated LNMO; and (J) rate performance of the bare $LiFePO_4$/carbon composite (LFP/C) and $Li_3V_2(PO_4)_3$ (LVP)-coated LFP/C electrodes. *(Part A–D, reprinted with permission from C.Z. Lu, J.M. Chen, Y. Da Cho, W.H. Hsu, P. Muralidharan, G.T.K. Fey, Electrochemical performance of $LiCoO_2$ cathodes by surface modification using lanthanum aluminum garnet, J. Power Sources 184 (2008) 392–401 [50], Copyright © 2008, Elsevier; part E–I, reprinted with permission from X. Hao, B.M. Bartlett, Improving the electrochemical stability of the high-voltage li-ion battery cathode $LiNi_{0.5}Mn_{1.5}O_4$ by titanate-based surface modification, J. Electrochem. Soc. 160 (2013) A3162–A3170 [52], Copyright © 2013, The Electrochemical Society; part J, reprinted with permission from Zhipeng, G. Shao, X. Wang, J. Song, G. Wang, Z. Ma, et al., $Li_3V_2(PO_4)_3$ modified $LiFePO_4/C$ cathode materials with improved high-rate and low-temperature properties, Ionics 19 (2013) 1861–1866 [49], Copyright © 2013, Springer).*

coatings, but this process is time-consuming and sometimes limited to large surface area by reactor. Sputtering and sol–gel can be employed easily with fast processing time at the cost of uniformity and thickness compared to ALD. Nanoscale surface coating techniques have been constantly improving with time, and extensively researched further to gain precise control on thickness, homogeneity, and uniformity to greatly improve the battery materials outcomes.

4 CONCLUSIONS

In conclusion, nanotechnology provides new roads to design, synthesize, and modify cathode materials to overcome capacity limitation and greatly enhance battery performance. We summarized different nanostructural design and synthesis of cathode materials for LIBs. Nanotemplate method offers us to synthesize nanostructure materials with complex morphology using a template in single step with good control of size and shape, which would otherwise be impossible to produce. Solvothermal/hydrothermal and coprecipitation synthesis methods are both widely used, simple, and low-cost routes to synthesize cathode nanostructures with high yield but lack control over the size and shape of the final nanostructure. Solid-state reaction methods can be preferred to produce polycrystalline cathode materials with high-throughput but lack good control over the final size, shape, and morphology of the material. Nanostructural synthesis of cathode materials resulted in enhanced performance of battery, which were attributed to their improved structure and reduced Li^+ diffusion path lengths facilitating fast Li^+ insertion/extraction reactions. Advancements in the nanosynthesis methods have been gaining precise control over size, shape, and reproducibility with low or medium processing costs, and the limitation gaps are minimizing rapidly. Nanoscale surface modification of the cathode materials using popular deposition techniques ALD, CVD, sputtering, and wet-coating/sol–gel methods were also discussed. The whole approach of surface coating on cathodes is to become resistant to electrolyte/electrode side reactions, show reduced self-discharge behavior, exhibit thermal and structural stability, suppress the impedance growth and further enhance the battery performance. ALD is the most competitive and cost effective technique, which offer highly conformal, pinhole free, and angstrom (Å)-level surface depositions with high throughput. But ALD is slow and time-consuming and sometimes limited to large surface area by reactor. CVD can provide up to nanoscale surface deposition, and sputtering and wet-coating methods are more preferred when depositing thicker submicron level coatings. Sputtering is a low cost and fast deposition method, which is also widely used. Wet coating/sol–gel is a low cost, low-temperature, and the most commercially employed surface deposition technique and offers high purity and high adhesion coatings. However, it requires high-temperature posttreatment to grow crystalline coatings. Sputtering and sol–gel can be employed easily with fast processing time at the cost of uniformity and thickness compared to ALD. In short, for nanoscale thickness ALD and CVD methods are recommended, and for thicker submicron to micron-level coating requirements, sputtering and sol–gel methods are advised. However, the improvement of surface modified cathodes electrochemical performance is dependent on the type, the amount, and the thickness of the coating material. Nanoscale surface-coating techniques have also been constantly improving with time, and extensively researched further to gain precise control over thickness, homogeneity, and uniformity to greatly improve the battery materials outcomes. Nanotechnology-driven development of cathode material is an essential part for the future LIBs evolution. With the progress of nanosynthesis and nanosurface modification techniques of cathode materials in recent years, the battery technology future seems very promising, thanks to nanotechnology.

GLOSSARY

3DOM/m	Three-dimensionally ordered macroporous and meso-/microporous
ALD	Atomic layer deposition
BET	Brunauer–Emmett–Teller
BLMO	Bare lithium manganese oxide
CNTs	Carbon nanotubes
CVD	Chemical vapor deposition
ECR-MOCVD	Electron cyclotron resonance–metal organic chemical vapor deposition
EDS	Energy dispersive spectroscopy
EIS	Electrochemical impedance spectra
FCC	Face-centered cubic
GLMO	Graphite-coated lithium manganese oxide
GLP	Glassy lithium phosphate
HRTEM	High-resolution transmission electron microscope
LAG	Lanthanum aluminum garnet ($3LaAlO_3$:Al_2O_3)
LCO	Lithium cobalt oxide ($LiCoO_2$)
LFP	Lithium iron phosphate ($LiFePO_4$)
LFP/C	Lithium iron phosphate ($LiFePO_4$)–carbon composite
LIBs	Lithium-ion batteries
LMO	Lithium manganese oxide ($LiMnO_4$)
LMO-NCM	$0.3Li_2MnO_3-0.7LiNi_{5/21}Co_{5/21}Mn_{11/21}O_2$
LMR-NMC	Lithium manganese rich–nickel manganese cobalt oxide [$xLi_2MnO_3 \cdot (1-x)LiMO_2$; M = nickel (Ni); manganese (Mn); cobalt (Co)]
LNMO	Lithium nickel–manganese oxide ($LiNi_{0.5}Mn_{1.5}O_4$)
LTO	Lithium titanate ($Li_4Ti_5O_{12}$)
LVP	Lithium vanadium phosphate [$Li_3V_2(PO_4)_3$]
MPCVD	Microwave plasma chemical vapor deposition
NCM	Lithium nickel–cobalt–manganese oxide ($LiNi_{0.6}Co_{0.2}Mn_{0.2}O_2$)
NCNT	Nitrogen-doped carbon nanotubes
NMC	Lithium nickel–manganese–cobalt oxide ($LiNi_{1/3}Co_{1/3}Mn_{1/3}O_2$)
NMP	*N*-2-methyl pyrrolidone
OLO	Overlithiated layered oxide ($Li_{1.17}Mn_{0.56}Co_{0.095}Ni_{0.175}O_2$)

PEDOT:PSS	Poly(3,4-ethylenedioxythiophene):polystyrene sulfonate
PEG	Poly(ethylene glycol)
PEO	Poly(ethylene oxide)
PMMA	Poly(methyl methacrylate)
PPO	Poly(propylene oxide)
PVD	Physical vapor deposition
PVDF	Polyvinylidene fluoride
RF	Radio frequency
RMS	Reactive magnetron sputtering
SAED	Selected area electron diffraction
SEM	Scanning electron microscope
STEM	Scanning transmission electron microscope
ST-NCM	Surface-treated $LiNi_{0.6}Co_{0.2}Mn_{0.2}O_2$
TEM	Transmission electron microscope
XRD	X-ray diffraction

REFERENCES

[1] C. Zhao, X. Wang, R. Liu, F. Xu, Q. Shen, β-MnO_2 sacrificial template synthesis of $Li_{1.2}Ni_{0.13}Co_{0.13}Mn_{0.54}O_2$ for lithium ion battery cathodes, RSC Adv. 4 (2014) 7154.

[2] G.G. Yadav, A. David, H. Zhu, J. Caruthers, Y. Wu, Microemulsion-based synthesis and electrochemical evaluation of different nanostructures of $LiCoO_2$ prepared through sacrificial nanowire templates, Nanoscale 6 (2014) 860–866.

[3] S. Qiu, Z. Chen, F. Pei, F. Wu, Y. Wu, X. Ai, et al. Synthesis of monoclinic $Li[Li_{0.2}Mn_{0.54}Ni_{0.13}Co_{0.13}]O_2$ nanoparticles by a layered-template route for high-performance Li-ion batteries, Eur. J. Inorg. Chem. 2013 (2013) 2887–2892.

[4] A. Vu, A. Stein, Multiconstituent synthesis of $LiFePO_4$/C composites with hierarchical porosity as cathode materials for lithium-ion batteries, Chem. Mater. 23 (2011) 3237–3245.

[5] X. Xie, D. Su, B. Sun, J. Zhang, C. Wang, G. Wang, Synthesis of single-crystalline spinel $LiMn_2O_4$ nanorods for lithium-ion batteries with high rate capability and long cycle life, Chem. A Eur. J. 20 (2014) 17125–17131.

[6] B.Y. Xia, P. Yang, Y. Sun, Y. Wu, B. Mayers, B. Gates, et al. One-dimensional nanostructures: synthesis, characterization, and applications, Adv. Mater. 15 (2003) 353–389.

[7] J. Zhu, J. Fiore, D. Li, N.M. Kinsinger, Q. Wang, E. DiMasi, et al. Solvothermal synthesis, development, and performance of $LiFePO_4$ nanostructures, Cryst. Growth Des. 13 (2013) 4659–4666.

[8] Z. Li, K. Zhu, J. Li, X. Wang, Morphological and orientational diversity of $LiFePO_4$ crystallites: remarkable reaction path dependence in hydrothermal/solvothermal syntheses, Cryst. Eng. Comm. 16 (2014) 10112–10122.

[9] S. Yang, X. Zhou, J. Zhang, Z. Liu, Morphology-controlled solvothermal synthesis of $LiFePO_4$ as a cathode material for lithium-ion batteries, J. Mater. Chem. 20 (2010) 8086.

[10] Y. Park, K.C. Roh, W. Shin, J. Lee, Novel morphology-controlled synthesis of homogeneous $LiFePO_4$ for Li-ion batteries using an organic phosphate source, RSC Adv. 3 (2013) 14263.

[11] G. Qin, S. Xue, Q. Ma, C. Wang, The morphology controlled synthesis of 3D networking $LiFePO_4$ with multiwalled-carbon nanotubes for Li-ion batteries, CrystEngComm. 16 (2014) 260–269.

[12] Z. Ma, G. Shao, Y. Fan, G. Wang, J. Song, T. Liu, Tunable morphology synthesis of $LiFePO_4$ nanoparticles as cathode materials for lithium ion batteries, ACS Appl. Mater. Interfaces 6 (2014) 9236–9244.

[13] J. Song, L. Wang, G. Shao, M. Shi, Z. Ma, G. Wang, et al. Controllable synthesis, morphology evolution and electrochemical properties of $LiFePO_4$ cathode materials for Li-ion batteries, Phys. Chem. Chem. Phys. 16 (2014) 7728.

[14] N. Zhou, E. Uchaker, H.-Y. Wang, M. Zhang, S.-Q. Liu, Y.-N. Liu, et al. Additive-free solvothermal synthesis of hierarchical flower-like $LiFePO_4$/C mesocrystal and its electrochemical performance, RSC Adv. 3 (2013) 19366.

[15] X. Lou, Y. Zhang, Synthesis of $LiFePO_4$/C cathode materials with both high-rate capability and high tap density for lithium-ion batteries, J. Mater. Chem. 21 (2011) 4156.
[16] M. Wang, Y. Yang, Y. Zhang, Synthesis of micro-nano hierarchical structured $LiFePO_4$/C composite with both superior high-rate performance and high tap density, Nanoscale 3 (2011) 4434.
[17] Y. Xue, Z. Wang, F. Yu, Y. Zhang, G. Yin, Ethanol-assisted hydrothermal synthesis of $LiNi_{0.5}Mn_{1.5}O_4$ with excellent long-term cyclability at high rate for lithium-ion batteries, J. Mater. Chem. A 2 (2014) 4185.
[18] L. Wang, G. Liu, W. Wu, D. Chen, G. Liang, Synthesis of porous peanut-like $LiNi_{0.5}Mn_{1.5}O_4$ cathode materials through an ethylene glycol-assisted hydrothermal method using urea as a precipitant, J. Mater. Chem. A 3 (2015) 19497–19506.
[19] K. Saravanan, P. Balaya, M.V. Reddy, B.V.R. Chowdari, J.J. Vittal, Morphology controlled synthesis of $LiFePO_4$/C nanoplates for Li-ion batteries, Energy Environ. Sci. 3 (2010) 457.
[20] L. Tan, Z. Luo, H. Liu, Y. Yu, Synthesis of novel high-voltage cathode material $LiCoPO_4$ via rheological phase method, J. Alloys Compd. 502 (2010) 407–410.
[21] L. Si, L. Yue, D. Jin, Solvothermal synthesis of flower-like lanthanum tartrate and lanthanum oxide microspheres in ethanol–water mixed system, Cryst. Res. Technol. 46 (2011) 1149–1154.
[22] H.-I. Chen, H.-Y. Chang, Homogeneous precipitation of cerium dioxide nanoparticles in alcohol/water mixed solvents, Coll. Surf. A 242 (2004) 61–69.
[23] Y. Zhang, Y. Li, X. Niu, D. Wang, D. Zhou, X. Wang, et al. A peanut-like hierarchical micro/nano-$Li_{1.2}Mn_{0.54}Ni_{0.18}Co_{0.08}O_2$ cathode material for lithium-ion batteries with enhanced electrochemical performance, J. Mater. Chem. A 3 (2015) 14291–14297.
[24] A.R. West, Solid State Chemistry and its Applications, Wiley, (2014).
[25] Q. Li, F. Zheng, Y. Huang, X. Zhang, Q. Wu, D. Fu, et al. Surfactants assisted synthesis of nano-$LiFePO_4$/C composite as cathode materials for lithium-ion batteries, J. Mater. Chem. A 3 (2015) 2025–2035.
[26] L. Zhou, D. Zhao, X. Lou, $LiNi_{0.5}Mn_{1.5}O_4$ hollow structures as high-performance cathodes for lithium-ion batteries, Angew. Chem. Int. Ed. 51 (2012) 239–241.
[27] C. Zhu, T. Akiyama, Designed synthesis of $LiNi_{0.5}Mn_{1.5}O_4$ hollow microspheres with superior electrochemical properties as high-voltage cathode materials for lithium-ion batteries, RSC Adv. 4 (2014) 10151.
[28] H. Luo, P. Nie, L. Shen, H. Li, H. Deng, Y. Zhu, et al. Synthesis of $LiNi_{0.5}Mn_{1.5}O_4$ hollow microspheres and their lithium-storage properties, ChemElectroChem 2 (2015) 127–133.
[29] P. Patnaik, Dean's Analytical Chemistry Handbook, McGraw-Hill, (2004).
[30] W. Hua, Y. Wang, Y. Zhong, G. Wang, B. Zhong, B. Fang, et al. An approach towards synthesis of nanoarchitectured $LiNi_{1/3}Co_{1/3}Mn_{1/3}O_2$ cathode material for lithium ion batteries, Chin. J. Chem. 33 (2015) 261–267.
[31] X. Yang, X. Wang, Q. Wei, H. Shu, L. Liu, S. Yang, et al. Synthesis and characterization of a Li-rich layered cathode material $Li_{1.15}[(Mn_{1/3}Ni_{1/3}Co_{1/3})_{0.5}(Ni_{1/4}Mn_{3/4})_{0.5}]_{0.85}O_2$ with spherical core–shell structure, J. Mater. Chem. 22 (2012) 19666.
[32] P. Remith, N. Kalaiselvi, $Li_{1.2}Mn_{0.6}Ni_{0.1}Co_{0.1}O_2$ microspheres constructed by hierarchically arranged nanoparticles as lithium battery cathode with enhanced electrochemical performance, Nanoscale 6 (2014) 14724–14732.
[33] X. Xiao, X. Liu, L. Wang, H. Zhao, Z. Hu, X. He, et al. $LiCoO_2$ nanoplates with exposed (001) planes and high rate capability for lithium-ion batteries, Nano Res. 5 (2012) 395–401.
[34] Y.-S.S. Jung, A.S. Cavanagh, A.C. Dillon, M.D. Groner, S.M. George, S.-H. Lee, Enhanced stability of $LiCoO_2$ cathodes in lithium-ion batteries using surface modification by atomic layer deposition, J. Electrochem. Soc. 157 (2010) A75.
[35] X. Li, J. Liu, X. Meng, Y. Tang, M.N. Banis, J. Yang, et al. Significant impact on cathode performance of lithium-ion batteries by precisely controlled metal oxide nanocoatings via atomic layer deposition, J. Power Sources 247 (2014) 57–69.
[36] J. Liu, B. Xiao, M.N. Banis, R. Li, T.-K. Sham, X. Sun, Atomic layer deposition of amorphous iron phosphates on carbon nanotubes as cathode materials for lithium-ion batteries, Electrochim. Acta 162 (2015) 275–281.
[37] R.L. Patel, H. Xie, J. Park, H.Y. Asl, A. Choudhury, X. Liang, Significant capacity and cycle-life improvement of lithium-ion batteries through ultrathin conductive film stabilized cathode particles, Adv. Mater. Interfaces 2 (2015) n/a.

[38] W. Chang, J.-W. Choi, J.-C. Im, J.K. Lee, Effects of ZnO coating on electrochemical performance and thermal stability of $LiCoO_2$ as cathode material for lithium-ion batteries, J. Power Sources 195 (2010) 320–326.

[39] J. Wang, Q. Zhang, X. Li, Z. Wang, H. Guo, D. Xu, et al. Sputtering graphite coating to improve the elevated-temperature cycling ability of the $LiMn_2O_4$ electrode, Phys. Chem. Chem. Phys. 16 (2014) 16021–16029.

[40] M.L. Marcinek, J.W. Wilcox, M.M. Doeff, R.M. Kostecki, Microwave plasma chemical vapor deposition of carbon coatings on $LiNi_{1/3}Co_{1/3}Mn_{1/3}O_2$ for Li-ion battery composite cathodes, J. Electrochem. Soc. 156 (2009) A48.

[41] J. Lee, W. Choi, Surface modification of over-lithiated layered oxides with PEDOT:PSS conducting polymer in lithium-ion batteries, J. Electrochem. Soc. 162 (2015) A743–A748.

[42] R.W. Johnson, A. Hultqvist, S.F. Bent, A brief review of atomic layer deposition: from fundamentals to applications, Mater. Today 17 (2014) 236–246.

[43] X. Wang, G. Yushin, Chemical vapor deposition and atomic layer deposition for advanced lithium ion batteries and supercapacitors, Energy Environ. Sci. 8 (2015) 1889–1904.

[44] Y. Lee, T.Y. Kim, D.-W. Kim, J.K. Lee, W. Choi, Coating of spinel $LiNi_{0.5}Mn_{1.5}O_4$ cathodes with SnO_2 by an electron cyclotron resonance metal–organic chemical vapor deposition method for high-voltage applications in lithium ion batteries, J. Electroanal. Chem. 736 (2015) 16–21.

[45] J.-H. Park, T.S. Sudarshan, Chemical Vapor Deposition, ASM International, OH, 2001. Available from: https://books.google.com/books?id=oQ3wW4cQdLYC&pgis=1

[46] B. Qiu, J. Wang, Y. Xia, Z. Wei, S. Han, Z. Liu, Enhanced electrochemical performance with surface coating by reactive magnetron sputtering on lithium-rich layered oxide electrodes, ACS Appl. Mater. Interfaces 6 (2014) 9185–9193.

[47] G. Tan, F. Wu, L. Li, R. Chen, S. Chen, Coralline glassy lithium phosphate-coated $LiFePO_4$ cathodes with improved power capability for lithium ion batteries, J. Phys. Chem. C 117 (2013) 6013–6021.

[48] H. Kim, M.G. Kim, H.Y. Jeong, H. Nam, J. Cho, A new coating method for alleviating surface degradation of $LiNi_{0.6}Co_{0.2}Mn_{0.2}O_2$ cathode material: nanoscale surface treatment of primary particles, Nano Lett. 15 (2015) 2111–2119.

[49] G. Zhipeng, X. Shao, J. Wang, G. Song, Z. Wang, Ma, et al. $Li_3V_2(PO_4)_3$ modified $LiFePO_4$/C cathode materials with improved high-rate and low-temperature properties, Ionics 19 (2013) 1861–1866.

[50] C.Z. Lu, J.M. Chen, Y. Da Cho, W.H. Hsu, P. Muralidharan, G.T.K. Fey, Electrochemical performance of $LiCoO_2$ cathodes by surface modification using lanthanum aluminum garnet, J. Power Sources 184 (2008) 392–401.

[51] C. Ghanty, P.P. Dahiya, R.N. Basu, J.-K. Chang, S.B. Majumder, Improvement of the electrochemical characteristics of lithium and manganese rich layered cathode materials: effect of surface coating, J. Electrochem. Soc. 162 (2015) A1957–A1965.

[52] X. Hao, B.M. Bartlett, Improving the electrochemical stability of the high-voltage Li-ion battery cathode $LiNi_{0.5}Mn_{1.5}O_4$ by titanate-based surface modification, J. Electrochem. Soc. 160 (2013) A3162–A3170.

[53] C.J. Brinker, G.W. Scherer, Sol–Gel Science: The Physics and Chemistry of Sol–Gel Processing, Academic Press Inc, (1990).

CHAPTER FOUR

Nanotechnology in Electrochemical Capacitors

E. Goikolea, R. Mysyk
CIC energiGUNE, Miñano, Spain

Contents

1 INTRODUCTION

Supercapacitors or ultracapacitors are one of the electrical energy storage technologies undergoing extensive developments in the last years. In the energy-power spectrum, supercapacitors take an intermediate place between batteries and dielectric capacitors, delivering higher power densities than the former and higher energy densities than the latter [1]. Typical values of energy and power densities of commercially available supercapacitors are in the range of 4–5 Wh/kg and 10–20 kW/kg, correspondingly, but recent developments suggest that these figures can be overcome in the foreseeable future. Although supercapacitors present lower energy densities than batteries, they possess numerous assets that batteries do not, primarily a virtually infinite lifetime. They also feature the ability for rapid charge (as distinct from battery materials), a wider temperature range, environmental friendliness, better safety, higher reliability, and maintenance-free operation [2,3].

As distinct from batteries, supercapacitors can mainly be viewed as pulse power devices with a typical discharge time ranging from a few seconds down to less than 1 s [4]. Nowadays there is proven demand for supercapacitors in transportation (hitting an approximately 70% supercapacitor market share in the United States), especially public transportation, such as buses and tramways, as both a stand-alone and complementary energy source. Supercapacitors are particularly useful in starting engines and are able to accumulate charge in regenerative energy braking. They also find applications in renewable energy (e.g., adjusting the pitch of a wind turbine's rotor blades), industrial machines (forklifts, cranes), electronics, and home appliances [5,6].

The first reference to supercapacitors dates back to 1957 with General Electric's patent [7] followed by the devices patented by SOHIO Corporation in 1960–70 [8,9]. Further developments were made by NEC, Panasonic (Japan), ECOND (Russia), and many other companies around the world. Particularly, Maxwell developed a supercapacitor reaching 20 kW/kg in the beginning of the 1990s under the Department of Energy's Ultracapacitor Development Program [10,11]. In Europe, Batscap (nowadays Blue Solutions) offered products with similar performance. Nowadays a nonexhaustive list of supercapacitor manufacturers from all over the world includes Yunasko, Maxwell, Nesscap, JM Energy, Ioxus, and Blue Solutions, to cite just a few.

This chapter aims at delineating the relation between nanoscience/nanotechnology and its impact on the performance of supercapacitors. In fact, apart from the progress made in the research of new electrode materials, nanostructuring offers a promising solution to fulfil the requirements of a high performing supercapacitor [12]. We start off with an introduction to the basics of supercapacitors, their operating principles, and the handiest electrochemical tools for evaluating supercapacitors in laboratory. We further consider parameters governing supercapacitor performance and major determining factors. The main body of the chapter is devoted to the relation between the properties of materials at nanoscale and the electrochemical response of supercapacitor electrodes. We focus mainly on the textural properties of porous carbons and nanosizing inorganic pseudocapacitive materials.

2 BASIC PRINCIPLES AND CLASSIFICATION OF ELECTROCHEMICAL CAPACITORS

2.1 Supercapacitor materials and cell configurations

Active electrode materials for electrochemical capacitors can be broken down into two main categories: materials for electric double-layer capacitors (EDLC, non-Faradic

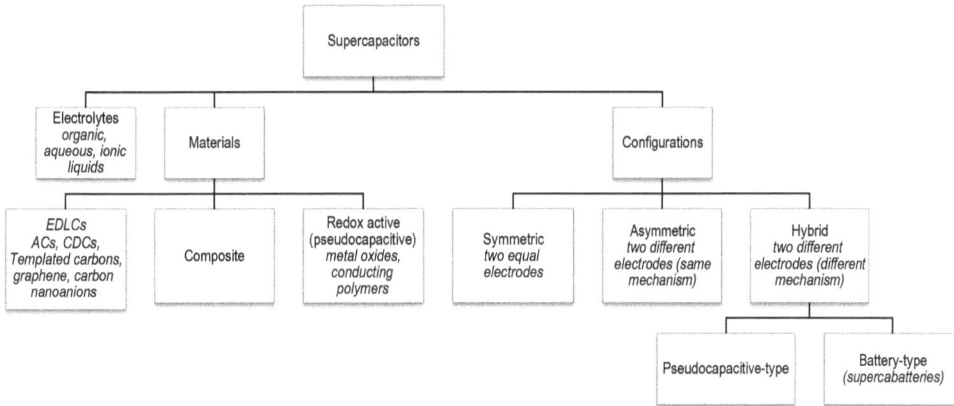

Figure 4.1 ***Taxonomy of materials, configurations, and electrolytes for supercapacitors [13].***

capacitors) and those for pseudocapacitors (Faradic capacitors). Combinations between the two types of materials or materials exhibiting both non-Faradic and Faradic activity also are widely used in the art, leading to various types of asymmetric (hybrid) systems (Fig. 4.1) [3,13].

A supercapacitor cell comprises two electrodes with each of them being stuck on a current collector providing a link to an electrical circuit, a separator between the electrodes, and an electrolyte (Fig. 4.2A). The electrodes can be identical (symmetric cells, Fig. 4.2B) or different (asymmetric cells, Fig. 4.2C). The separator allows minimizing the volume of a supercapacitor cell while preventing the short circuit between the electrodes and enabling ionic charge transfer by the electrolyte ions. Each electrode can be represented as a separate capacitor, therefore both electrodes can be considered as series-connected capacitors with their characteristic capacitance value. The electrodes are termed according to the change in electric potential when they are charged. As long as both electrodes build up charge, the potential of the *negative electrodes* decreases whereas that of the *positive electrode* increases. The terms *negative electrode* and *positive electrode* are used for supercapacitors more often than *anode* and *cathode*. Therefore, the total capacitance of a two-electrode cell can be calculated as

$$C = C_+C_-/(C_+ + C_-) \tag{4.1}$$

where C_+ and C_- are the capacitance of the positive and the negative electrodes in Farads (F), correspondingly.

In gravimetric terms,

$$C_g = \frac{C_{g,+}m_+C_{g,-}m_-}{(C_{g,+}m_+ + C_{g,-}m_-)(m_+ + m_-)} \tag{4.2}$$

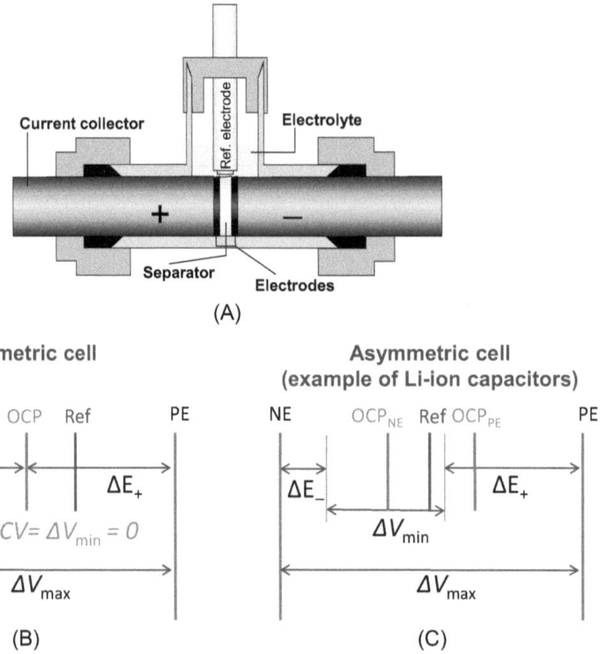

Figure 4.2 (A) Schematics of supercapacitor cells and the voltage and potential breakdown in (B) symmetric and (C) asymmetric cells. *OCP,* the initial potentials of the negative, positive electrodes in an asymmetric cell, and both electrodes in a symmetric cell; *OCV,* the zero current voltage; *NE,* the negative electrode; *PE,* the positive electrodes; *Ref,* the reference electrode. The operating potential range of each electrode in an asymmetric system is determined by the corresponding electrode processes (Li-ion capacitor is given as an example; for other types of asymmetric systems ΔE_-, ΔE_+ are different) and the voltage limits of the cell.

where C_g is the gravimetric cell capacitance, $C_{g,+}$ and $C_{g,-}$ the gravimetric capacitances of the positive and negative electrodes, accordingly, m_+ and m_- the masses of the corresponding electrodes. In a symmetric cell with equal electrode masses, this formula becomes

$$C_g = \frac{C_{g,+}C_{g,-}}{2(C_{g,+} + C_{g,-})} \tag{4.3}$$

The capacitance of each electrode is determined as

$$C_{+(-)} = Q/_{\Delta E_{+(-)}} \tag{4.4}$$

where Q is the charge in Coulombs and $\Delta E_{+(-)}$ is the electric potential difference in the corresponding electrode. In practice, the values of ΔE_+ and ΔE_- are not known for a two-electrode cell since only the total potential difference (cell voltage) can be measured, $\Delta U = \Delta E_+ + \Delta E_-$, that is, only C_g can be calculated. To evaluate the capacitance of a single electrode ($C_{g,e}$), it has become common to assume that the gravimetric capacitances of positive and negative electrodes are equal. Therefore, for a symmetric cell with equal electrode masses

$$C_g = C_{g,e} / 4 \tag{4.5}$$

The exact values of $C_{g,+}$ and $C_{g,-}$ can be determined by introducing a constant-potential reference electrode (Fig. 4.2A). Knowing $C_{g,+}$ and $C_{g,-}$ is important in asymmetric full (two-electrode) cells since capacitance values can be quite different [Eq. (4.5) is only approximate], and the masses of electrodes should be balanced according to the equivalence of charges.

$$C_{g,+(-)} = Q/_{m_{+(-)}} \Delta E_{+(-)} \tag{4.6}$$

A practically important value is a zero-current potential (OCP, also referred to as the *equilibrium* potential or the rest potential), which can be the same if an identical material is employed in both electrodes (the zero current voltage, OCV, of such a cell is therefore 0). OCP depends on the impurities in a material and the previous cycling (such as in situ formed functional groups). Noteworthy, a common feature of materials for both non-Faradic (EDLC) and Faradic (pseudo) capacitors is the potential difference gradually increasing with charge build-up (the state of charge) whereas the charge storage mechanisms are completely different (electrosorption vs. electrochemical reaction, as detailed in what follows) [14].

Asymmetric configurations can be of various types involving combinations, such as two non-Faradic electrodes with different capacitance values, capacitive and pseudocapacitive electrodes, capacitive and battery type electrodes, two composite electrodes with each involving both capacitive and battery type processes, and so forth [15]. Asymmetric cells are usually designed to provide higher energy than that of symmetric cells [16]. A particular type of supercapacitor cells are the so-called battery-type hybrids that combine non-Faradic and Faradic battery-type behavior in a single cell, and can be classified as internal series hybrids (ISHs) and internal parallel hybrids (IPHs) [17]. An ISH contains one electrode, which is exclusively a battery-type electrode, ideally working at a constant potential whereas the other is a capacitive electrode exhibiting a tilted linear potential profile typical of a supercapacitor under constant-current discharge. As distinct, an IPH involves battery and capacitive materials as part of each electrode, providing higher energy at low current owing to the use of high-capacity battery materials, but also being able to operate at high current through non-Faradic materials when battery materials fail [18]. Some hybrid systems are nowadays commercially available (e.g., lithium-ion capacitors also referred to as LICs) [19].

The electric energy stored in an EDLC capacitor is given by

$$E_{sc} = \int_{Q_0}^{Q_1} U\,dq = \frac{Q^2}{2C} = \frac{1}{2}QU = \frac{1}{2}CU^2 \tag{4.7}$$

In other words, energy is determined by the capacitance and the cell voltage, with more impact from the latter.

EDLCs are the most widespread type of supercapacitors, operating by electrosorptive redistribution of electrolyte ions onto high-surface area electrodes [20] without any electrochemical charge transfer process. Electrostatic charge build-up onto the surface of active materials is provided by an electrical stimulus from an external energy source whereas discharge is associated with reverse ion redistribution to the initial state, releasing energy. In a typical two-electrode EDLC cell, the charging process drives the negative electrode material to an excess of negative charge, stimulating a predominance of cations in the vicinity of the electrode surface on the solution side, forming a so-called "electrical double layer." An opposite process takes places on the positive electrode, in which the electrode is positively charged with a concomitant preponderance of anions on the solution side.

High capacitance values of 100 F/g of active material are attainable with EDLCs owing to a high specific surface area (SSA) of electrode materials and a very small charge separation between the surface atoms and the ions from the electrolyte (typically below 1 nm) [20]. This can be approximately quantified with a parallel plate capacitor model proposed first by Helmholtz [21]:

$$C_g = \frac{\varepsilon_0 \varepsilon_r A}{d} \tag{4.8}$$

where ε_r is the effective dielectric permitivity, ε_0 the electric constant, A the SSA, and d the charge separation, that is, the distance between electronic and ionic charges.

Such enormous gravimetric capacitance forms the basis for the term "supercapacitor". Various carbon nanomaterials have been explored as EDLC electrodes, in particular, template carbons [22], carbon onions [23], graphene [24], and others, but microporous carbons remain the material of choice owing to their moderate cost, widespread availability, and the most suitable textural properties. Carbons with pores narrower than 2 nm (microporous carbons according to the IUPAC nomenclature) usually provide the highest SSA (>1000 m^2/g) among various carbon nanomaterials and are the most suitable non-Faradic materials for traditional supercapacitor applications [25]. These will be detailed in the corresponding subchapter.

Another type of super capacitor materials is the so-called pseudocapacitive materials, which are inorganic materials exhibiting surface or near-surface fast and reversible redox reactions with the charge build-up being proportional to the potential difference, that is, their electrical response is ideally identical to that of EDLCs [26]. These materials will be discussed in greater detail in Section 5.

2.2 Electrolytes for supercapacitors

Electrolytes for supercapacitors can be broken down into aqueous and organic. Typical aqueous electrolytes are 1 M H_2SO_4 or 6 M KOH. Aqueous basic or acidic electrolytes provide a practical operational voltage of 1 V, which is limited by a maximum of 1.23 V (thermodynamic water stability window) if equal potential windows are available with respect to the OCP potential. Neutral aqueous electrolytes can provide higher practical stability limits, 1.6–2.2 V according to laboratory studies [27]. Standard organic electrolytes [28] (1 M tetraethylammonium tetrafluoroborate or triethylmethylammonium tetrafluroborate dissolved in acentonitrile or propylene carbonate) give rise to the cell voltage of 2.7 V, which can be slightly extended in an asymmetric configuration. However, they exhibit drawbacks such as the need for cell assembly in inert atmosphere, environmental pollution on cell recycling or failure, flammability, and high cost [29]. On the other hand, aqueous electrolytes are easy to manipulate, environmentally friendly, nonflammable, and low-cost solutions. However, the main factor determining the current industrial landscape is energy density, that is, the operating voltage. Therefore, organic electrolytes are used predominantly in industry since they provide higher energy density according to Eq. (4.7). In the last years, organic electrolytes using spiro-structure quaternary ammonium salts have been introduced to commercial supercapacitors [30]. They show better high-voltage and high-temperature resistances, and some of them show higher capacitance values than electrolytes based on tetraalkylammonium salts.

Solvent-free room temperature ionic liquid (RTIL) electrolytes have also been proposed with an operating voltage of above 2.7V [31]. However, they require an expensive drying process for activated carbons (ACs), have low inherent electrical conductivity and poor compatibility with microporous carbons (bulky organic ions have difficulty accessing microporous surface). In practical terms, the energy density of RTIL-based systems can be limited by their low capacitance arising due to low accessible SSA, but they can be used for extending the temperature range of supercapacitors over typical limits [32]. Internal hybrids use conventional electrolytes of metal ion batteries, such as lithium hexafluorophosphate solution in ethylene carbonate and diethyl carbonate, 1.0 M $LiPF_6$.

The main properties of the most common electrolytes used in supercapacitors are summarized in Table 4.1.

2.3 Electroanalytical methods for studying supercapacitors: cyclic voltammetry, galvanostatic cycling, impedance spectroscopy

Cyclic voltammetry (CV) is a widely used electroanalytical technique in supercapacitor research. CV consists in recording the current flowing through an electrochemical cell

Table 4.1 Summary of the main properties of aqueous, organic, and ionic liquid electrolytes used in supercapacitors

	Aqueous	Organic	Ionic liquids
Voltage of a symmetric cell, V	0.7–1.0	2.7	3–5
Maximum capacitance, F/g (F/cm^3)	up to 300 (120–160)[a] ~1000 (>1000)[b]	up to 200 (60–80)	up to 100–120 (<60)
Temperature range, °C	Room temperature	−45 + 60	0 + 100 −50 +100[e]
Electrical conductivity, mS/cm	400–800[c] 100–200[d]	10–60	10–20

[a] Double layer capacitance with carbon materials
[b] Pseudocapacitance with electroactive materials
[c] Acidic and basic electrolytes
[d] Neutral aqueous electrolytes
[e] Mixed ionic liquids [32].

when the potential difference is increased/decreased linearly with a constant potential (voltage) sweep rate. A pair of sweeps with an increasing and decreasing potential difference is referred to as a "cycle." Importantly, CV may be run using two-electrode and three-electrode cell configurations.

A two-electrode configuration is the laboratory prototype of a supercapacitor device, in which the voltage sweep rate is set between the electrodes, but the growth/decay of potential difference at each electrode is not controllable and is governed by the corresponding electrode processes. For example, in a configuration including two non-Faradic electrodes, the potential differences are determined by the capacitance and the masses of electrodes according to Eq. (4.6). Ideal cyclic voltammograms of supercapacitors can be easily deduced on the basis of Eq. (4.4). It can be written for a two-electrode supercapacitor:

$$c = \frac{dQ}{dU} = i\frac{dt}{dU} = \frac{i}{v} \tag{4.9}$$

where c is the differential capacitance, i is the instant current, dt is the time corresponding to the voltage change, dU, and $v = dU/dt$ is the sweep (scan) rate. Therefore, if the capacitance is constant throughout ΔU, the voltammogram will present just a straight line in the i-t plot with a constant $i = \pm I$ value depending on the capacitance [Fig. 4.3A, Eq. (4.9)]. The most common representation is a current (capacitance)-voltage plot, giving rise to a rectangular graph (Fig. 4.3B).

Integral capacitance can be defined as

$$C = \frac{Q}{\Delta U} = \int i\frac{dt}{\Delta U} = \frac{I}{v} \tag{4.10}$$

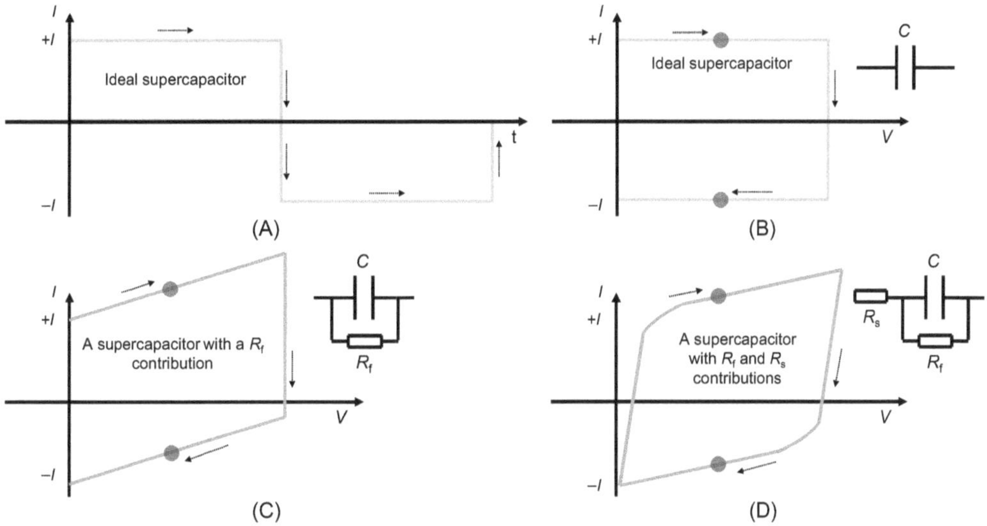

Figure 4.3 ***Voltammetry response of a two-electrode supercapacitor with constant capacitance.*** (A) *i–t* plot, (B) the *i–V* plot, (C) the *i–V* plot of a supercapacitor with a series resistance, and (D) the *i–V* plot of a supercapacitor with series and leak resistances.

where $\int i\,dt$ is the charge accumulated over ΔU. In gravimetric terms, cell capacitance can be calculated as

$$C_g = \frac{C}{m_+ + m_-} = \int \frac{i\,dt}{\Delta U(m_+ + m_-)} \tag{4.11}$$

Therefore, the average capacitance per electrode can be calculated according to Eq. (4.5)

$$C_{g,e} = 4C_g = 4\int \frac{i\,dt}{\Delta U(m_+ + m_-)} = 4\frac{I}{v(m_+ + m_-)} \tag{4.12}$$

For electrodes with equal masses, $m_+ = m_- = m_e$, this translates into

$$C_{g,e} = 2\frac{I}{vm_e} \tag{4.13}$$

where I is the average current over the time of the run.

A real supercapacitor cell exhibits deviations from ideal CV shape. Fig. 4.3C shows the deformation of a CV if a leak resistance, R_f, is presented in parallel to capacitance. Apart from R_f, a real supercapacitor also contains a series resistance, R_s, that leads to some distortion of voltammetry curves at the beginning of charge and discharge (Fig. 4.3D). This distortion makes it difficult to precisely determine capacitance, especially at high sweep rates where the contribution of resistance to the shape of CV curves is nonnegligible. Therefore, CV can be used for knowing the values of capacitance only approximately. As distinct from real devices, in laboratory practice it is often advantageous to use a two-electrode set-up additionally (but optionally) involving a reference electrode, which only serves for monitoring potential evolution at separate electrodes, with the scan rate being set between the positive and negative electrodes (Fig. 4.2A). A typical voltammogram of two electrode cells is reproduced in Fig. 4.4A, showing that additional information can be extracted on the potential evolution of each electrode using a reference electrode with a fixed potential value (the upper x-axis). The capacitance of each separate electrode can therefore be calculated according to Eq. (4.6) instead of the average capacitance per electrode.

The three-electrode set-up allows an isolated electrode material to be studied, not a prototype device. Three-electrode measurements mainly serve two purposes: (1) measuring the capacitance of a material in a given electrolyte and (2) establishing the range of electric potential differences (the potential stability window, ΔE_{max}) providing for a stable and reversible charge/discharge process. The three electrodes are the working electrode, WE (the material of interest as the main constituent), the reference electrode, RE (an electrode with a constant electrochemical potential), and the counter electrode, CE (an electrode used to complete the circuit). It is important to ensure that the CE is properly chosen to exclude side electrode processes (such as electrolyte decomposition) able to pollute the electrolyte and impact the behavior of the WE. In a three-electrode configuration, while the current flows between the WE and the CE the sweep rate is set between the WE and the RE, and not between the WE and the CE as in a two-electrode

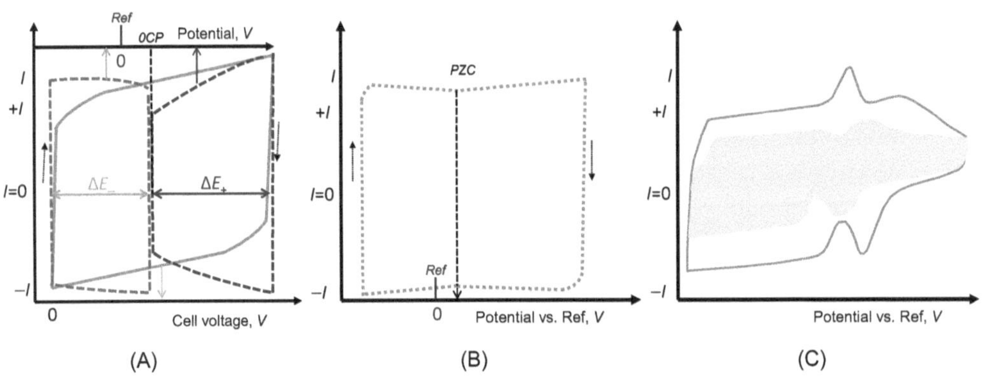

Figure 4.4 (A) Cyclic voltammetry response of a two-electrode supercapacitor and (B) of an isolated material in a three-electrode configuration without any pseudocapacitive activity. (C) Separation of capacitive contribution *(filled area)* from overall charge accumulation also involving diffusion-limited according to Eq. (4.15). *PZC*, point of zero charge; *Ref*, the potential of the reference electrode; *OCP*, zero current potential of electrodes in a symmetric two-electrode cell with the same material at both electrode (ideally constant); ΔE_{+}, ΔE_{-} the operating potential ranges of negative and positive electrodes, correspondingly, in a two-electrode cell.

set-up. As the equivalence of charges should be maintained according to Eq. (4.6) between the WE and the CE, the mass of the CE should be selected properly to make sure that the potential difference at the CE is kept within the electric potential limits of the electrolyte stability, that is, the charge passing through the WE is balanced by a reversible capacitive process at the CE. In practice of supercapacitor research, the CE is often chosen to be an AC electrode with a mass exceeding by far that of the studied WE, so, ΔE at the CE can be kept within narrow limits, not going beyond the electrolyte breakdown potential.

In a three-electrode configuration, the differential capacitance of a material can be calculated as

$$c = \frac{dQ}{dE m_e} = i\frac{dt}{dE m_e} = \frac{i}{\nu m_e} \tag{4.14}$$

where dE is the differential potential of WE and m_e is the mass of WE.

Fig. 4.4B shows a schematic voltammetry response of a carbon material scanned with a fixed scan rate versus a reference electrode over the potential stability range. A distinct minimum of the CV curve not involving any Faradic contribution (usually the case for organic electrolytes) can be observed for carbon materials with a limited amount of defects, which corresponds to the minimum of the density of electronic states at the Fermi level [33–35]. This value is the potential of zero charge (PZC), that is, the electrode potential when the surface density of electrical charge is zero. It can also be experimentally measured, for example, as the immersion potential of a vacuum-kept material introduced in deaerated electrolyte solution, and may be shifted depending on the surface chemistry by almost 1 V [36].

Three-electrode CV measurements are also useful for kinetic analysis. The capacitive or pseudocapacitive current should be linearly scaled with the sweep rate according to Eq. (4.9). This point delineates the commonality of capacitive and pseudocapacitive processes and also differentiates them from slower diffusion-limited redox processes typical of battery-type electrodes. Voltammogramms of battery-type electrodes contain peaks with the peak current being proportional to the square route of the scan rate. For some materials exhibiting pseudocapacitive activity, there is a mixed contribution of pseudocapacitive and diffusion-limited kinetic responses that can be separated using the "mapping" procedure. The measured current can be fitted according to the power law $i = a^{\nu b}$, where both a and b are adjustable parameters [37]. When $b = 1$ the reaction is not controlled by mass diffusion and can often be associated with a (pseudo)capacitive process [according to Eq. (4.9)], whereas $b = 0.5$ corresponds to a diffusion-controlled Faradic process. It was proposed to evaluate the relative contribution of pseudocapacitive and diffusion-limited processes by fitting the total current according to Eq. (4.15) at various potential values for a range of scan rates [38,39].

$$i = k_1\nu + k_2\nu^{1/2} \tag{4.15}$$

Galvanostatic charge/discharge (GA) is another useful electroanalytical technique consisting in applying a constant current to a supercapacitor cell while simultaneously imposing a voltage window. Typically, voltage increases and decreases linearly during charge/discharge for systems with constant capacitance. The linearity of voltage-time graphs follows simply from the definition of capacitance [Eq. (4.10)]. Thus, the current and the reciprocal of the slope of a ΔU–t discharge line determine the capacitance. For comparison with battery technologies, it is sometimes useful to calculate capacity (charge), not capacitance, Q, which is most often presented as mAh/g (1 mAh = 3.6 Coulombs). However, as is the case for CV, the GA plot of a real supercapacitor exhibits different resistance components.

Fig. 4.5 depicts the endurance test (sometimes referred to as "floating" or "the float-voltage" test), which consists in maintaining a supercapacitor at the maximum (rated) voltage at an elevated temperature (usually 60°C) and periodically measuring its capacitance values using GA [40,41]. The close-up of the discharge curve demonstrates that the beginning of discharge (decreasing cell voltage) is a vertical line, further taken over by a curvy portion before the tilted linear part sets in finally. The vertical part is usually referred to as equivalent series resistance (ESR) whereas both the curvy and vertical parts altogether are termed equivalent distributed resistance (EDR). ESR is influenced by resistive components, such as the inherent resistivity of active material, the resistance of electrolyte and separator(s), and so forth. The curvy part is mainly associated with charge redistribution process due to the inhomogeneity of the porous network [1] since the voltage of an electrode is measured at the external electrode surface. Switching to open circuit leads to charge redistribution, which equalizes potential all over the electrode, resulting in a loss in potential. ESR and EDR also differ in that the former is

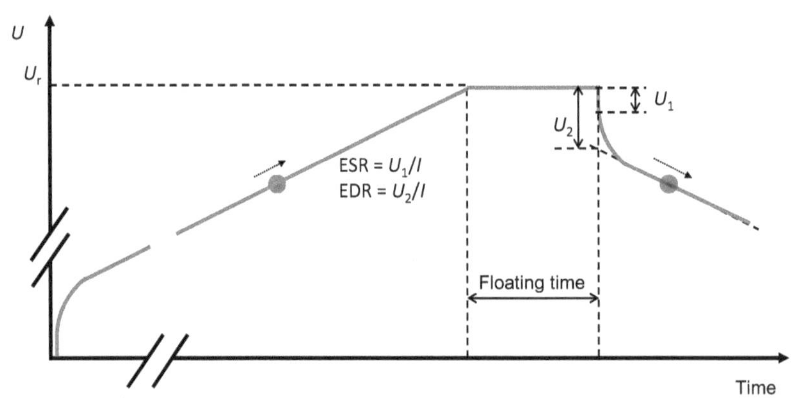

Figure 4.5 ***Endurance test consisting in galvanostatic charge, constant voltage, and galvanostatic discharge.***

independent of the current value whereas the latter is usually impacted by the current value, being higher for low current densities [41].

GA charge/discharge is also used for cycling ability tests, which consist in repeated charge/discharge cycles until the capacitor fails. In practice, such cyclability tests are common in the field of batteries, but it is quite difficult to reach failure in a reasonable testing time under repeated GA tests for EDLCs since the cycle ability of a well-assembled cell is almost infinite due to the highly reversible character of electrosorption. For this reason, floating tests are preferred for EDLCs since they provide harsher conditions that can induce cell failure in an acceptable testing time. The tests are performed after cell assembly and after selected floating times until end-of-life criteria are met. The end-of-life criteria are usually set as a double increase of resistance and 20–30 % loss of capacitance [40]. For asymmetric (hybrid) cells working at least partially through electrochemical processes, cyclability tests are more useful than for EDLCs.

A common use of GA testing is to submit a capacitor to different current densities, determining the capacitance and resistance components for each current density, to establish capacitance retention as a function of current density: the flatter the $C = f(I)$ plot, the higher the power capability [42].

Electrochemical impedance spectroscopy is another widely employed electrochemical technique that can also be a useful tool for evaluating supercapacitors. It consists in setting voltage (or current) to some fixed value and applying a small sinusoidal signal (±10–20 mV for voltage) over the set value. Correspondingly, the response of current (or voltage) is measured. A sinusoidal signal is applied consecutively for different frequencies since the response of different processes occurring in the cell exhibits frequency dependence. The impedance can be defined as complex resistance on the basis of ac current and voltage [43]:

$$U = U_0 e^{-j\omega t} \tag{4.16}$$

$$I = I_0 e^{-j\omega t+\varphi} \tag{4.17}$$

$$Z = \frac{U_0}{I_0} I_0 e^{-j\varphi} \tag{4.18}$$

where $\omega = 2\pi f$.

The most common representation of impedance response is a so-called Nyquist plot (also referred to as the Cole-Cole plot or the Complex Impedance Plane plot), which shows the absolute value of the imaginary impedance component as a function of the real impedance component. As with other technologies, the impedance spectra of a supercapacitor can be modelled with an equivalent circuit, which is a way of simulating cell impedance using both classical electrical components, such as resistors and capacitors and some specialized electrochemical elements.

However, equivalent circuits should be maximally assimilated to processes likely happening in the cell; otherwise, a too complex circuit can provide a satisfactory description of an impedance spectrum, but is useless for decrypting real phenomena occurring in an electrochemical capacitor cell. In what follows we will address the physical significance of the experimental impedance response of supercapacitors rather than giving a comprehensive overview of proposed equivalent circuits. The ideal plot of a capacitor in-series connected to resistance (R_s) is given in Fig. 4.6A. The capacitive component is represented by a vertical line crossing the x-axis at the value of series resistance. However, this kind of response is not observed with porous electrodes whose schematic impedance spectrum is exemplified in Fig. 4.6B. Roughly 3 regions can be separated: (1) the high-frequency region where the tilted line intersects the x-axis. The intersection point corresponds to the resistance in series to the capacitance; (2) the tilted line corresponds to the ionic resistance inside the pores. The theoretical background for this line was given by de Levie [44,45] who proposed to model this phenomenon with a transmission line circuit; (3) the quasivertical line intersecting the tilted line at what is referred to as "the knee frequency" is indicative of mostly capacitive behavior. Important treatments considering various pore shapes were also proposed to explain impedance spectra [46]. The deviations from the vertical line (the ideal capacitor) can be described using a combination of so-called constant phase elements (CPE, nonideal capacitance). A CPE is given by the formula:

$$Z = \frac{1}{Y(j\omega)^{\alpha}} \tag{4.19}$$

where $0 < \alpha < 1$ and $\omega = 2\pi f$.

A supercapacitor cell (Fig. 4.6C) also can additionally contain a semicircle at a high frequency, which is an undesirable resistive contribution increasing the overall resistance. Mathematically, such a semicircle can be modelled with a resistance (R_p) connected

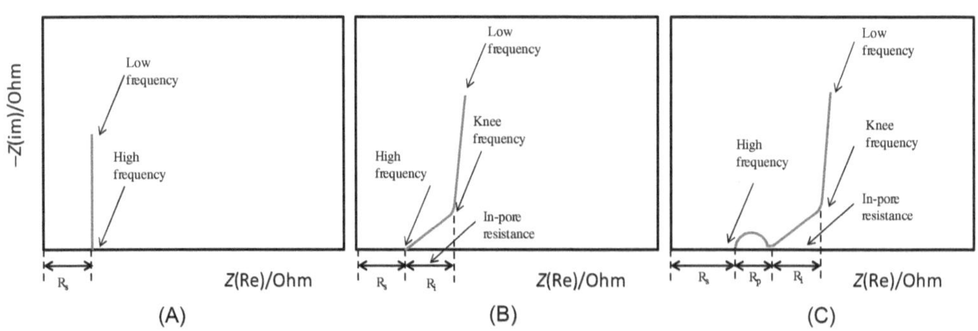

Figure 4.6 ***Nyquist plots.*** (A) An ideal capacitor with a series resistance, (B) a capacitor using porous electrodes, and (C) a capacitor using porous electrodes with a parallel leak resistance.

in parallel to capacitance. Several hypotheses were put forward to extract the physical origin of this component: intergranular electronic resistance among the AC particles, AC–current collector interface, and the resistance of the passive oxide layer developed on current collectors' surface. It was shown that a pretreatment of current collectors' surface helps in eliminating such a semicircle, and it is commonly believed that a well-assembled cell should display a semicircle-free Nyquist plot [47]. The Nyquist plot has a drawback of hiding the frequency values corresponding to the impedance components, that is, additional plots, such as the Bode plot (phase angle vs. frequency) are sometimes useful, especially if a supercapacitor is considered for electronic applications.

Impedance data can also be treated in terms of a complex capacitance [48]:

$$C(\omega) = C'(\omega) + jC''(\omega) \tag{4.20}$$

with

$$C'(\omega) = \frac{Z''}{|Z|^2 \omega} \tag{4.21}$$

$$C''(\omega) = \frac{Z'}{|Z|^2 \omega} \tag{4.22}$$

The maximum of $C''(\omega)$ allows extracting critical frequency (f_k) representing the threshold value for the ratio of stored-to-lost energy, which is above 50% for $f < f_k$. For a given material, the critical frequency can depend on the electrolyte and the electrode thickness.

3 PARAMETERS GOVERNING SUPERCAPACITOR PERFORMANCE

3.1 Energy and power density of supercapacitors

The energy density (E_{sc}) of a supercapacitor can be defined in the same way as for other electrochemical energy sources [Eq. (4.7)]. The main difference of a supercapacitor material from an ideal battery material is that the electrode potential changes continuously with Q whereas a battery material ideally has an equilibrium potential, which is constant as long as the reduced and oxidized species coexist [20]. Therefore, the energy of an ideal battery containing two ideal battery electrodes would be

$$E_B = \int_{Q_0}^{Q_t} U\,dq = QU \tag{4.23}$$

where U is the cell voltage, which is constant independently of Q as the potential of each electrode stays constant during charge build-up. Real battery materials often exhibit deviations from such signature, for example, due to kinetic limitations, consecutive chemical reactions, and so forth, but still the charge/discharge profile is a necessary (but not sufficient) diagnostic feature allowing one to distinguish between battery-type and supercapacitor-type behavior.

This shows that the energy of a battery will be twice that of a supercapacitor for the same capacity (Coulombs or mAh/g). However, in practice the capacity of a battery material exceeds that of a supercapacitor. This can be understood on the basis of specific capacitance of carbon electrodes, which ranges experimentally between 100 and slightly above 200 F/g. For the capacitive electrode in a symmetric organic–based cell, the working potential stability range is usually about 1.35 V. Therefore, capacitive electrodes have the maximum estimated capacity of about 75 mAh/g. In aqueous electrolytes, capacitive electrodes usually work in the potential range of $\pm$ 0.4 to $\pm$ 0.8V (the latter for neutral aqueous electrolytes). Therefore, the capacity of such an electrode in aqueous media is about 30–40 mAh/g. For comparison, the capacity of a battery material in most advanced batteries is usually above 100 mAh/g. This limitation of capacitive materials arises because of the surface capacitance (about 5–10 $\mu F/cm^2$ only) of carbon electrodes, which is lower than for metal electrodes. When multiplied by a SSA of 1000–2000 m^2/g, this leads to a maximum capacitance of 100–200 F/g depending on the electrolyte, which is the known experimental upper limit. This is a fundamental limitation that needs more detailed insight. The areal capacitance of a non-Faradic electrode can be represented as a sum of three components [49]:

$$\frac{1}{C_{A,exp}} = \frac{1}{C_q} + \frac{1}{C_H} + \frac{1}{C_{dl}} \tag{4.24}$$

where $C_{A,exp}$ is the experimental area-normalized capacitance value, C_q the quantum capacitance for low-dimensional materials or space-charge capacitance for 3D carbons, C_H the capacitance of the compact double layer (the Stern model), C_{dl} the capacitance of the diffuse ionic layer (all capacitance values measured in $\mu F/cm^2$). It is evident from this formula that $C_{A,exp}$ can only be lower than the lowest of the three contributions. $C_{A,exp}$ is difficult to know precisely for porous carbons due to the limitations of methods available for calculating accessible surface area (we will consider this further when discussing the electrical double layer in nanoporous materials). However, for nonporous carbons, for example, highly oriented pyrolitic graphite (HOPG), the aforementioned value is about 5 $\mu F/cm^2$ [33], limited by C_q being low because of carbons' electronic structure. C_{dl} is $>$100 $\mu F/cm^2$ in concentrated electrolytes, and can be excluded from consideration as a nonlimiting factor. C_q is not precisely known for porous carbons, but estimations using literature data for nonporous carbons, such as graphene and graphite yield the values of 5–15 $\mu F/cm^2$ for basal graphene plane and $>$70 $\mu F/cm^2$ for edge carbon atoms (10–20 % of all C atoms) [50]. However, edge carbon atoms contain unsatisfied valence (dangling bonds) being prone to undesirable electrochemical electrolyte decomposition, which is detrimental to cycle life.

Thus, increasing capacitance by increasing C_q (e.g., by doping carbons as explained further) or using pseudocapacitive materials with higher than double capacitance is required to push energy density above the current limits.

The second parameter for energy density is the maximum cell voltage, which is determined by electrolyte breakdown potential at one of the electrodes as explained in Section 2.2. Thus, the highest energy density can be ensured with organic electrolytes, and for this reason they have become the most widespread type used in commercial supercapacitors.

In practice, the utilization of supercapacitors is most often restricted between U_{max} and $U_{max}/2$ for the compatibility with other components of the electronic circuitry such as boost DC/DC converters employed to provide constant output voltage. It is also clear that the major part of energy can also be extracted in this voltage range. Therefore, it is recommended to calculate capacitance and energy using a voltage range of U_{max} down to $U_{max}/2$. In other words, available energy for commercial devices (E_{avail} in Wh kg):

$$E_{avail} = \frac{1}{2}C_g\left[U_{max}^2 - (U_{max}/2)^2\right] = \frac{3}{8}C_g U_{max}^2 = \frac{3}{32}C_{g,e}U_{max}^2 = \frac{3}{8}\frac{CU_{max}^2}{M} \quad (4.25)$$

where M is the total mass of all the components (active electrode material, current collectors, separators, electrolyte, casing, etc.), not only the mass of active material [41]. For practical calculations, the value of (U_{max} − EDR) is used instead of U_{max}.

Power density of supercapacitors can be defined in two different ways: maximum power, P_{max}, and real power output, P. The first depends only on the resistance and the maximum operating voltage, U_{max}.

$$P_{max} = {}^{U_{max}^2}\!/_{4R} \quad (4.26)$$

The second is defined as the rate of energy delivery that is useful for presenting the Ragone plot.

$$P = {}^{E}\!/_{\Delta t} \quad (4.27)$$

The typical power of supercapacitors based on ACs is usually limited to 10–15 kW/kg. Apart from gravimetric values, high volumetric energy and power values are more practically important since transportation is one of the main application areas [51], that is, more energy should be packed into a unit volume. The optimization of materials and electrode processing parameters (particle size, adhesion to current collectors, etc.) is needed to enable high volumetric capacitance, energy, and power densities.

3.2 Other relevant metrics: cost, cycle life, temperature range, safety

Energy density is commonly considered as the main target for electrochemical energy storage technologies. In this sense, supercapacitors do not seem to be competitive with batteries, which usually provide higher energy density (e.g., 10–20 times higher energy density for Li-ion batteries if they are taken for comparison). However, each particular application put more requirements for an energy storage system than energy density only. Primarily, those requirements can be cost, operational temperature range, cycle life, and safety. A supercapacitor cost is estimated to be 20,000 $/kWh versus somewhat below 1000 $/kWh for lithium ion batteries [52]. However, the cycle life of a supercapacitor exceeds 1 million cycles whereas that of a battery can be a few thousand only. Thus, supercapacitors' cost can be estimated to be 0.02 $/kWh per cycle whereas batteries become more expensive in the long term with a cost of 1 $/kWh per cycle. Thus, whenever a supercapacitor or a battery can be equally chosen for a given application, the initial investment is higher for supercapacitors, but also the higher the lifetime of the system and the lower the cost per cycle. For this reason, supercapacitors are used in regenerative energy breaking recovery in industrial equipment and public transportation vehicles because they are simply the lowest-cost solutions [52]. This also demonstrates that the cycle life of EDLCs is their major practical advantage, being 3 orders of magnitude higher than that of batteries. Another critical parameter is safety. Supercapacitor cells are usually safer than batteries because they are free of metals, such as lithium and associated side chemistries (e.g., lithium plating) that can provoke, for example, thermal runaway or another destructive process. For this reason, EDLCs are much more tolerant to harsh usage conditions. Furthermore, supercapacitors inherently provide maintenance-free operation. Finally, supercapacitors feature a wider operation temperature range than batteries. Although such a comparison underlines the advantages and limitations of each technology, supercapacitors should not be viewed as competitors to batteries; both technologies are rather complementary, with supercapacitors being mostly suitable for pulse rather than continuous energy delivery.

4 NANOTECHNOLOGY IN ELECTRICAL DOUBLE LAYER CAPACITORS

4.1 Electrical double layer: nanopores versus planar surface

Several theories were proposed to describe the electrical double layer, but most of them were developed for planar electrodes. As noted earlier, practical EDLCs use microporous carbon materials where the state of electrosorbed species can be essentially different from what occurs on plane surface. Further we briefly describe the development of electrical double layer theories, focusing mainly on electrosorption into microporous carbons.

The Helmholtz model is the simplest approximation to modeling spatial charge distribution at double layer interfaces. The charge of a solid electronic conductor is neutralized by opposite sign ions at distance *d* from the solid. This theory considers rigid layers counterbalancing the charges from the solid, to which a simple parallel plate capacitor model applies [53,54].

The Gouy-Chapman or diffuse model suggests that the same amount of opposite ionic charge concentrates in a liquid surrounding a charged solid, but ions are not rigidly attached to surface [55,56]. The ions in solution tend to diffuse into the liquid phase until the counter potential set up by their departure restricts this tendency. The kinetic energy of the ions will partially determine the thickness of the diffuse layer, and ion concentration in the diffuse layer follows the Boltzmann distribution. This model fails for highly charged double layers. Experimentally, the measured thickness of double layers is greater than the calculated. The Gouy-Chapman model takes a better approach to reality than the Helmholtz model, but still has limited quantitative validity. It assumes that ions are the point charges that can unrestrictedly approach electrode surface.

The Stern model presents further advance, assuming that there can be specifically surface-adsorbed ions in the vicinity of electrode surface. The ions are strongly adsorbed by the electrode within a so-called compact layer. In the compact layer there are specifically adsorbed ions (forming the inner Helmholtz plane, IHP), and nonspecifically adsorbed ions (forming the outer Helmholtz plane, OHP) [57]. In the aggregate, the Stern model includes the internal Stern layer (IHP and OHP) and an outer diffuse layer (e.g., the Gouy-Chapman layer).

Although the aforementioned models are commonly accepted for the electrical double layer on plane surface, they have limited applicability to the real behavior of porous electrodes employed in supercapacitors. The peculiarities of ion electrosorption in porous media make the process of charge storage extremely difficult, and there is still a lack of complete understanding of ions' behavior in micropores sized similarly to ions. A major problem still under debate is how pore size impacts capacitance. This is directly connected to fundamental questions, such as how ions are accommodated in pores, whether they are solvated (and if so, to what extent), what the mechanism of charging in micropores is, and so forth. Establishing both ion size and pore size is not a trivial task. Ion size depends on factors, such as ion geometry, flexibility, and solvation, all of them complicating quantitative estimations. Pore size depends greatly on the method used for its calculation and the effective pore size should also be calculated with respect to ion size [58]. Furthermore, pore size distribution and its dispersion have impact on capacitance. Eliad et al. found that the value of areal capacitance exhibits a maximum whenever pore size and ion size are best matched, with the values of C_{exp} of up to ~15 mF/cm^2 [59]. The authors also found using a series of porous carbons with controlled porosity that ions are electrosorbed differently from aqueous electrolyte: cations are hydrated whereas monovalent anions are dehydrated. By contrast, bivalent sulfate anions are hydrated in the pores. In organic electrolyte, $TEABF_4$/ACN, it was suggested using both carbide-derived carbons (CDCs) [60] and ACs [61] that areal or volumetric capacitance increases substantially if carbons are made up of pores in the subnanometer pore size range. As the solvated ions of $TEABF_4$/ACN have a diameter superior to 1 nm (1.2 nm for TEA^+ and 1.04 nm for BF_4^-), it was suggested that they have to be desolvated to access subnanometer pores (fully desolvated size 0.68 nm for TEA^+ and 0.48 nm for BF_4^-) [60]. Thus, the results of those studies go in line with the earlier work for aqueous electrolytes [59]. Further developments suggested that the solvation sheath was probably not fully stripped off the ions inside the pores [62]. To further confirm the capacitance increase in subnanometer pores, solvent-free RTILs were used with ion size below 1 nm with a series of carbons with different porosity, showing the same trend as the solution-based electrolyte [63]. These experimental findings were corroborated theoretically, and several formulae were proposed to explain those results [64,65]. In particular, those are the cylinder-in-wire model for microporous carbons, the double cylinder model for mesopores, with the parallel plate Helmholtz model applying to macropores [66].

However, pore shape is an important factor, and the most likely approximation to pore shape in microporous carbons is a slit, not cylinder, which can be supported, for instance, by preferred stereoselective elecrosorption of planar NO_3^- ions [67] over bulky anions of the same longest dimension. Therefore, the sandwich capacitor model was later proposed [66]:

$$C_H = \frac{\varepsilon_0 \varepsilon_r}{d_{eff}} = \frac{\varepsilon_0 \varepsilon_r}{(b - a_0)} \tag{4.28}$$

where ε_r is the effective dielectric permittivity, ε_0 the electric constant, d_{eff} the effective distance between electronic and ionic charges, b the pore half-width, and a_0 the ion radius.

It is necessary to note that matching ion and pore sizes may lead to the situation where the average pore size becomes smaller than the ion size. This leads to the emergence of an ion-sieving effect [59] consisting in the deformation of cyclic voltammograms as shown in Fig. 4.7, with the current decreasing gradually as the potential goes to the extreme values [68]. The current does not drop sharply, but progressively, which is connected with the presence of pore size distribution rather than a fixed pore size value. In a two-electrode cell, such behavior may depend on the maximum voltage applied to the cell. For example, some carbons show quite rectangular voltammograms in organic electrolytes in the narrow voltage window (0–1 V) with the decay in capacitive current

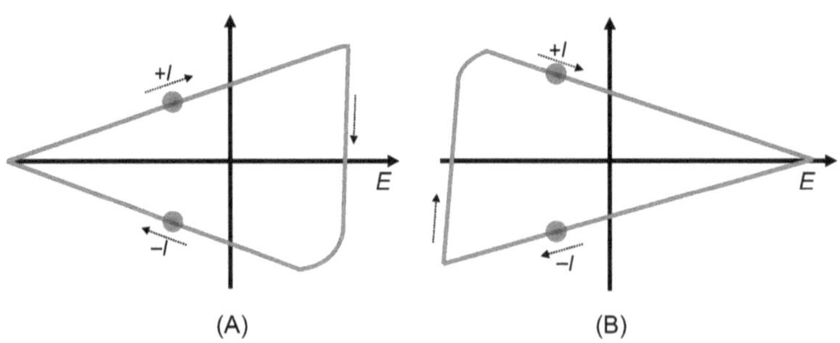

Figure 4.7 Schematic CV response of materials exhibiting (A) cation- and (B) anion-sieving effects.

appearing only at the higher voltage. By only considering the ion-accessible part of SSA (not considering SSA of pores smaller than the size of bare TEA^+ ions, 0.68 nm), it was evidenced that the maximum charge that can be stored in the accessible porosity coincides virtually with the experimentally measured one. Therefore, this led to the hypothesis that SSA of microporous carbons materials can be saturated with ions within operating cell voltage due to the higher areal capacitance than the typical value of 10 $\mu F/cm^2$ that can be calculated for capacitive electrodes displaying rectangular voltammogramms [68,69].

Noteworthy, relating capacitance to pore size requires experimental determination of parameters, such as SSA and average pore size, which depend substantially on the selected method. It was suggested that the most widely used Brunauer–Emmett-Teller (BET) method might underrate SSA for ultramicroporous carbons (pore size below 0.7 nm) or overrate SSA when carbons preferentially contain large micropores (pore size between 1 and 2 nm) [70,71]. Other methods used for the analysis of micropores (the Dubinin-Radushkevich equation, Kaneko's comparison plot as well as immersion calorimetry [72]) often provide SSA values somewhat distant from the BET SSA. On the basis of those methods it is suggested that areal capacitance can eventually be about constant, independently of the pore size. Recent studies suggest that ε_r increases with the pore size [73,74], which could mitigate the dependence of capacitance on the pore size [75]. To conclude, the methods available for evaluating the textural properties of carbons still lack the precision needed for providing unambiguous conclusion on the dependence of capacitance on pore size, which leaves room for alternative views.

The assessment of porosity can also be altered by the binder used in the preparation of electrodes; N_2 adsorption/desorption is usually conducted on powdered materials whereas electrochemical measurements are made using binder-containing electrodes where part of carbon porosity can be obstructed [76]. For this reason, the most recent studies are based on carbon monoliths [77]. It also needs to be recalled that too close size match between pores and ions can be detrimental for high-current capacitance as evidenced by the stronger capacitance decay for microporous carbons in comparison to their mesoporous counterparts [78]. Therefore, the optimum should be sought not only in terms of maximum capacitance, but also (and mostly) with respect to its value at high current densities. Furthermore, as industry looks primarily for high volumetric capacitance, it becomes clear that too narrow pores are electrolyte-inaccessible whereas too wide pores contain some pore volume unused for electrosorption. Both features decrease the density of electrodes and are thus detrimental to volume-based capacitance. Thus, there are no doubts that tuning pore size is needed [60] in practical terms to maximize volume-based performance.

Apart from pore size, parameters, such as pore shape, length and tortuosity as well as the chemistry of carbon surface (heteroatoms, defects, etc.) are known to impact capacitance. They cannot be easily controlled and the extent to which they influence capacitive charge storage needs further clarification. Thus, understanding of electrosorption in nanopores is still insufficient for a more rational design of better performing materials.

Much theoretical and experimental effort is pursued on the dynamics of charge storage, particularly in situ studies using electrochemical quartz microbalance (EQCM) [79,80], nuclear magnetic resonance (NMR) [81], and small-angle X-ray spectroscopy (SAXS) [82,83].

The EQCM technique was proposed as an extremely useful tool for in situ monitoring ionic fluxes in nanoporous carbons [80]. It allows direct gravimetric measurements of changes in an electrode's mass with the state-of-charge. The interpretation of EQCM data consists particularly in comparing the theoretical mass build-up corresponding to the state-of-charge by electrosorption of bare ions with the experimental changes of electrode mass during electrochemical experiments. In particular, it allows demonstrating such phenomenon as mixed ionic fluxes in the vicinity of PZC. It is also useful to evaluate the solvation of particular ions as a function of potential, and establish the ions responsible for charge storage. Most importantly, EQCM studies demonstrate that the charging process consists in the redistribution of ions in pores rather than only adsorption of counter ions under corresponding polarization.

NMR has also recently received extensive use for studying electrosorption in pores. In particular, an NMR study [84] concludes that TEA^+ ions have more affinity toward intergraphene pore space than BF_4^- ions, but the degree of such affinity depends on the carbon origin. For standard organic electrolyte, $TEA^+BF_4^-$/ACN and standard industrial carbon for supercapacitors, a recent NMR work suggests that the preferred mechanism of charge build-up in microporous carbons is cation adsorption under negative polarization whereas positive polarization first initiates anion adsorption changing for ion swapping at the higher potential [85]. Ion swapping also was established as the main feature of ion dynamics for aqueous neutral electrolytes using in situ SAXS measurements [83].

To conclude, all of the cited literature demonstrates unambiguously that tuning nanoporous carbons to capacitive charge storage is an important yet complicated nanotechnology task to move forward supercapacitor technology.

4.2 Tuning nanoporous carbons to optimum capacitive charge storage

The synthesis of porous carbon materials consists in physical and chemical activation of carbonaceous matter (e.g., fossils) or an amorphous carbon char obtained by pyrolysis of carbon-rich biomass in inert atmosphere. This is achieved by a physical or chemical activation of carbons [86]. Physical activation is conducted at high temperature under a mixed inert/oxidizing atmosphere (e.g., CO_2 or H_2O + N_2 or Ar.) [87,88], whereas chemical activation is performed on amorphous carbons previously mixed with chemicals, such as alkalis, carbonates, chlorides, or acids (e.g., KOH, K_2CO_3, H_3PO_4, $ZnCl_2$...). The result of activation processes is formation of porous networks in carbons with high specific SSA. Activation processes can be tuned to provide the desired pore size range: for example, chemical activation is believed to be more suitable than physical activation for pore size tuning in carbons from fossil charcoals [89]. However, such tuning is limited: chemical activation gives rise to significant ultramicropore volume (pores below 0.7 nm), which is not accessible even to almost fully desolvated ions of organic electrolytes. Also, higher SSA values are usually reached with chemical activation versus physical activation. For such carbons, the maximum gravimetric capacitance values in organic electrolyte are about 200 F/g and slightly over 60 F/cm^3 [90]. Some commercial ACs for organic-based supercapacitors are produced using steam activation with good reproducibility, mostly using coconut shell as the carbon precursor [91]. Physical activation is preferred industrially as it is less expensive.

In the academic domain, more advanced but expensive carbons have been extensively researched. CDCs are touted as materials providing high control over pore size distribution owing to the availability of varied distribution of carbon atoms in carbide precursors. CDCs are prepared by high-temperature extraction of metals from carbides serving as precursors. CDCs are commonly produced by high-temperature chlorination [92,93] and vacuum decomposition [94,95] of TiC, SiC, and VC among others, giving capacitance values up to 220 F/g and 126 F/cm^3 in KOH and 120 F/g and 72 F/cm^3 in organic electrolyte.

Kyotani's group developed templated carbon based on FAU (faujasite) zeolite with a porosity that consists of a diamond-like framework of spherical cavities [22]. The pore walls of such zeolite are converted into three-dimensionally arrayed and mutually connected 1.2 nm nanopores. This carbon exhibits extremely uniform pore size distribution, remarkably not containing inaccessible ultramicropores (as in chemically ACs) and a high BET SSA (3000 m^2/g). Such optimum properties lead to both very high gravimetric (140–190 F/g) and volumetric (75–83 F/cm^3) capacitance in organic electrolyte (Fig. 4.8).

Graphene-like carbons have recently become popular but typical commercial graphene powders suffer from the restacking of graphene layers (e.g., low SSA and low capacitance) and the low density (i.e., the low volumetric capacitance). Ruoff et al. claim that chemical activation with KOH can be applied to exfoliated graphene oxide for producing activated graphene with a BET SSA of above 3000 m^2/g and a broad pore size distribution (between 0.6 and 5 nm). When coupled with a propylene-carbonate-diluted IL, activated graphene delivers a gravimetric capacitance of 140 F/g and 60 F/cm^3 in a symmetric cell [96]. In another work, the preparation of compactly interlinked graphene nanosheets was conducted by an evaporation-induced drying of a graphene hydrogel, with the highest volumetric capacitance of up to 376 F/cm^3 in 6 M KOH aqueous electrolyte (Fig. 4.9). Such graphene material has a SSA of 370 m^2/g only, losing a great part of its capacitance after heat-treatment at 800°C alongside the associated increase in SSA of up to 700 m^2/g. This led the authors to think that the exceptional volumetric capacitance should be mostly attributed to pseudocapacitance rather than non-Faradic capacitance [97].

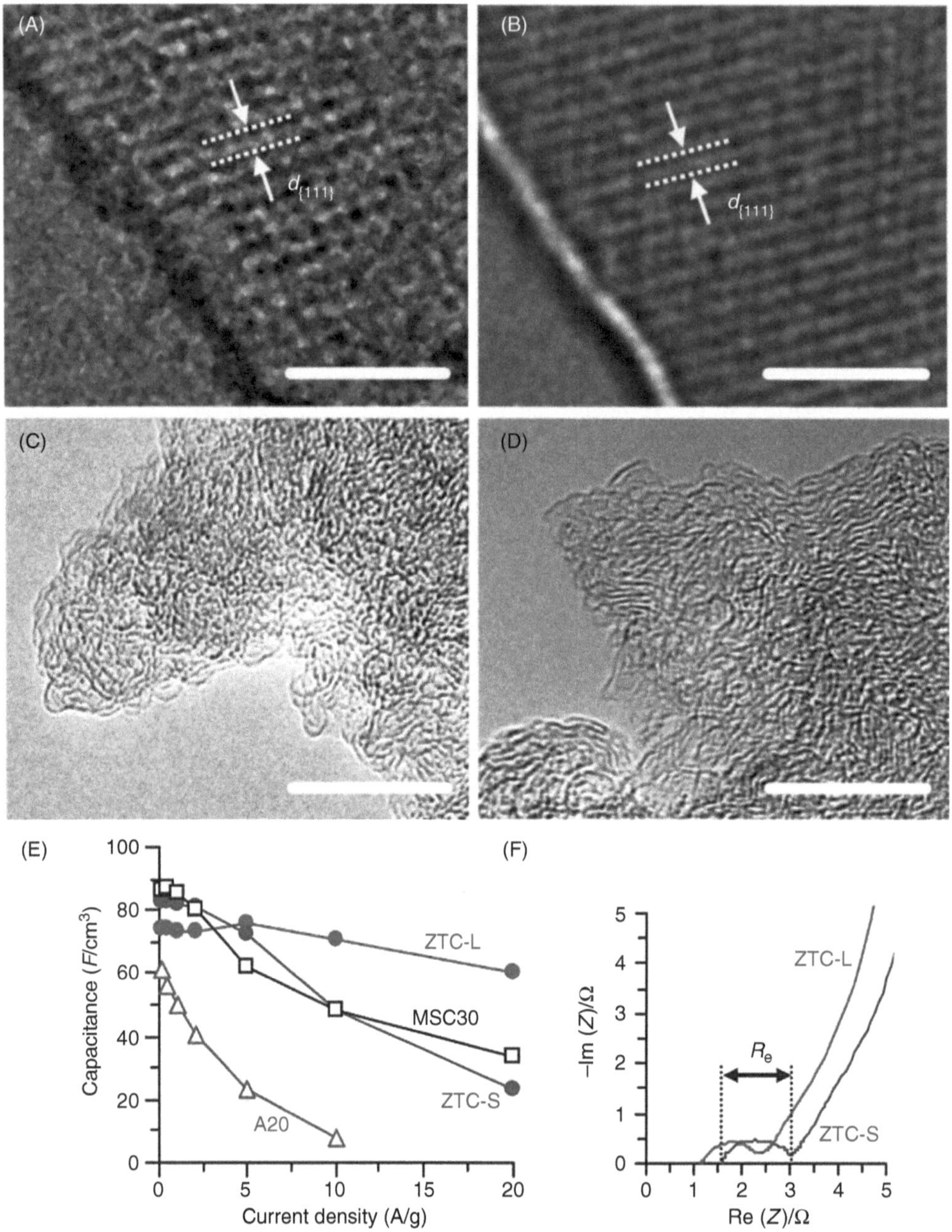

Figure 4.8 TEM images of (A, B) zeolite-templated carbons and (C, D) activated carbons. (A) ZTC-L, (B) ZTC-S, (C) MSC30, and (D) A20. Scale bars are 10 nm. (E) Volumetric capacitance (*Cv*) versus current density. (F) Nyquist plots for the two ZTCs. The size of the semicircle (*Re*) is shown for ZTC-S. *(Reprinted (adapted) with permission from H. Itoi, H. Nishihara, T. Kogure, T. Kyotani, Three-dimensionally arrayed and mutually connected 1.2-nm nanopores for high-performance electric double layer capacitor, J. Am. Chem. Soc. 133 (5) (2011) 1165–1167 [22]. Copyright © 2011 American Chemical Society).*

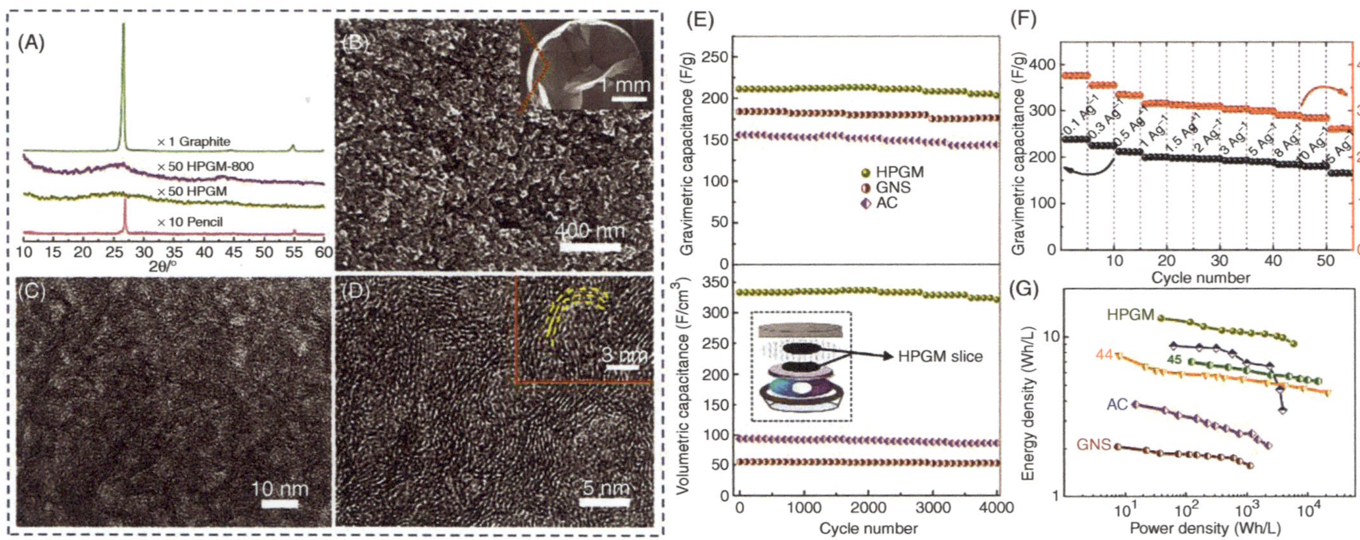

Figure 4.9 ***Structure and morphological characterization of HPGM.*** (A) XRD patterns of graphite, soft pencil lead, HPGM, and HPGM-800. (B) Cross-sectional SEM images of HPGM, with the inset showing the cross-section of HPGM at low magnification. (C, D) TEM images of HPGM-800. (E) Cycle performance of HPGM, GNS, and AC at a current density of 0.5 A/g: the inset is a diagram of the two-electrode measurement device. (D) Rate performance of HPGM. (F) Volumetric Ragone plots comparing HPGM with AC, GNS, and carbon materials reported in the reference number indicated. *(Reprinted (adapted) by permission from Macmillan Publishers Ltd: Scientific Reports, Y. Tao, X. Xie, W. Lv, D.-M. Tang, D. Kong, Z. Huang, H. Nishihara, T. Ishii, B. Li, D. Golberg, F. Kang, T. Kyotani, Q.-H. Yang, Towards ultrahigh volumetric capacitance: graphene derived highly dense but porous carbons for supercapacitors, Sci. Rep. 3 (2013) 1–8 [97], Copyright 2013).*

The capillary compression of graphene gel films in the presence of nonvolatile liquid electrolyte was reported to enable densely packed porous graphene electrodes with a varied density of up to ~1.33 g/cm^3 and maximum volumetric capacitance of about 200 F/cm^3 [98]. To achieve this, chemically converted graphene dispersion was exchanged with a miscible mixture of volatile and nonvolatile liquids (the last serving further as electrolyte) with subsequent removal of the volatile liquid by vacuum evaporation.

Graphene has also been proposed for unconventional EDLC applications, such as 120 Hz *ac* line filtering (e.g., parallel graphene sheets plasma-grown on current collectors [99]). In this case, the main figure of merit is the phase angle rather than capacitance. It is necessary to note that graphene work is still in the early development stage, that is, the often-claimed extraordinary performance still needs multiple cross-checks and further verification to reach the level of maturity sufficient for industry.

Other open-surface carbon nanomaterials, such as mesoporous carbons, nanotubes, and carbon onions provide lower capacitance values than microporous carbons because of their lower SSA, but they have been demonstrated as promising for high-power demand, since ion transport is not hindered as is the case for micropores [100]. An original approach is to use such carbons with eutectic mixtures of ILs, extending the operating temperature range to −50 to 100°C [32]. Apart from porous structure, the surface chemistry of carbons (functional groups, defects, and heteroatoms) is a major factor contributing to capacitance through the following parameters: (1) quantum capacitance (C_q); (2) electrical conductivity (resistance and rate capability); (3) wettability of carbons with electrolytes, (the utilization of available surface for charge storage); (4) extra Faradic (often pseudocapacitive) contribution.

The doping of carbons with heteroatoms is suggested to be a way to heighten C_q, hence, the overall capacitance [101]. For example, an increase in $C_{A,exp}$ from 6 to 22 $\mu F/cm^2$ was achieved by incorporating 2.3 at.% of nitrogen into graphene [102]. More recent work shows an exceptional performance of nitrogen-enriched carbons in neutral aqueous electrolytes, especially in terms of high-current capacitance.

5 PSEUDOCAPACITIVE MATERIALS

Pseudocapacitive materials exhibit an electrochemical response qualitatively comparable to that of EDLC materials, but the mechanism responsible for it is different in nature. While in EDLC materials the charge is electroadsorbed onto a high surface material, here the charge storage mechanism is essentially Faradic. Similar to batteries, pseudocapacitive materials undergo redox reactions, however, these are very fast, reversible, and in general they are assumed to occur at the surface and not in the bulk of the material. Consequently, although there is a charge passage across the double layer, the derivative dq/dU is constant in the whole potential window, being capacitance the reason between both parameters [20] [Eq. (4.9)].

The values commonly ascribed to a pure electrostatic capacitive process are in the range of tens of $\mu F/cm^2$ whereas pseudocapacitive charge transfer can give values one or two orders of magnitude higher [20]. Thus, one of the most exploited strategies to increase the energy density of supercapacitors is the use of pseudocapacitive materials, both as the sole active material of one of the electrodes or in combination with a purely EDLC material, such as AC. Nevertheless, for the real implementation of pseudocapacitors there are still several drawbacks to overcome. They suffer from lower power density values than EDLC materials due to the slower kinetics of chemical reactions compared to an electrostatic process and, moreover, as a result of these redox reactions the cyclability or stability upon cycling is much poorer. Unlike carbon materials frequently featuring a similar electrochemical potential difference between the OCP potential and the electrolyte breakdown potential under both positive and negative polarization, pseudocapacitive materials are usually operational only in the limited operating potential difference on one side of an OCP potential. In most cases, a pseudocapacitive material requires coupling with another (pseudo)capacitive material working under opposite polarization to provide an asymmetric electrode configuration with cell voltage competitive to that of non-Faradic systems [103].

5.1 Pseudocapacitance in carbon nanomaterials: charge storage by carbon functionalities and reversible hydrogen electrosorption

The charge storage capacity of nanoporous carbons in aqueous media can be enhanced through additional Faradic contribution, in particular, reversible hydrogen adsorption and additional pseudocapacitance owing to oxygen- and nitrogen-containing functional groups being either formed on the surface of carbon materials during synthesis and storage or deliberately introduced. For example, the hydroquinone/benzoquinone group often emerges on carbon surface, showing usable electrochemical activity in acidic media close to OCP potential [104]. Nitrogen can be incorporated through the ammoxidation of carbon surface or pyrolysis/physical activation of nitrogen-rich precursors, such as melamine [105,106]. Whereas in-plane nitrogen atoms enhance quantum capacitance, edge-bound pyridine nitrogen can accept/release hydrogen in acidic media through a redox event [107]. Additionally to Faradic contribution, nitrogen also allows extending the practical operating voltage range of acidic symmetric cells up to 1.2V. An interesting example is zeolite-template N-enriched carbon based on acrylonitrile, which exhibits capacitance values above 300 F/g in 1 M H_2SO_4 [108] Another avenue is provided by some naturally abundant materials, such as seaweeds, which can be pyrolized, producing porosity and retaining surface functionality while activation can be skipped [109]. The derived carbons show high capacitance values of about 200 F/g at moderate SSA (about 500 m^2/g) due to addition/removal of hydrogen in the quinone/hydroquinone transformation and pyrone groups, but the batch-to-batch reproducibility of such carbons can depend on the nonconstant precursor composition. As a result, the electrochemical response of resultant carbons can be altered.

Additional Faradic contribution can also be provided by reversible adsorption of atomic (nascent) hydrogen, which is formed in situ on the surface of negative electrodes [110]. Under negative polarization, reduction of water produces atomic hydrogen, which is weakly chemisorbed in the pores of the negative carbon electrode. In neutral aqueous electrolytes, this phenomenon leads to the excess of OH^- ions in the negative electrode, pushing the potential of water reduction to lower values according to the Nernst equation [111]. This leads to the increase in usable cell voltage up to 1.6 V and sometimes higher values over the termodymanic limit of water stability (1.23 V). Reversible hydrogen storage also occurs in aqueous asymmetric systems using a battery-type [112] or pseudocapacitive [113] material as the positive electrode. In this case, the aqueous system can stably work at about 2 V. In both symmetric and asymmetric cells with neutral aqueous electrolytes, charge storage in the negative electrode is a combined contribution from electrosorption and hydrogen storage.

5.2 Nanosizing in pseudocapacitive inorganic materials: oxide supercapacitors

Conducting polymers (CPs) and metal oxides are the best known pseudocapicitive materials. Compared to metal oxides, CPs offer a much more affordable option. They are environmentally friendly, they have conductivity values up to 500 S/cm in the doped state and the cell voltage can be as high as 3V in nonaqueous electrolytes. However, their usage as electrode materials is questioned due to their very low stability upon cycling. During the charge and discharge ions are inserted into the polymer matrix triggering the swelling and shrinking of the compound and consequently the cycling performance very rarely exceeds 1000 cycles of charge and discharge. Taking into account the limited application of these systems, in the present chapter we will focus on the use of inorganic compounds.

Among the major requirements that a metal oxide should fulfill in supercapacitor applications, high electronical conductivity and accessible oxidation states over a range of potentials with no irreversible transformations are of great importance [114]. Regarding these properties and the resulting electrochemical performance, RuO_2 remains the best choice. This oxide was first studied as electrode material for electrochemical capacitors in the late 70s and early 80s, exhibiting capacitance values around 150–200 $\mu F/cm^2$, that is, about 10 times higher than those obtained for carbonaceous materials [115,116].

Depending on the synthetic route, the water content and the crystallinity of the compound will be very much different as well as the pseudocapacitance level. In this sense, anhydrous RuO_2 working in acidic media exhibited capacitance values of 720 F/g [116] while hydrous ruthenium oxide (RuO_2–$0.5H_2O$) showed a capacitance as high as 900 F/g [117]. A high crystallinity degree hinders the permeability of electrons in the bulk of the material and, thus, fast and reversible reactions are limited to the surface level.

Amorphous compounds can better diffuse electrons as they better tolerate structural expansion/contractions and, consequently, in these cases the capacitive response does not only come from the surface but also from the bulk of the material [118].In the case of hydrated compounds, this capacitance is ascribed to the easy pathway that hydrous structures exhibit for both electron and proton transport [119].

In any case, pseudocapacitive charge storage still takes place mostly at the surface level, and thus, the greater the SSA the larger the capacitance in F/g. In this sense, nanostructuring of pseudocapacitive materials is of great importance for the improvement of the energy density of the final device. Moreover, going down to the nanoscale not only provides larger SSA but also shortens diffusion path lengths and facilitates proton transport, which is involved in the charge storage reaction. For example, for nanostructured RuO_2 shown in Fig. 4.10 specific capacitance values up to 1300 F/g have been reported, very close to the theoretical 1358 F/g [120,121].

Nevertheless, the high cost of ruthenium limits the use of RuO_2 to military, spatial, or medical applications where the budget is not an issue. In the search of a synergy between cost, safety, and electrochemical performance other oxides, such as IrO_2 [122], NiO [123], Co_3O_4 [124], SnO_2 [125], V_2O_5 [126], MoO_x [127], and especially MnO_2 [128–131] have turned out to be good candidates. Note that although all of these compounds have been typically tested for supercapacitor applications, in many cases the extent of the chemical reaction is not a continuous function of the potential U, and therefore, the derivative dQ/dU, that is, capacitance, is not constant. In those cases, the oxides should not be considered pseudocapacitive but instead battery-type materials [14]. For instance, nickel and cobalt oxides have been largely tested for supercapacitor applications; however, their CVs exhibit pronounced redox peaks at a big voltage difference and GA curves display distinct plateaus [132]. Thus, capacitance cannot be considered a constant parameter in this type of compounds and as such they should not be named pseudocapacitive. However, going down to the nanoscale offers another option to convert a battery material into pseudocapacitive. As mentioned earlier, transport paths are shortened when particle size is decreased, and thus, the response of the material in terms of power increases. Moreover, taking into consideration that in a particle of ~ 2 nm 63% of the atoms are located at the surface, then, most of the Li^+ storage (or Na^+ or other chemistry type battery materials) will predominantly occur at the surface region. For example, V_2O_5 anchored to carbon nanotubes (CNTs) performs similarly to EDLC supercapacitors in terms of energy and power [133], although cyclability still remains an issue.

MnO_2 polymorphs have turned out to be the best alternatives to RuO_2 [128–131], among which the layered birnessite has emerged as a good compromise between electrochemical performance and sample preparation [134]. In any case, hindered by its poor conductivity, the specific capacitance of MnO_2 is far from the theoretical 1230 F/g (0.9 V) and instead it rarely exceeds 400 F/g. The large fluctuations in specific capacitance values that can be found in the literature come principally from the very diverse

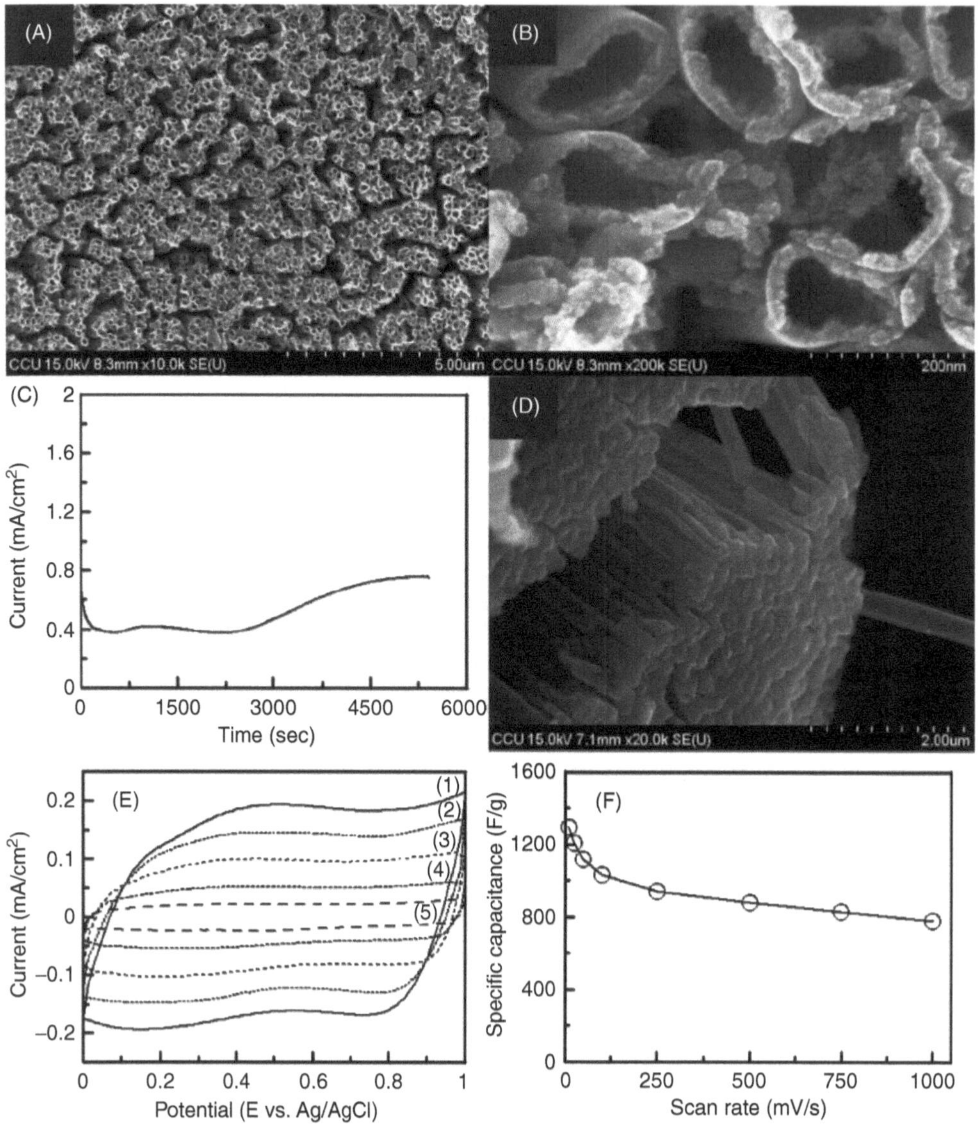

Figure 4.10 (A, B) SEM top images of an $RuO_2{\cdot}xH_2O$ NTs arrayed electrode. (C) The *i–t* curve of anodic deposition measured at 1.0 V from a 10 mM $RuCl_3{\cdot}xH_2O$ + 0.1 M CH_3COONa solution. (D) A SEM image showing the onset-overfilled morphology of a $RuO_2{\cdot}xH_2O$ NT arrayed electrode. All NTs were deposited at 1.0 V from a 10 mM $RuCl_3{\cdot}xH_2O$ + 0.1 M CH_3COONa solution. (E) Cyclic voltammograms measured at (1) 1000, (2) 750, (3) 500, (4) 250, and (5) 100 mV/s, and (F) dependence of the capacitance loss on the scan rate of CV from 10 to 1000 mV/s for an annealed $RuO_2{\cdot}xH_2O$ NT arrayed electrode (0.19 ± 0.01 mg/cm^2). *(Reprinted (adapted) with permission from C.-C. Hu, K.-H. Chang, M.-C. Lin, Y.-T. Wu, Design and tailoring of the nanotubular arrayed architecture of hydrous RuO_2 for next generation supercapacitors, Nano Lett. 6 (12) (2006) 2690–2695 [121]. Copyright © 2006 American Chemical Society).*

synthetic routes followed, which affects the chemical composition, crystallinity, morphology, and particle size.

The pseudocapacitance of manganese oxides is attributed to the fast and reversible redox reactions with protons or cations from the electrolyte that implicate Mn(VI)/Mn(IV), Mn(IV)/Mn(III), and Mn(III)/Mn(II) oxidation states. Despite the clearly redox nature of the energy storage mechanism, MnO_2 electrodes exhibit typical rectangular-shaped CV curves, comparable to RuO_2 and even high surface carbon materials. In general, the thicker the electrode, the lower the specific capacitance due to the low electrical conductivity of MnO_2. Moreover, in MnO_2 thin films the distances for the solid-state transport of insertion cations are also reduced, which is also favorable for capacitive applications. For instance, in thin films of MnO_2 supported on platinum (<5 μm) capacitance values up to 1380 F/g have been measured [130].

5.3 Pseudocapacitive charge storage by composites between nanocarbons and inorganic materials

Besides their higher specific capacitance values and hence higher energy density, oxide supercapacitors may not be employed alone as supercapacitor electrodes for practical purposes. Apart from RuO_2, the conductivity of most metal oxides is very low. The high resistivity of metal oxides increases both the sheet resistance and the charge transfer resistance of the electrode, and especially causes a large IR loss at high current densities. Thus, the power density and the rate capability are poor, which hinders practical application. Very recently, transition metal nitrides have attained great attention due to their high conductivity and chemical stability and their electrochemical behavior, which is close to that of an ideal EDLC [135,136]. Nevertheless, the practical use of metal nitrides is still at a very early stage. The high cycle stability is one of the strongest assets of electrochemical capacitors; however, the strain developed in metal oxides during charge and discharge induces the mechanical failure of the electrode, leading to poor long-term stability. Moreover, surface area and porosity are difficult to tailor in metal oxides and they could contribute through non-Faradic processes to increase the total capacitance. Therefore, the development of composite electrodes containing both carbon and transition metal oxides as supercapacitor electrodes emerges as a natural solution. In such nanostructured carbon/metal oxide composite, the carbon substrate serves a high-surface-area 3D current collector, and defines the internal pore structure of the electrode, facilitating the infiltration and rapid transport of the electrolyte to the nanoscopic oxide phase. In this case, metal oxides will be the main contributors to the high specific capacitance and high energy density values of the composite electrodes [137].

Recently, the use of graphene family materials (graphene, graphene oxide, rGO) as support of metal oxides has emerged as a practical solution to combine synergistically the best of both types of materials [138]. For example, RuO_2 nanoparticles anchored on the surface of rGO have shown higher specific capacitance than that of either RuO_2 or rGO. In this system oxide nanoparticles prevent the restacking of rGO, increasing the capacitance of the carbonaceous material while rGO avoids the agglomeration of RuO_2 nanoparticles and offers a conductive pathway [139]. A similar result was also observed in the MnO_2 nanowires–rGO composite material [140].

6 CONCLUSIONS AND PERSPECTIVES

The trends in energy consumption and production are changing toward a more sustainable model, and within this new framework, the development of improved energy storage systems is essential for the consolidation of different renewable energy sources. On top of that, both the amount and variety of portable electronic devices are increasing, starting from consumer electronics and reaching the hybrid electric vehicle (HEV) stage, and electrochemical capacitors or supercapacitors have found their niche as a promising solution for fast charging and regenerative energy acquisition.

In this chapter we have aimed at introducing the basics of supercapacitors and summarizing the optimum properties required from materials in order to better understand the trends in materials development. Most of the research efforts in the field of materials are directed toward the energy density improvement, and in order to get to this goal nanostructuring is of paramount importance. (1) Capacitance is proportional to the number of active sites (both Faradic and non-Faradic), and thus, high SSA materials are required as electrode materials. (2) Special care must be taken to provide a good electrode–electrolyte match in order to achieve good capacitance ratings. Suitable pore-size distributions (in the range of ~1 nm), pore network, and pore lengths are needed for good ion diffusion at high rates. (3) Low internal resistance or short ion/electron diffusion path lengths need to be provided, and (4) Better electrochemical and mechanical stability is required for good cycling performance.

Apart from the materials, the design and optimization of new cell configurations is a growing field of opportunities for developing hybrid battery/supercapacitor systems. Such systems will be in great demand in those applications where a battery or supercapacitor alone cannot meet specifications. Equally important to the materials and cell development is the processing of electrodes and cell assembly, which should also be optimized to maximize the performance.

GLOSSARY

A20	A20 activated carbon from Unitika, Ltd. (from Ref. [22])
AC	Activated carbon
ACN	Acetonitrile
BET	Brunauer–Emmett–Teller
CDC	Carbide-derived carbon
CE	Counter electrode

CNTs	Carbon nanotubes
CP	Conducting polymer
CPE	Constant phase element
CV	Cyclic voltammetry
DOE	Department of Energy (US)
EDLC	Electric double-layer capacitor
EDR	Equivalent distributed resistance
EQCM	Electrochemical quartz microbalance
ESR	Equivalent series resistance
FAU	Faujasite
GA	Galvanostatic charge/discharge
GNS	Graphene nanosheets, Powdered graphene (from Ref. [96])
HEV	Hybrid electric vehicle
HPGM	High density porous graphene macroform (from Ref. [96])
HPGM-800	High density porous graphene macroform annealed at 800°C (from Ref. [96])
IL	Ionic liquid
IPH	Internal parallel hybrids
ISH	Internal series hybrids
IUPAC	International Union of Pure and Applied Chemistry
LIC	Lithium-ion capacitor
MSC30	MSC30 activated carbon from Kansai Coke and Chemicals Co., Ltd. (from Ref. [22])
NE	Negative electrode
NT	Nanotube
NMR	Nuclear magnetic resonance
OCP	Zero-current potential or open circuit potential
OCV	Zero-current voltage or open-circuit voltage
PE	Positive electrode
PZC	Potential of zero charge
RE	Reference electrode
rGO	reduced graphene oxide
RTIL	Room-temperature ionic liquid
SAXS	Small-angle X-ray spectroscopy
SEM	Scanning electron microscopy
SSA	Specific surface area
TEA^+	Tetraethylammonium cation
$TEABF_4$	Tetraethylammonium tetrafluoroborate (also referred as NEt_4BF_4)
TEM	Transmission electron microscopy
WE	Working electrode
XRD	X-ray diffraction
ZTC-L	Zeolite-templated carbon with large carbon frameworks (from Ref. [22])
ZTC-S	Zeolite-templated carbon with small carbon frameworks (from Ref. [22])

REFERENCES

[1] R. Kötz, M. Carlen, Principles and applications of electrochemical capacitors, Electrochim. Acta 45 (15–16) (2000) 2483–2498.

[2] P. Simon, Y. Gogotsi, Materials for electrochemical capacitors, Nat. Mater. 7 (2008) 845–854.

[3] A. González, E. Goikolea, J.A. Barrena, R. Mysyk, Review on supercapacitors: technologies and materials, Renew. Sust. Energ. Rev. 58 (2016) 1189–1206.

[4] A.F. Burke, Ultracapacitors: why, how, and where is the technology?, J. Power Sources 91 (2000) 37–50.

[5] J.R. Miller, A.F. Burke, Electrochemical capacitors: challenges and opportunities for real-world applications, Electrochem. Soc. Interf. 17 (2008) 53–57.
[6] J.R. Miller, P. Simon, Electrochemical capacitors for energy management, Science 321 (2008) 651–652.
[7] H. Becker, "Low voltage electrolytic capacitor." US Patent 2,800,616, 1957.
[8] R.A. Rightmire, "Electrical energy storage apparatus." US Patent 3288641A, 1966.
[9] D.L. Boos, J.E. Metcalfe, Electrolytic capacitor employing paste electrodes. US Patent 3,536,963A, 1972.
[10] T.C. Murphy, W.E. Kramer, Proceedings of the Fourth International Seminar on Doublelayer capacitor and similar energy storage devices, in Florida Educational Seminars, 1994.
[11] IEA, Report on the first phase of the IEA implementing agreement for hybrid electric vehicle technology and programmes, Paris, 1993–1999.
[12] Z. Yu, L. Tetard, L. Zhai, J. Thomas, Supercapacitor electrode materials: nanostructures from 0 to 3 dimensions, Energ. Environ. Sci. 8 (2015) 702–730.
[13] R. Berenguer, Trends and research challenges in supercapacitors, Boletín del Grupo Español del Carbón (37) (2015) 9–13.
[14] P. Simon, Y. Gogotsi, B. Dunn, Where do batteries end and supercapacitors begin?, Science 343 (6176) (2014) 1210–1211.
[15] F. Béguin, E. Frackowiak, Supercapacitors: Materials, Systems and Applications, Wiley-VCH Verlag GmbH & Co. KGaA, Weinheim, Germany, (2013).
[16] W.G. Pell, B.E. Conway, Peculiarities and requirements of asymmetric capacitor devices based on combination of capacitor and battery-type electrodes, J. Power Sources 136 (2004) 334–345.
[17] D. Cericola, R. Kötz, Hybridization of rechargeable batteries and electro-chemical capacitors: principles and limits, Electrochim. Acta 72 (2012) 1–17.
[18] D. Cericola, R. Kötz, Hybridization of electrochemical capacitors and rechargeable batteries: An experimental analysis of the different possible approaches utilizing activated, J. Power Sources 196 (2011) 10305–10313.
[19] JM Energy Corporation, [Online]. Available from: http://www.jmenergy.co.jp/en/
[20] B.E. Conway, Electrochemical Supercapacitors. Scientific Fundamentals and Technological Applications, Kluwer Academics/Plenum Publishers, New York, (1999).
[21] H. Helmholtz, Studien über electrische Grenzschichten, Annalen der Physiks und Chemie 243 (7) (1879) 337–382.
[22] H. Itoi, H. Nishihara, T. Kogure, T. Kyotani, Three-dimensionally arrayed and mutually connected 1.2-nm nanopores for high-performance electric double layer capacitor, J. Am. Chem. Soc. 133 (5) (2011) 1165–1167.
[23] J.K. McDonough, A.I. Frolov, V. Presser, J. Niu, C.H. Miller, T. Ubieto, M.V. Federov, Y. Gogotsy, Influence of the structure of carbon onions on their electrochemical performance in supercapacitor electrodes, Carbon 50 (9) (2012) 3298–3309.
[24] X. Zhang, Z. Haitao, L. Chen, K. Wang, X. Sun, Y. Ma, Recent advances in porous graphene materials for supercapacitor applications, RSC Adv. 4 (2014) 45862–45884.
[25] H. Shi, Activated carbons and double layer capacitance, Electrochim. Acta 41 (10) (1996) 1633–1639.
[26] B.E. Conway, V. Birss, J. Wojtowicz, The role and utilization of pseudocapacitance for energy storage by supercapacitors, J. Power Sources 66 (1–2) (1997) 1–14.
[27] Q. Abbas, P. Ratajczak, P. Babuchowska, A. Le Comte, D. Bélanger, T. Brousse, F. Béguin, Strategies to improve the performance of carbon/carbon capacitors in salt aqueous electrolytes, J. Electrochem. Soc. 162 (5) (2015) A5148–A5157.
[28] M. Ue, Chemical capacitors and quaternary ammonium salts electrochemistry, Electrochem. 75 (8) (2007) 565–572.
[29] F. Béguin, V. Presser, A. Balducci, E. Frackowiak, Carbons and electrolytes for advanced supercapacitors, Adv. Mater. 26 (14) (2014) 2219–2251.
[30] D. Ruan, F. Zuo, High voltage performance of spiro-type quaternary ammonium-salt-based electrolytes in commercial large supercapacitors, Electrochem. 83 (11) (2015) 997–999.
[31] M. Ue, Double layer capacitors, in: H. Ohno (Ed.), Electrochemical Aspects of Ionic Liquids, second ed., John Wiley & Sons, 2011, pp. 205–223.

[32] R. Lin, P.-L. Taberna, S. Fantini, V. Presser, C.R. Pérez, F. Malbosc, N.L. Rupesinghe, K.B.K. Teo, Y. Gogotsi, P. Simon, Capacitive energy storage from −50 to 100°C using an ionic liquid electrolyte, J. Phys. Chem. Lett. 2 (19) (2011) 2396–2401.
[33] H. Gerischer, An interpretation of the double layer capacity of graphite electrodes in relation to the density of states at the Fermi level, J. Phys. Chem. 89 (11) (1985) 4249–4251.
[34] M. Hahn, M. Baertschi, O. Barbieri, J.-C. Sauter, R. Kötz, R. Gallay, Interfacial capacitance and electronic conductance of activated carbon double-layer electrodes, Electrochem. Solid St. 7 (2) (2004) A33–A36.
[35] O. Barbieri, M. Hahn, A. Herzog, R. Kötz, Capacitance limits of high surface area activated carbons for double-layer capacitors, Carbon 43 (6) (2005) 1303–1310.
[36] H. Tobias, A. Soffer, The immersion potential of high-surface electrodes, J. Electroanal. Chem. 148 (1983) 221–232.
[37] A.J. Bard, L.R. Faulkner, Electrochemical Methods: Fundamentals and Applications, John Wiley & Sons, New York, (2001) 233.
[38] J. Wang, J. Polleux, J. Lim, B. Dunn, Pseudocapacitive contributions to electrochemical energy storage in TiO_2 (anatase) nanoparticles, J. Phys. Chem. C 111 (2007) 14925–14931.
[39] T. Brezesinski, J. Wang, J. Polleux, B. Dunn, S.H. Tolbert, Templated nanocrystal-based porous TiO_2 films, J. Am. Chem. Soc. 131 (2009) 1802–1809.
[40] D. Weingarth, A. Foelske-Schmitz, R. Kötz, Cycle versus voltage hold—which is the better stability test for electrochemical double-layer capacitors?, J. Power Sources 225 (2013) 84–88.
[41] Idaho National Laboratory Report DOE/NE-ID-11173, FreedomCAR Ultracapacitor Test Manual, 21 September, 2004.
[42] W.G. Pell, B.E. Conway, Analysis of power limitations at porous supercapacitor electrodes under cyclic voltammetry modulation and dc charge, J. Power Sources 96 (1) (2001) 57–67.
[43] B.E. Conway, Impedance Behavior of Electrochemical Supercapacitors and Porous Electrodes, in: E. Barsoukov and J.R. Mcdonald, (Eds.), Impedance Spectroscopy. Theory, Experiment and Applications, second ed., John Wiley & Sons, p. 469.
[44] R. de Levie, Porous electrodes in electrolyte solutions I, Electrochim. Acta 8 (1963) 751–780.
[45] R. de Levie, Porous electrodes in electrolyte solutions II, Electrochim. Acta 9 (1964) 1231–1245.
[46] H. Keiser, K.D. Beccu, M.A. Gutjahr, Estimation of the pore structure of porous electrodes by impedance measurements, Electrochim. Acta 21 (1976) 539–543.
[47] C. Portet, P.-L. Taberna, P. Simon, C. Laberty-Robert, Modification of Al current collector surface by sol–gel deposit for carbon–carbon supercapacitor applications, Electrochim. Acta 49 (2004) 905–912.
[48] P.-L. Taberna, P. Simon, J.F. Fauvarque, Electrochemical characteristics and impedance spectroscopy studies of carbon-carbon supercapacitors, J. Electrochem. Soc. 150 (3) (2003) A292–A300.
[49] D. Qu, Studies of the activated carbons used in double-layer, J. Power Sources 109 (2) (2002) 403–411.
[50] L.R. Radovic, Surface chemistry of activated carbon materials: state of the art and implications for adsorption, Surfaces of Nanoparticles and Porous Materials, CRC Press, New York, USA, 1999, p. 529.
[51] P. Simon, Y. Gogotsi, Nanostructured activated carbons from natural precursors for electrical double-layer capacitors, Nat. Mater. 7 (11) (2008) 845–854.
[52] J.R. Miller, Engineering electrochemical capacitor applications, J. Power Sources 326 (2016) 726–735.
[53] H. Helmholtz, Ueber einige Gesetze der Vertheilung elektrischer Ströme in körperlichen Leitern mit Anwendung auf die thierisch-elektrischen Versuche, Annalen der Physik und Chemie 165 (6) (1853) 211–233.
[54] D. Qu, H. Shi, Studies of activated carbons used in double-layer capacitors, J. Power Sources 74 (1) (1998) 99–107.
[55] L.G. Gouy, Sur la constitution de la charge électrique à la surface d'un electrolyte, Comptes Rendus de l'Académie des sciences 149 (1909) 654–657.
[56] D.L. Chapman, A contribution to the theory of electrocapillarity, Philos. Mag. 25 (1913) 475–481.
[57] O. Stern, The theory of the electrolytic double shift, Zeitschrift fur Elektrochimie und Angewandte Physikalische Chemie 30 (1924) 508–516.
[58] F. Stoeckli, T.A. Centeno, Optimization of the characterization of porous carbons for supercapacitors, J. Mater. Chem. A 1 (23) (2013) 6865–6873.

[59] L. Eliad, G. Salitra, A. Soffer, D. Aurbach, Ion sieving effects in the electrical double layer of porous carbon electrodes: estimating effective ion size in electrolytic solutions, J. Phys. Chem. B 105 (29) (2001) 6880–6887.
[60] J. Chmiola, G. Yushin, Y. Gogotsi, C. Portet, P. Simon, P.-L. Taberna, Anomalous increase in carbon capacitance at pore sizes less than 1 nanometer, Science 313 (5794) (2006) 1760–1763.
[61] C. Vix-Guterl, E. Frackowiak, K. Jurewicz, M. Friebe, J. Parmentier, F. Béguin, Electrochemical energy storage in ordered porous carbon materials, Carbon 43 (6) (2005) 1293–1302.
[62] J. Chmiola, C. Largeot, P.-L. Taberna, P. Simon, Y. Gogotsi, Desolvation of ions in subnanometer pores and its effect on capacitance and double-layer theory, Angewandte Chemie International Edition 47 (18) (2008) 8020–8024.
[63] C. Largeot, C. Portet, P.-L. Taberna, Y. Gogotsi, P. Simon, Relation between the ion size and pore size for an electric double-layer capacitor, J. Am. Chem. Soc. 130 (2008) 2730–2731.
[64] J. Huang, B. Sumpter, V. Meunier, Theoretical model for nanoporous carbon supercapacitors, Angewandte Chimie International Edition 120 (3) (2008) 530–534.
[65] J. Huang, B.G. Sumpter, V. Meunier, A universal model for nanoporous carbon supercapacitors applicable to diverse pore regimes, carbon materials, and electrolytes, Chem.-Eur. J. 14 (22) (2008) 6614–6626.
[66] G. Feng, R. Qiao, J. Huang, B.G. Sumpter, V. Meunier, Ion distribution in electrified micropores and its role in the anomalous enhancement of capacitance, ACS Nano 4 (4) (2010) 2382–2390.
[67] M. Noked, E. Avraham, Y. Bohadana, A. Soffer, D. Aurbach, Development of anion stereoselective, activated carbon molecular sieve electrodes prepared by chemical vapor deposition, J. Phys. Chem. C 7316–7321 (17) (2009) 113.
[68] R. Mysyk, E. Raymundo-Piñero, F. Béguin, Saturation of subnanometer pores in an electric double-layer capacitor, Electrochem. Comm. 11 (2009) 554–556.
[69] R. Mysyk, E. Raymundo-Piñero, J. Pernak, F. Béguin, Confinement of symmetric tetraalkylammonium ions in nanoporous carbon electrodes of electric double-layer capacitors, J. Phys. Chem. C 113 (30) (2009) 13443–13449.
[70] T.A. Centeno, F. Stoeckli, The assessment of surface areas in porous carbons by two model-independent techniques, the DR equation and DFT, Carbon 48 (9) (2010) 2478–2486.
[71] F. Stoeckli, T.A. Centeno, Pore size distribution and capacitance in microporous carbons, Phys. Chem. Chem. Phys. 14 (33) (2012) 11589–11591.
[72] T.A. Centeno, O. Sereda, F. Stoeckli, Capacitance in carbon pores of 0.7 to 15 nm: a regular pattern, Phys. Chem. Chem. Phys. 27 (13) (2011) 12403–12406.
[73] A. Garcia-Gomez, V. Barranco, G. Moreno-Fernandez, J. Ibañez, T.A. Centeno, Correlation between capacitance and porosity in microporous carbon monoliths, J. Phys. Chem. C 118 (10) (2014) 5134–5141.
[74] E. Redondo, J. Carretero-González, E. Goikolea, J. Ségalini, R. Mysyk, Effect of pore texture on performance of activated carbon supercapacitor electrodes derived from olive pits, Electrochim. Acta 160 (2015) 178–184.
[75] S. Kondrat, A. Kornyshev, F. Stoeckli, T.A. Centeno, The effect of dielectric permittivity on the capacitance of nanoporous electrodes, Electrochem. Comm. 348–350 (2013) 34.
[76] Q. Abbas, D. Pajak, E. Frackowiak, F. Béguin, Effect of binder on the performance of carbon/carbon symmetriccapacitors in salt aqueous electrolyte, Electrochim. Acta 140 (2014) 132–138.
[77] A. García-Gómez, G. Moreno-Fernández, B. Lobato, T.A. Centeno, Constant capacitance in nanopores of carbon monoliths, Phys. Chem. Chem. Phys. 17 (24) (2015) 15687–15690.
[78] H. Tamai, M. Kouzu, M. Morita, H. Yasuda, Highly mesoporous carbon electrodes for electric double-layer capacitors, Electrochem. Solid St. 6 (10) (2003) A214–A217.
[79] M.D. Levi, S. Sigalov, D. Aurbach, L. Daikhin, Electrochemical quartz crystal admittance methodology for tracking compositional and mechanical changes in porous carbon electrodes, J. Phys. Chem. C 117 (29) (2013) 14876–14889.
[80] M.D. Levi, G. Salitra, N. Levy, D. Aurbach, J. Maier, Application of a quartz-crystal microbalance to measure ionic fluxes in microporous carbons for energy storage, Nat. Mater. 8 (2009) 872–875.
[81] J.M. Griffin, A.C. Forse, W.-Y. Tsai, P.-L. Taberna, P. Simon, C.P. Grey, In situ NMR and electrochemical quartz crystal microbalance techniques reveal the structure of the electrical double layer in supercapacitors, Nat. Mater. 14 (2015) 812–819.

[82] S. Boukhalfa, D. Gordon, L. He, Y. Melnichenko, N. Nitta, A. Magasinski, G. Yushin, In situ small angle neutron scattering revealing ion sorption in microporous carbon electrical double layer capacitors, ACS Nano 8 (3) (2014) 2495–2503.

[83] C. Prehal, D. Weingarth, E. Perre, R.T. Lechner, H. Amenitsch, O. Paris, V. Presser, Tracking the structural arrangement of ions in carbon supercapacitor nanopores using in situ small-angle X-ray scattering, Energ. Environ. Sci. 8 (2015) 1725–1735.

[84] M. Deschamps, E. Gilbert, P. Azais, E. Raymundo-Piñero, M.R. Ammar, P. Simon, D. Massiot, F. Béguin, Exploring electrolyte organization in supercapacitor electrodes with solid-state NMR, Nat. Mater. 12 (2013) 351–358.

[85] J.M. Griffin, A.C. Forse, H. Wang, N.M. Trease, P.-L. Taberna, P. Simon, C.P. Grey, Ion counting in supercapacitor electrodes using NMR spectroscopy, Faraday Discuss. 176 (2014) 49–68.

[86] F. Rodriguez-Reinoso, M. Molina-Sabio, Activated carbons from lignocellulosic materials by chemical and/or physical activation: an overview, Carbon 30 (1992) 1111–1118.

[87] F. Rodriguez-Reinoso, M. Molina-Sabio, M.T. Gonzalez, The use of steam and CO_2 as activating agents in the preparation of activated carbons, Carbon 33 (1) (1995) 15–23.

[88] M. Molina-Sabio, M.T. Gonzalez, F. Rodriguez-Reinoso, A. Sepúlveda-Escribano, Effect of steam and carbon dioxide activation in the micropore size distribution of activated carbon, Carbon 34 (4) (1996) 505–509.

[89] A. Linares-Solano, M.A. Lillo-Rodenas, J.P. Marko-Lozar, M. Kunowsky, A.J. Romero-Anaya, NaOH and KOH for preparing activated carbons used in energy an environmental applications, Int. J. Energ. Environ. Econ. 20 (4) (2012) 59–91.

[90] G. Dobele, T. Dizhbite, M.V. Gil, A. Volperts, T.A. Centeno, Production of nanoporous carbons from wood processing wastes and their use in supercapacitors and CO_2 capture, Biomass Bioenerg. 46 (2012) 145–154.

[91] Kuraray Chemical Co., LTD, [Online]. Available from: http://www.kuraraychemical.com/products/sc/capacitor.htm

[92] D.A. Ersoy, M.J. McNallan, Y. Gogotsi, Carbon coatings produced by high temperature chlorination of silicon carbide ceramics, Mater. Res. Innov. 5 (2001) 55–62.

[93] Y.G. Gogotsi, I.-D. Jeon, M.J. McNallan, Carbon coatings on silicon carbide by reaction with chlorine-containing gases, J. Mater. Chem. 7 (1997) 1841–1848.

[94] Z.G. Cambaz, G.N. Yushin, K.L. Vyshnyakova, L.N. Pereselentseva, Y. Gogotsi, Formation of carbide-derived carbon on beta-silicon carbide whiskers, J. Am. Ceram. Soc. 89 (2006) 509–514.

[95] A. Janes, T. Thomberg, E. Lust, Synthesis and characterisation of nanoporous carbide-derived carbon by chlorination of vanadium carbide, Carbon 45 (14) (2007) 2717–2722.

[96] Y. Zhu, S. Murali, M.D. Stoller, K.J. Ganesh, W. Cai, P.J. Ferreira, A. Pirkle, R.M. Wallace, K.A. Cychosz, M. Thommes, D. Su, E.A. Stach, R.S. Ruoff, Carbon-based supercapacitors produced by activation of graphene, Science 332 (6037) (2011) 1537–1541.

[97] Y. Tao, X. Xie, W. Lv, D.-M. Tang, D. Kong, Z. Huang, H. Nishihara, T. Ishii, B. Li, D. Golberg, F. Kang, T. Kyotani, Q.-H. Yang, Towards ultrahigh volumetric capacitance: graphene derived highly dense but porous carbons for supercapacitors, Sci. Rep. 3 (2013) 1–8.

[98] X. Yang, C. Cheng, Y. Wang, L. Qiu, D. Li, Liquid-mediated dense integration of graphene materials for compact capacitive energy storage, Science 341 (2013) 534–537.

[99] J.R. Miller, R.A. Outlaw, B.C. Holloway, Graphene double-layer capacitor with ac line-filtering performance, Science 329 (2010) 1637–1639.

[100] J.K. McDonough, A.I. Frolov, V. Presser, J. Niu, C.H. Miller, T. Ubieto, M.V. Federov, Y. Gogotsi, Influence of the structure of carbon onions on their electrochemical performance in supercapacitor electrodes, Carbon 50 (9) (2012) 3298–3309.

[101] H. Ji, X. Zhao, Z. Qiao, J. Jung, Y. Zhu, Y. Lu, L.L. Zhang, A.H. McDonald, R.S. Ruoff, Capacitance of carbon-based electrical double-layer capacitors, Nat. Comm. 5 (2013) 1–7.

[102] L.L. Zhang, X. Zhao, H. Ji, M.D. Stoller, L. Lai, S. Murali, S. Mcdonell, B. Cleveger, R.M. Wallace, R.S. Ruoff, Nitrogen doping of graphene and its effect on quantum capacitance, and a new insight on the enhanced capacitance of N-doped carbon, Energ. Environ. Sci. 5 (2012) 9618–9625.

[103] T. Brousse, D. Bélanger, A Hybrid Fe_3O_4–MnO_2 Capacitor in Mild Aqueous Electrolyte, Electrochem. Solid St. 6 (11) (2003) A244–A248.

[104] S. Roldán, C. Blanco, M. Granda, R. Menéndez, R. Santamaría, Towards a further generation of high-energy carbon-based capacitors by using redox-active electrolytes, Angewandte Chemie International Edition 50 (7) (2011) 1699–1701.

[105] D. Hulicova, J. Yamashita, Y. Soneda, H. Hatori, M. Kodama, Supercapacitors prepared from melamine-based carbon, Chem. Mater. 17 (5) (2005) 1241–1247.

[106] K. Jurewicz, R. Pietrzak, P. Nowicki, H. Wachowska, Capacitance behavior of brown coal based active carbon modified through chemical reaction with urea, Electrochim. Acta 53 (2008) 5469–5475.

[107] G. Lota, K. Lota, E. Frackowiak, Nanotubes based composites rich in nitrogen for supercapacitor application, Electrochem. Comm. 9 (7) (2007) 1828–1832.

[108] C.O. Ania, V. Khomenko, E. Raymundo-Piñero, J.B. Parra, F. Béguin, The large electrochemical capacitance of microporous doped carbon obtained by using a zeolite template, Adv. Funct. Mater. 17 (2007) 1828–1836.

[109] M.P. Bichat, E. Raymundo-Piñero, F. Béguin, High-voltage supercapacitor built with seaweed carbons in neutral aqueous electrolyte, Carbon 48 (2010) 4351–4361.

[110] B. Fang, H. Zhou, I. Honma, Ordered porous carbon with tailored pore size for electrochemical hydrogen storage application, J. Phys. Chem. B 110 (2006) 4875–4880.

[111] K. Jurewicz, E. Frackowiak, F. Béguin, Towards the mechanism of electrochemical hydrogen storage in nanostructured carbon materials, Appl. Phys. A 78 (7) (2004) 981–987.

[112] J. Niu, W. Pell, B. Conway, Requirements for performance characterization of C double-layer supercapacitors: Applications to a high specific-area C-cloth material, J. Power Sources 156 (2) (2006) 725–740.

[113] V. Khomenko, E. Raymundo-Piñero, F. Béguin, Optimisation of an asymmetric manganese oxide/activated carbon capacitor working at 2 V in aqueous medium, J. Power Sources 153 (1) (2006) 183–190.

[114] G. Wang, L. Zhang, J. Zhang, A review of electrode materials for electrochemical supercapacitors, Chem. Soc. Rev. 41 (2012) 797–828.

[115] S. Hadzi-Jordanov, H. Angerstein-Kozlowska, M. Vucovic, B.E. Conway, Reversibility and growth behavior of surface oxide films at ruthenium electrodes, J. Electrochem. Soc. 125 (1978) 1471–1480.

[116] J.P. Zheng, P.J. Cygan, T.R. Jow, Hydrous ruthenium oxide as an electrode material for electrochemical capacitors, J. Electrochem. Soc. 142 (8) (1995) 1699–1703.

[117] J.W. Long, K.E. Swider, C.I. Merzbacher, D.R. Rolison, Voltammetric characterization of ruthenium oxide-based aerogels and other RuO_2 solids: The nature of capacitance in nanostructured materials, Langmuir 15 (1999) 780–785.

[118] J.P. Zheng, Y. Xin, Characterization of $RuO_2{\cdot}xH2O$ with various water contents, J. Power Sources 110 (1) (2002) 86–90.

[119] D.A. McKeown, P.L. Hagans, L.P.L. Carette, A.E. Russell, K.E. Swider, D.R. Rolison, Structure of hydrous ruthenium oxides: implications for charge storage, J. Phys. Chem. B 103 (1999) 4825–4832.

[120] W. Sugimoto, H. Iwata, Y. Yasunaga, Y. Murakami, Y. Takasu, Preparation of ruthenic acid nanosheets and utilization of its interlayer surface for electrochemical energy storage, Angewandte Chemie International Edition 42 (34) (2003) 4092–4096.

[121] C.-C. Hu, K.-H. Chang, M.-C. Lin, Y.-T. Wu, Design and tailoring of the nanotubular arrayed architecture of hydrous RuO_2 for next generation supercapacitors, Nano Lett. 6 (12) (2006) 2690–2695.

[122] Y.M. Chen, J.H. Cai, Y.S. Huang, K.Y. Lee, D.S. Tsai, Preparation and characterization of iridium dioxide-carbon nanotube nanocomposites for supercapacitors, Nanotechnol. 22 (11) (2011) 115706.

[123] C.Z. Yuan, X.C. Zhang, L.H. Su, B. Gao, L.F. Shen, Facile synthesis and self-assembly of hierarchical porous NiO nano/micro spherical superstructures for high performance supercapacitors, J. Mater. Chem. 19 (32) (2009) 5772–5777.

[124] R.B. Rakhi, W. Chen, D.Y. Cha, H.N. Alshareef, Substrate dependent self-organization of mesoporous cobalt oxide nanowires with remarkable pseudocapacitance, Nano Lett. 12 (5) (2012) 2559–2567.

[125] S.W. Hwang, S.H. Hyun, Synthesis and characterization of tin oxide/carbon aerogel composite electrodes for electrochemical supercapacitors, J. Power Sources 172 (1) (2007) 451–459.

[126] C.-C. Hu, C.-M. Huang, K.-H. Chang, Anodic deposition of porous vanadium oxide network, J. Power Sources 185 (2008) 1594–1597.

[127] T. Brezesinski, J. Wang, S.H. Tolbert, B. Dunn, Ordered mesoporous alpha-MoO_3 with iso-oriented nanocrystalline walls for thin-film pseudocapacitors, Nat. Mater. 9 (2010) 146–151.
[128] Q. Feng, H. Kanoh, K. Ooi, Manganese oxide porous crystals, J. Mater. Chem. 9 (1999) 319–333.
[129] H.Y. Lee, J.B. Goodenough, Supercapacitor behavior with KCl electrolyte, J. Solid State Chem. 144 (1999) 220–223.
[130] M. Toupin, T. Brousse, D. Bélanger, Charge storage mechanism of MnO_2 electrode used in aqueous electrochemical capacitor, Chem. Mater. 16 (16) (2004) 3184–3190.
[131] W.F. Wei, X.W. Cui, W.X. Chen, D.G. Ivey, Manganese oxide-based materials as electrochemical supercapacitor electrodes, Chem. Soc. Rev. 40 (3) (2011) 1697–1721.
[132] G. Hu, C. Tang, C. Li, H. Li, Y. Wang, H. Gong, The sol-gel-derived nickel-cobalt oxides with high supercapacitor performances, J. Electrochem. Soc. 158 (6) (2011) A695–A699.
[133] M. Sathiya, A.S. Prakash, K. Ramesha, J.-M. Tarascon, A.K. Shukla, V2O5-anchored carbon nanotubes for enhanced electrochemical energy storage, J. Am. Chem. Soc. 133 (2011) 16291–16299.
[134] O. Ghodbane, J.-L. Pascal, F. Favier, Microstructural effects on charge-storage properties in MnO_2-based electrochemical capacitors, ACS Appl. Mater. Interfaces 1 (5) (2009) 1130–1139.
[135] D. Choi, G.E. Blomgren, P.N. Kumta, Fast and reversible surface redox reaction in nanocrystalline vanadium nitride supercapacitors, Adv. Mater. 18 (2006) 1178–1182.
[136] O. Bondarchuk, A. Morel, D. Bélanger, E. Goikolea, T. Brousse, R. Mysyk, Thin films of pure vanadium nitride: evidence for anomalous non-Faradaic capacitance, J. Power Sources 324 (2016) 439–446.
[137] M. Zhi, C. Xiang, J. Li, M. Li, N. Wu, Nanostructured carbon–metal oxide composite electrodes for supercapacitors: a review, Nanoscale 5 (2013) 72–88.
[138] Z.-S. Wu, G. Zhou, L.-C. Yin, W. Ren, F. Li, H.-M. Cheng, Graphene/metal oxide composite electrode materials for energy storage, Nano Energy 1 (1) (2012) 107–131.
[139] Z.-S. Wu, D.-W. Wang, W. Ren, J. Zhao, G. Zhou, F. Li, H.-M. Cheng, Anchoring hydrous RuO2 on graphene sheets for high-performance electrochemical capacitors, Adv. Funct. Mater. 20 (20) (2010) 3595–3602.
[140] Q. Cheng, J. Tang, J. Ma, H. Zhang, N. Shinya, L.-C. Qin, Graphene and nanostructured MnO_2 composite electrodes for supercapacitors, Carbon 49 (9) (2011) 2917–2925.

CHAPTER FIVE

Characterization of Nanomaterials for Energy Storage

A.M. Bittner*,, V. Koroteev†**
*The Foundation for Science, Bilbao, Spain
**CIC nanoGUNE, Donostia-San Sebastián, Spain
†Novosibirsk State University, Novosibirsk, Russia

Contents

Nanomaterials are designed or chosen to contain nanoscale features because these features bestow new or enhanced properties. One should, however, keep in mind that nanomaterials are first and foremost materials, which means they require suitable characterization [1–6]. "Suitable" may or not involve the nanoscale. As for a definition of "nano" for characterization methods, it is common practice to define lengths scales, here roughly separated into macro (>mm), micro (mm to μm), nano (μm to nm), and atomic (<nm). For the development of most energy storage applications, a standard macroscale characterization is absolutely necessary, and quite often this on its own suffices completely. True nanoscale methods, as introduced here, require very small probes, such as particles or focused radiation (photons) (Fig. 5.1). They are typically applied in research, while development and quality control require well-established standards, which are still not common on the nanoscale.

Emerging Nanotechnologies in Rechargeable Energy Storage Systems

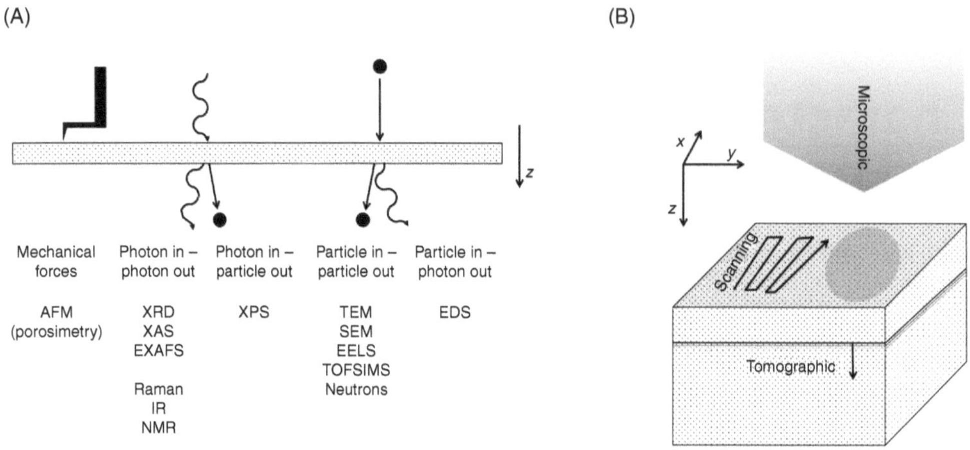

Figure 5.1 (A) Comparison of sample excitation and information collection methods; (B) scanning and imaging microscopy modes.

Is nanoscale resolution indeed required at all for characterization? Generally the answer is "no"; the standard macroscale characterization methods work with nanomaterials. There are three main "yes" cases: (1) if the material is porous, one might want to image the pores (which in fact is a huge challenge for any pore below 5 nm size); (2) if there are typical nanostructural elements, such as crystallites or surface patterns, imaging is helpful or even required; and (3) if there are chemical inhomogeneities (e.g., in material blends), they often occur on small scales. Imaging is carried out with a beam of particles (please recall that also photons are particles), which is either focused and read out by special optics (imaging in the true sense), or scanned point by point (scanning methods), which requires a much narrower focus. In contrast, scanning probe methods are carried out with a solid probe in very close proximity to the sample. The nanoscale resolution in imaging depends on the beam-sample orientation: Specular, parallel to the probe beam/path of the probe particles direction (Fig. 5.1A); or perpendicular to the beam, which in many cases means lateral nanoscale resolution "2D" on the probed surface (or in a probed slice of the sample volume, see Fig. 5.1B). Both combined, for example, by a scanning method for the perpendicular orientation as shown in Fig. 5.1B, would correspond to true nanoscale "3D" imaging. However, it is important to distinguish nanoscale imaging (or scanning or mapping) from macroscale indirect methods (e.g., scattering), which often are technically much simpler, provide good sample averages, and are able to deliver the desired information.

In the following, macro- and microscale characterization methods will be compared (Section 1), and in situ analysis will be distinguished from ex situ (Section 2). Structural and chemical analysis (Sections 3 and 4) list nonimaging methods that give indirect information on the atomic scale; they mainly fall into diffraction and spectroscopy, respectively. The nanoscale methods themselves are discussed in two sections: a list of suitable techniques (Section 5) and electron microscopy, which is in widespread use and offers so many options that it merits an own section (Section 6). The future is notoriously difficult to predict, but some trends are beginning to surface (Section 7).

1 MACRO- AND MICROSCALE CHARACTERIZATION

Macroscale analyses are standard physicochemical methods, and do not differ in neither application nor data evaluation when nanomaterials are used. Examples [5,6] comprise elemental analysis (combustion), mass spectroscopy, gravimetry, analysis of the water content (Karl Fischer titration); furthermore, calorimetry [7], and electrical/electrochemical characterization (impedance, voltammetry) [8] are very typical. The sample preparation is a key issue to obtain suitable data; it is ideally put in the context of in situ and ex situ characterization (Section 2). The microscale can be of great relevance, and in fact complicates many methods significantly, for example, by undesired light scattering for photon methods, by focus and contrast problems in electron-based methods; scanning probe methods can usually not even be applied on microrough surfaces. Still, most techniques allow selecting microscale areas, and thus a comparison between various parts of a sample (Fig. 5.2). The results should always be checked for their compatibility with the average values, measured on the macroscale. In other words, a single microscale area is not always representative for the complete sample. The micromethods require either new equipment, or add-ons to an existing nonimaging or macroscale imaging tool. Examples are:

- UV–vis and Raman spectroscopy [1,2,9] combined with an optical microscope;
- IR microscopy combined with a FTIR spectrometer [2];
- Focus methods for X-rays, usually installed at synchrotron sources, and combined with XAS (including EXAFS) and XRD;

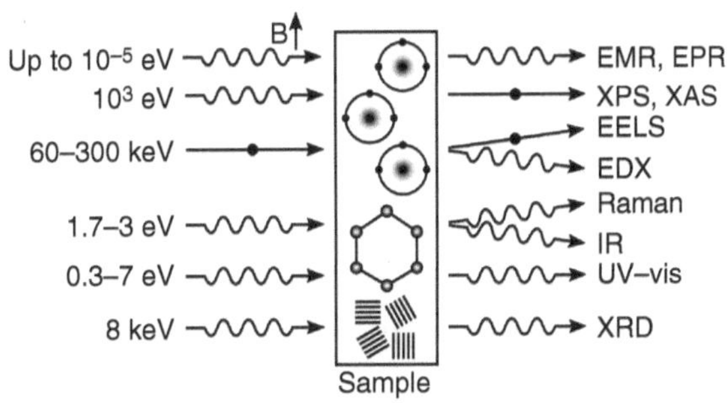

Figure 5.2 ***Various methods, which are usually applied on the macroscale, that is, without spatial resolution, that can be modified to provide information of milli- and micrometer-sized sample parts.***

- XPS with microimaging, where the obtained spectral data correspond to standard macroscale data.

Especially for the X-ray techniques, the application of microimaging can be complex. However, the microscopy options are very impressive, and allow imaging details and defects [10–12]; scaling down to the nanoscale is well underway, so these methods will be discussed in detail in the section on nanomethods (Section 5).

For a complete discussion of scales, one should not forget the atomic (or molecular scale). Only a very few methods allow researchers to address it directly (TEM or AFM at high resolution), but many methods provide structural data of utmost precision (XRD), or spectroscopic identification of chemical groups (IR) [2]. These methods will be discussed alongside the related nanoscale methods, without specific reference to the atomic scale, but mainly under the heading of chemical analysis (Section 4).

2 EX SITU, "POSTMORTEM" ANALYSIS VERSUS IN SITU ELECTROCHEMISTRY

The analysis of a material before or after use is called ex situ—this is the usual mode of material analysis. It has the huge advantage of allowing for sample preparation, cutting, thinning, polishing, dissolution, evaporation, and so forth, which is arguably most advanced in electron microscopy (a practically complete manual is Ref. [13]). It is furthermore of special relevance for development, when devices are tested after use (and disassembly); in passing we note that such an analysis should always include "before use" samples for comparison. Although the device is not necessarily destroyed, "postmortem" has somehow been established as description. The big advantage of "postmortem" analysis is that a real device is probed. It can, however, be outweighed by the disadvantage that nearly any sample preparation, including disassembly, draining of electrolyte, rinsing, drying, will change some properties. In principle, it is always an advantage to work in operando, which means—strictly speaking—with a realistic energy storage device. This is only rarely possible, and instead in situ techniques are applied to materials during use, meaning in a properly functioning electrochemical cell. This cell is usually an experimental device, used for research purposes only, practically never identical with the real device, and rarely approaching its performance. Unfortunately, with some very rare exceptions, complete devices cannot be analyzed, hence a part of the device or model has to be constructed. The art of in situ measurements is the marriage between a potentially complex tool with an electrochemical cell that mimics the full device [1]. Solid electrolytes without evaporating compound can be characterized in high vacuum, which offers the complete range of the powerful particle methods, especially electron microscopy [14].

Not surprisingly, especially successful methods are photon in/photon out techniques since radiation can penetrate matter very effectively: XRD [15] and XAS [1,15] require an X-ray-transparent window, which can be as simple as a polymer foil. Raman microscopy [9] requires a transparent window, such as quartz or glass. But also neutrons penetrate matter very easily, so neutron techniques are generally suitable for in situ cells [16]. However, some materials have very high absorption cross sections, such as ^{6}Li (a minority isotope) in lithium ion batteries (LiBs). All light elements scatter neutrons very effectively; hence hydrogen in polymers or in the electrolyte can be a problem. Electron techniques of any type (electron in/electron out in EM, electron in/photon out in EDS, photon in/electron out in XPS) pose huge challenges because of the low penetration depth. In particular low energy electrons cannot escape an electrochemical cell, which means a big challenge for all electron out methods, such as XPS. Scanning probe methods require even more: The scanning tip has to be inserted into the cell, which is only possible for rather simple models in cells open to the environment (which can be inert gas). Moreover, the surface roughness has to be below the microscale, which is rarely the case for technical electrodes.

3 STRUCTURAL ANALYSIS

Traditional structure determination on the atomic scale is based on *diffraction*, which requires crystallinity. XRD is carried out with X-ray tube sources in diffractometers or with synchrotron radiation in specialized beamlines [1,15]. The technique is applicable to nearly all materials, though the precision is much higher for heavy atoms. The main method is based on detection of X-ray intensity as a function of a single angle of sample orientation if the sample is a powder. The resulting diffractogram, which contains multiple peaks located at angles typical for lattice spacings, is usually compared to literature data. Moreover, the peak width contains information about the size of crystallites. A huge advantage of XRD is a very high compatibility with various environments, such that liquid cells can be employed, and in situ studies are straightforward. It is also applicable to oriented samples (e.g., fibers), when they are rotated around another angle, to provide the required orientational average.

A similar technique, again for the atomic scale, is *neutron diffraction* (Fig. 5.3) [16]. Neutrons require large installations like a nuclear reactor or a spallation source for production of the required intense beams. These particles do not interact much with matter, and thus need specialized beam guiding and focusing, and sensitive detectors. Neutrons are radioactive and induce radioactivity, hence careful material selection and shielding is a must. Given all this, neutron diffraction is not a standard technique. But while X-rays can only probe volumes close to the external casing of a real device, the interaction of neutrons with matter is so weak that complete cells are penetrated with great ease. Compared to XRD, the sensitivity favors light elements, which can be a big advantage for lithium-based structures and for carbon materials (Fig. 5.3), which is unique. An intense beam and a large area detector facilitate fast recording during actual operations, such as current cycling, giving

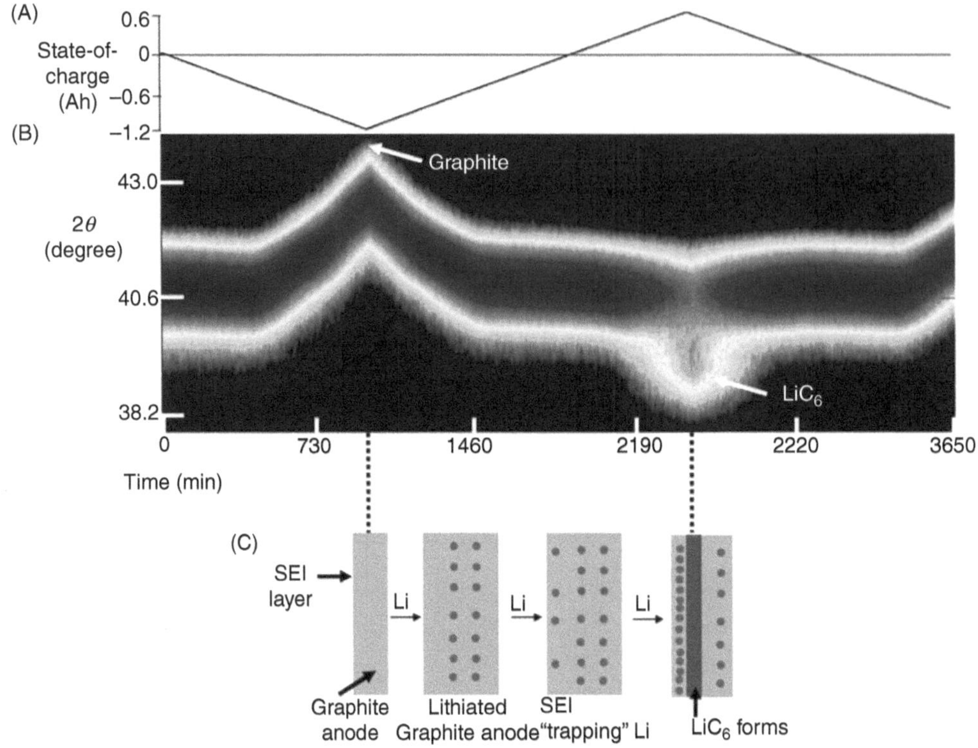

Figure 5.3 State-of-charge (A) and neutron diffraction patterns of the (002) graphite reflection (B) from the anode in a commercial LiB cell during cycling. The *gray scale* in the pattern represents the reflection intensity (except for the middle part around 41 degree in black, which corresponds to maximum intensity), hence the image shows diffractograms (intensity vs. angle 2θ) as function of time. Stoichiometric LiC_6 forms at full charging. Pure graphite (delithiated) exists where labeled. The proposed mechanism of Li insertion is sketched as function of charging (C). In the discharged state, pure graphite and SEI layer make up the anode. With charging, Li is inserted into graphite, followed by insertion into the SEI layer, and finally by formation of LiC_6. An area detector allows to synchronously record patterns also from the $LiCoO_2$ cathode. *(Reproduced from N. Sharma, V.K. Peterson, M.M. Elcombe, M. Avdeev, A.J. Studer, N. Blagojevic, R. Yusoff, N. Kamarulzaman, Structural changes in a commercial lithium-ion battery during electrochemical cycling: an in situ neutron diffraction study, J. Power Sources 195 (2010) 8258–8266 [16] with permission from Elsevier, Copyright © 2010).*

access to multiple in situ information. It is worth noting that the diffraction of neutrons can contain additional information since neutrons have a magnetic moment.

Gas (mainly nitrogen) *porosimetry* is a much more indirect method, which can give valuable information for any porous structure, whether crystalline or completely amorphous [17]. It accesses meso- (<50 nm), micro- (<2 nm), and ultramicropores (<0.7 nm)—these assignments are standardized, and widely used in technology, but the smaller ones are hard to access even with electron microscopy (Section 6). Porosimetry is based on the simple principle of establishing pressure equilibria at a fixed temperature (usually 77K). Volume–pressure curves (isotherms) show how nitrogen progressively fills all pores. The method requires careful precleaning by heating and evacuation. The isotherms can be interpreted by their shape, or quantitatively (adsorbed gas volume), which provides one of the most important measurements for porous solids, the BET value (internal surface area per mass). The actual value usually does not exactly correspond to the geometrical surface, but the value is of utmost importance in a technology context to classify materials. More advanced analysis requires modeling the adsorption with empirical potentials and/or constructing density functionals. The modeling finally yields an average pore size or a pore size distribution (often with some artifacts). Hence any interpretation requires great care, and ideally comparison with related methods [17].

4 CHEMICAL ANALYSIS (SPECTROSCOPIC TECHNIQUES)

Methods that are able to analyze the chemical composition can be addressed as (indirect) atomic scale methods, whenever they allow determining the relative arrangement of atoms [6,18]. All these methods are very much standard analysis techniques, without nanoscale resolution, that is, without the ability to determine absolute positions on the nanoscale.

UV–vis spectroscopy is based on the absorption of light in the visible and UV range, usually in transmission through a thin sample [18]. For in situ methods, reflection is possible, but it is usually hampered by too thick layers (of the energy storage material), or too low reflectivity. This limits the practical use to powders or dissolved materials. Ex situ, the most useful information to be gained is from the position and intensity of absorption bands of transition metal cations, caused by excitations of electronic levels. The absorption can easily be quantified, and furthermore it can specify the arrangement of ligands, for example, oxide and sulfide. However, the method is of limited use when many species are present, such as in energy storage materials.

NMR spectroscopy [1,18] excites transitions between electronic levels, split up by a strong magnetic field and the magnetic moment of atomic nuclei. This requires a complex, but standardized setup that combines the field and pulsed radiowaves. NMR is usually only employed for the (liquid) electrolyte, and molecules in solution (or in pores) can be analyzed with great ease, and their composition can be quantified, also the configuration of atoms. However, recent advances allowed an extension to the solid state that has become of great value for many, especially amorphous, materials [4,19,20]. Moreover, it also provides information on dynamics.

Infrared absorption spectra [18,21] arise from the excitation of molecular vibrations or of phonons. They can provide quantification and identification of chemical groups in molecules, or in mixtures/blends, including solids. Some information on the intermolecular interaction can be obtained, but the real advantage lies in the plethora of absorption bands, which are very characteristic for each material (fingerprint). This is generally true for solids, too, but the broad vibrational bands and the very broad phonons are less characteristic [21], for example, in supercapacitors [22]. In the most simple setup, the material in question is placed in the infrared beam, in front of the detector. Solids and in situ techniques require a reflection setup, which asks for highly reflective electrodes, such as gold.

X-ray absorption spectra (XAS or in more details XANES) provide elemental analysis and a detailed picture of the direct chemical environment (absorption edge) for each atomic species in a sample. Usually, a synchrotron source is required, and either absorption or emission (fluorescence) of X-rays provides the signal. Most important, the oxidation state is directly reflected in the exact position of the atomic absorption edges [6]. A Fourier transform procedure of the weak spectral features close to the edge (EXAFS) gives the local geometry of neighboring atoms [6]. EXAFS can easily give artifacts, so the interpretation is more difficult than for XAS, but it is chemically specific, and provides interatomic distances even for amorphous material.

Mössbauer spectroscopy is based on the emission of gamma rays from radioactive nuclei in solids, and it can only be carried out in specialized tools. There are not many applicable elements, but iron is the best known, and it is of relevance for some battery electrodes [3,23]. The detailed information on the local electronic environment is used for characterization, but it could possibly also track chemical reactions.

5 NANOSCALE CHARACTERIZATION

The most important method, or rather family of methods, is electron microscopy (EM; Section 6). The current section introduces non-EM techniques with nanoscale resolution, starting with methods that provide images with nanoscale resolution in 3D.

5.1 Nanoscale resolution in 3D

Scanning probe methods (Fig. 5.1) record some signal, for example, the topography z, as function of a lateral position (x,y) on a flat sample surface. The local topography can be detected by application and measurement of forces on a cantilever, or from mechanical properties of an oscillating cantilever. While these methods provide excellent resolution, the requirement for sample preparation (flatness) is so strict, that the experiment is totally dependent on developing a good model system [24,25]. Hence, despite the highest possible lateral and especially vertical (z) resolution, scanning probe is as yet very rarely applied. Examples are carbon-based materials in electrolytic double layer capacitors (EDLCs) and in LiB and Li-S electrodes: They can be modeled by pure graphite [24,25]. Lateral and z resolution of *atomic force microscopy* (AFM) are here in the low nm range, hence thin layers, such as the SEI can be visualized, but no chemical information can be determined. This means that even in situ information is of limited value because real electrodes have a complex microstructure. A way to introduce chemical information to

AFM is *scanning electrochemical microscopy* (SECM), for example, used for characterizing the SEI layers formed at the electrode–electrolyte interface [26]. Here the mechanism is not a force, but a local electrochemical current, which depends on the local redox chemistry. Again, due to the preparation of a flat electrode, results cannot easily be applied to the technical systems.

The most important probes are, however, photons. While infrared microscopy offers very low resolution (not on the nanoscale), visible wavelengths are more attractive. Optical microscopy is a standard, but it is usually not combined with UV–vis spectroscopy. *Raman spectroscopy* has become widely used [1,2,9] (Fig. 5.4). Here the focused beam can be confocally confined to a region of several hundreds of nm in diameter [6]. The scattered light is selected for a lower wavelength (with very low background), such that the energy difference corresponds to vibrations or phonons excited in the material (or in the electrolyte). The energy of vibrations in molecules (and phonons in solids) depends strongly on the local chemical environment, so Raman spectra are used to detect chemical species, and to follow their reactions in situ. The method is useful for most compounds, especially because one can select a strong resonance (large scattering cross section), which is found in most materials (carbon, polymers, oxides, such as $LiCoO_2$). Higher incident energies result in higher lateral resolution down to ca. 200 nm, but the trade-off can be excitation of unwanted fluorescence.

In the same spirit, X-rays are extremely useful, especially for in situ studies. However, in contrast to standard spectroscopic methods, or methods with a slight X-ray focus, microscopy on the nanoscale requires very high intensities and special focusing optics, hence not only a synchrotron, but a dedicated setup. The big advantage is that the properties of the X-rays allow researchers to combine a very high resolution (10 nm) with the very powerful chemical information from *X-ray absorption spectroscopy* (XAS): tomographic reconstructions (Fig. 5.5) offer views in 3D, and XAS at absorption edges gives a high selectivity for elements [10–12]. The edge energy, and features close to the edge [X-ray near-edge spectroscopy (XANES)] provide information on the oxidation state of the atom. A disadvantage is that X-rays interact little with elements of low atomic number, so carbon materials (in EDLCs) and lithium (in batteries) are not ideal.

5.2 Nanoscale resolution in lower dimensions (on a surface or in a slab of material)

Secondary ion mass spectrometry (SIMS) is usually analyzed with TOF (time of flight) techniques. TOFSIMS is as yet not very popular, although it is one of the simplest ways to move layer by layer through a solid sample. It requires vacuum, but offers a mass analysis of atoms and small clusters of material sputtered off by an ion beam. It thus allows for depth profiling of a large range of elements and molecules in parallel, offering a unique, but of course destructive view inside a solid electrode, as exemplified by recording Li,

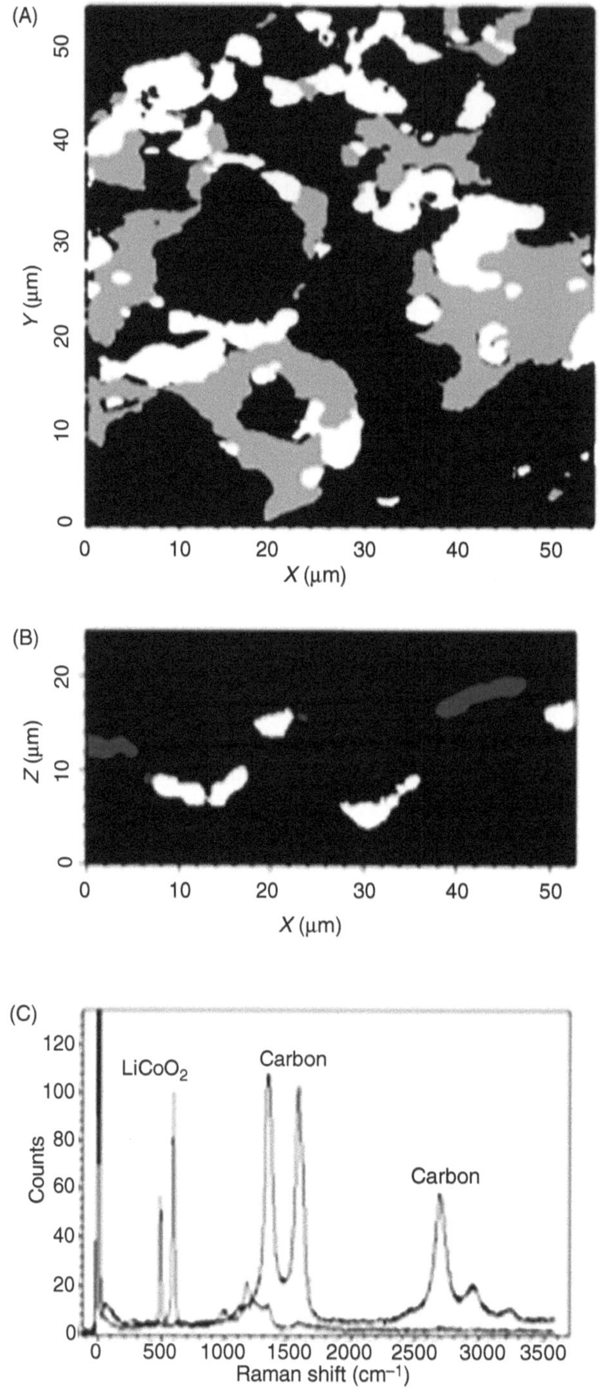

Figure 5.4 ***Raman spectromicroscopy of a carbon/$LiCoO_2$ electrode.*** *Gray* features correspond to carbon, *bright* ones to $LiCoO_2$. (A) The sample plane (*xy*), (B) an *xz* slice recorded from confocal microscopy (laser beam in *z*), and (C) two Raman spectra, from which the strongest lines were chosen for imaging. *(Reproduced from Tokyo Instruments (Japan). Available from: http://www.tokyoinst.co.jp/en/products/detail/NF02/index.html [27] with permission from The Royal Society of Chemistry. Courtesy of Tokyo instruments; sample courtesy of Kanamura Laboratory, Faculty of Urban Environmental Sciences, Tokyo Metropolitan University).*

(A)

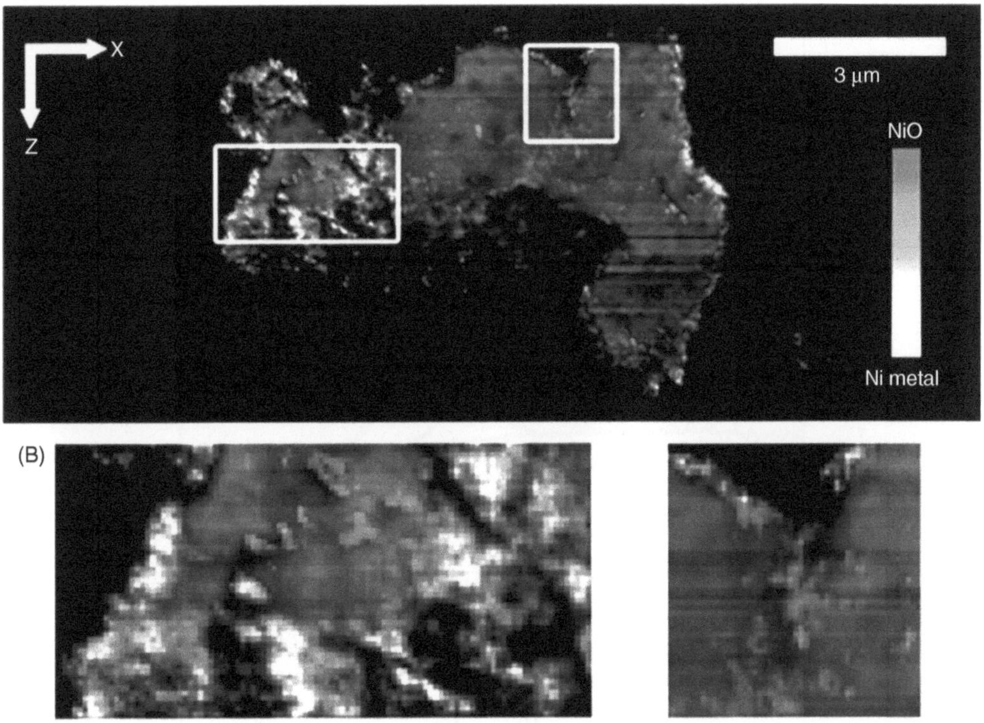

Figure 5.5 Reconstructed 3D XANES tomography data of a partially reduced Li-ion NiO electrode (A). *Bright* features correspond to metallic Ni, *dark* ones to NiO. The regions marked by *white rectangles* are also shown zoomed in (B). The voxels in these regions correspond to the lateral resolution limit of 50 nm. *(Reproduced from J.C. Andrews, B.M. Weckhuysen, Hard X-ray spectroscopic nano-imaging of hierarchical functional materials at work, ChemPhysChem 14 (2013) 3655–3666 [10] with permission from John Wiley & Sons Inc., Copyright © 2013).*

LiO, C, O, Cl, Ni, and Sn from an alloy anode, as a function of sputter time (electrode thickness) and cycling procedures [28].

X-ray photoelectron spectroscopy (XPS) is one of the most popular chemical analysis methods [it is also called electron spectroscopy for chemical analysis (ESCA)] [6]. It usually requires vacuum, and is very sensitive only to the surface—the emitted electrons have very low energies down to the eV range. Since surfaces transferred from experiments in cells always contaminate in air, a few nm (or more) of material are first removed by ion sputtering. The spectra are thus interpreted as stemming from the freshly exposed surface of the sample. A rough overview spectrum with all element-specific emission peaks is recorded, usually followed by detailed scans over each emission peak, which provides evidence for presence or absence of compounds. The peaks are evaluated by fitting curves. This procedure is crucial since excellent fits are nearly unique, and allow for quantitative detection and chemical analysis in terms of oxidation states (Fig. 5.6).

Neutron scattering or reflectometry [6] records structural changes in 1D (specular, z axis, Fig. 5.1). It can be used to track film formation and volume changes, usually in batteries

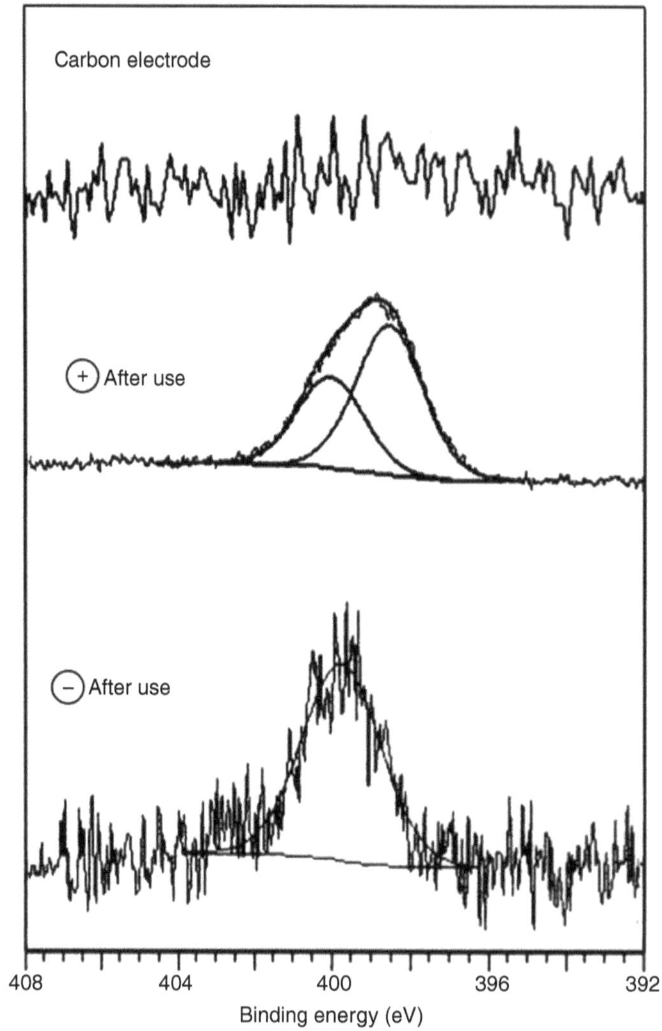

Figure 5.6 ***XPS of nitrogen species formed by long-term "aging" processes in an EDLC (postmortem analysis).*** Before use, the carbon electrode shows only noise (no nitrogen). After use as negative electrode, traces of nitrogen (above the noise level) are found, assigned to amides (400.2 eV), which form in the electrolyte. At the positive electrode, in addition to amides, also amines (398.8 eV) are detected, formed from the oxidized electrolyte cation (tetraethyl ammonia). The electrodes were carefully rinsed, so the nitrogen is covalently linked. *(Reproduced from A.M. Bittner, M. Zhu, Y. Yang, H.F. Waibel, M. Konuma, U. Starke, C.J. Weber, Ageing of electrochemical double layer capacitors, J. Power Sources 203 (2012) 262–273 [22] with permission from Elsevier, Copyright © 2012).*

[3]. The requirements are similarly exigent as for neutron diffraction (given earlier), and the in situ compatibility is very high.

6 ELECTRON MICROSCOPY

Electron microscopy is the most important electron in/electron out method (in addition X-ray emission can be detected for EDS/WDS), working on a surface, and in combination with ion beam sputtering also in 3D. It is used to characterize morphology, and to determine particle size or shape. The technique is extremely widespread (it was the first real nanoscale method), and it has become a standard procedure for a huge range of samples, from viruses to construction materials. It offers ultimate resolution down to the very low nm range and to single atoms. A typical tool includes electron beam, electron column with electromagnetic lenses, specimen holder, and a set of detectors. Electron microscopy methods usually require a good vacuum, and thus quite often special sample preparation. There are two fundamentally different geometries: transmission and scanning microscopies (Fig. 5.7) [29,30].

6.1 SEM

Scanning electron microscopes (SEM) have an electron beam of 1–40 keV, focused by one or two e-column lenses to a spot less than 1 nm in diameter [30]. The beam passes a scanning system, which deflects the beam in x and y directions, scanning the surface of the sample (Figs. 5.1 and 5.7B).

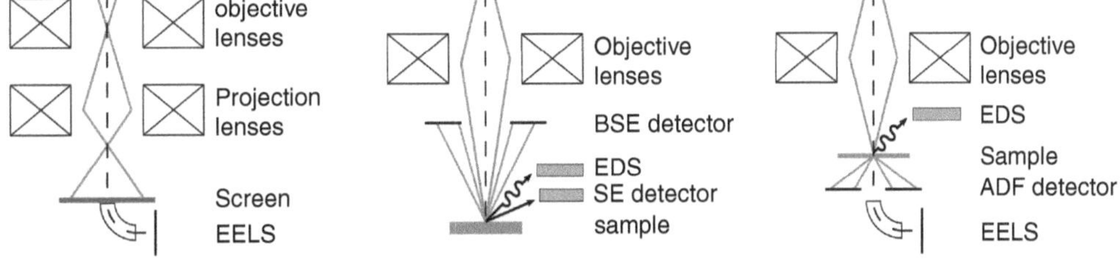

Figure 5.7 ***Schematic representation of different electron microscopy geometries.*** (A) Transmission electron microscopy (TEM), (B) scanning electron microscopy (SEM), and (C) transmission scanning electron microscopy (STEM).

When the primary electron beam enters the sample, the electrons are scattered and absorbed in a region called "interaction volume," which has a droplet shape and a typical size of ~1 μm. Interaction of the primary beam with the sample results in electron elastic scattering, emission of secondary electrons by inelastic scattering, and X-ray photon emission, each of which can be detected by specialized detectors. Beam scanning is synchronized with an image display; so, the resulting image is a distribution of the intensity of the signal, emitted from the sample. SEM has, strictly speaking, no magnification because it is a scanning technique, so field of view and resulting image are controlled by beam size and positioning precision on the sample. The magnification can be determined as the ratio of displayed image size to scanned sample area.

Modern SEM setups often come with different features, such as:

- focused ion beam columns, using the beams of Ga or He ions, providing higher depth of focus and ability to cut the sample in a precise way;
- backscattered electron detector, which increases the contrast for heavy atoms;
- energy-dispersive X-ray spectroscopy (EDS) and/or wavelength dispersive X-ray spectroscopy (WDS) detectors, which allow identification of chemical elements in the sample;
- electron backscattered diffraction (EBSD) add-on, which allows for crystal phase identification and works on the diffraction principle;
- scanning transmission electron spectroscopy (STEM) detector (given later), which enables a microscope to work in the transmitting regime;
- environmental regime, which allows for electron microscopy at higher pressures, which is important for in situ measurements and nonconducting samples studies.

The narrow beam allows SEM to identify chemical elements and phases on the nm scale, enabling mapping on the samples and investigation of local composition or inhomogeneity.

6.2 TEM

The *transmission electron microscope* (TEM) works on the same principle as an optical one, but, owing to the small de Broglie wavelength of electrons, the magnification is significantly higher, however, the resolution is limited by different aberration processes [29]. Recently, TEM resolution was greatly enhanced, with the development of chromatic aberration correctors, which enable an operator to examine fine details, as small as single molecules, atoms, or vacancies [so-called high resolution (HR) TEM]. TEM has a typical electron energy range from 40 up to 300 keV. The beam diameter is controlled by e-column lenses and apertures (Fig. 5.7) and determines the area observed in the image. Magnification is determined by projection lenses located below the sample. The resulting image is formed on a phosphorous screen or a charge-coupled detector. The technique requires a lot of sample preparation, which normally includes either dispersion of the sample or its thinning with an ion beam in SEM. With the progress of membrane preparation (including graphene), different techniques, which enable electrochemical processes with in situ TEM, are being developed [31].

Most of the TEMs can also work in the *scanning transmission electron microscopy mode* (STEM). In this mode, the beam is focused to a narrow diameter and scans over the sample, and the transmitted signal is collected. Due to the small diameter of the beam, this mode enables chemical element mapping either with EDS or with electron energy loss spectroscopy (EELS).

6.3 Application of SEM to materials characterization

In battery research, SEM, as a method of materials characterization, gives information on the level of individual particles, thus the method is used to study particle morphology and size distribution, investigation of films or coatings (e.g., SEI) formation, uniformity and smoothness, dendrite formation, electrode microstructure, impurities, ion diffusion, contamination, and so forth.

Intensive development in the field of Li-ion batteries requires reliable and safe products with high energy density. The most promising and dense anode is metallic lithium. But dendrite growth, which is an issue for anode materials, is more pronounced for the materials containing elemental metal [14]. Dendrite growth threatens the safety of batteries by piercing an ion-transporting separator between the cathode and anode. The problem of dendrite growth is widely studied with SEM. Various approaches were reported to inhibit dendrite formation, involving coatings, producing uniform Li^+ flux over the surface with tailored separators or 3D patterning of Li anode surfaces. All of these approaches require characterization of electrode and separator morphology and composition, done with SEM.

The morphology of materials also has a high impact on its electrical performance. Particle size influences electric conductivity, Li^+ diffusion rate and material intercalation performance. Typically, nanosized materials show excellent high C-rate capability, but can have worse long-term stability, while micron-sized particles show higher long-term stability with lower C-rate capacity [32]. The approach to obtain better stability and capacity with the material morphology tuning requires characterization of single nanoparticle shape, which is usually done with SEM and TEM.

In situ SEM microscopy is typically used in battery research to study coating formation and SEI growth. There are commercial cells available on the market. A typical in situ SEM cell probe requires either environmental SEM or ionic liquid electrolyte, which do not evaporate even in ultrahigh vacuum [33,34]. Since imaging is done with an electron beam under high electron surface density, it can cause high charging in regions under observation, so it is important to select regions of high electronic conductivity, for example, near the current collector.

SEM is highly demanded for metal–oxygen battery research (e.g., Li-O_2, Zn-O_2), since porosity, morphology, and microstructure of the electrode materials determine penetration rate and availability of oxygen molecules, which are reversibly transformed in the battery.

6.4 Application of TEM to materials characterization

TEM is a routine method to characterize morphology and crystallinity of the energy storage materials, such as cathodes, and to study thin layer coatings, applied on the surface of electrode materials.

Thin coatings play a significant role in passivation of electrode material surfaces, preventing capacity loss with cycling (AlF_3, Al_2O_3, TiO_2), and increase of electronic conductivity (carbon). Concentration gradient structures have also been reported to effectively prevent degradation. TEM is very useful to study local phase transitions and reactions, occurring on the coating interface, and understanding its influence on the properties of the resulting material. In situ TEM studies of phase transformations under Li intercalation involving not only imaging, but also ED, EDS, and EELS measurements, can reveal phase and morphology changes, chemical reactions, occurring during charge–discharge process and allow identification of the reaction products [35].

While there are some works on (S)TEM measurement of pore structure, these methods are mostly applicable for ceramics materials, such as silica and silicates. In this case, the pores are referred either as a part of crystal structure of ordered mesoporous materials [36], or as imperfections [37]. The utilization of (S)TEM images for porosity measurements in carbon-based materials is quite demanding, since the microscopy is a very local 2D technique, the study should involve more than one analysis method, and the result strongly depends on image postprocessing [38]. The reliability of such techniques is still ambiguous, since it is a well-known fact that carbon materials show damage and volatility under electron beams [39].

High performance electrode material design requires the knowledge of active or reactive sites and homogeneity of the samples, on different stages of the charge–discharge process. The location of such sites can be studied with TEM. For example, one could imagine that Li should penetrate inside carbon nanotubes from their ends, but in situ TEM studies show that Li intercalation results in bead-like structures, meaning that Li atoms diffuse through the carbon nanotube walls [40]. The samples before and after intercalation have similar morphologies and diameter distributions, meaning that the process of Li intercalation into carbon nanotubes is fully reversible.

Kinetics of the processes occurring in electrochemical cells can be also studied with in situ TEM. Many in situ experiments are targeting the SEI: the process of its formation, its position and its transformation with the number of cycles all play key roles for the development of new electrode materials, such as silicon. A special design of electrode materials can result in SEI formation on the surface of material coatings, but not on the active phase itself. The example of a pomegranate-like material, made from silicon encapsulated into carbon spheres, is presented in Fig. 5.8. In situ TEM studies of such types of materials provide information on SEI formation, lithiation kinetics and spare volume left inside the carbon shell [41].

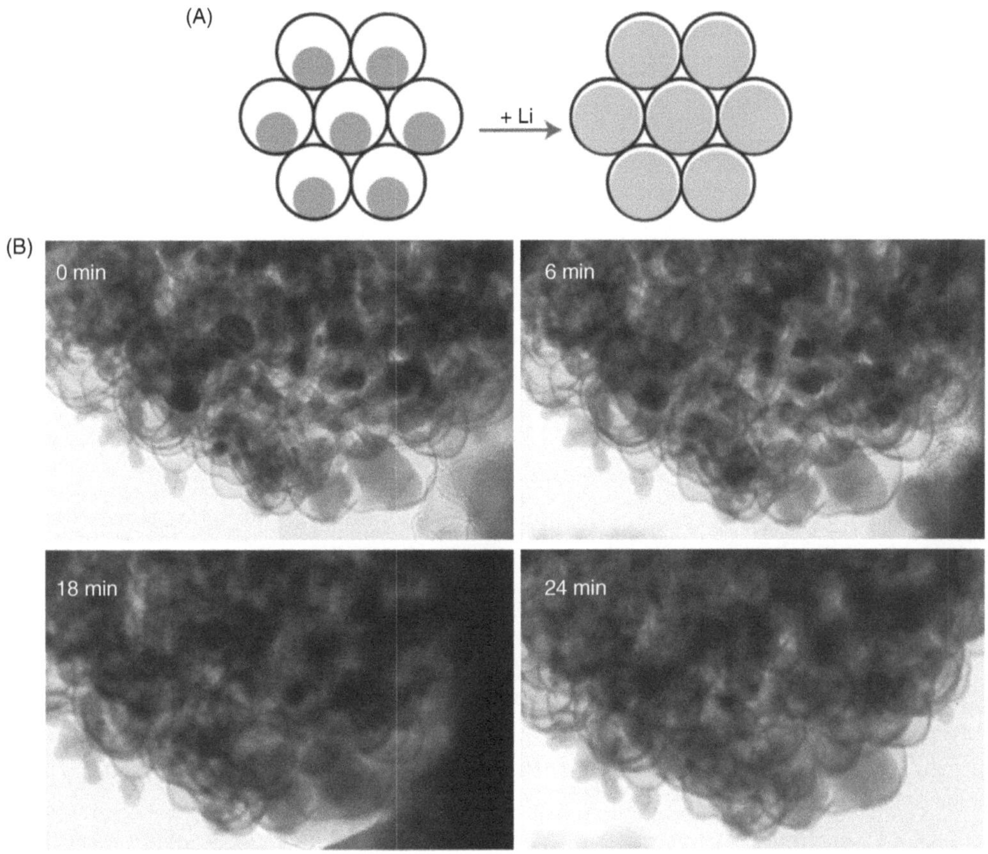

Figure 5.8 (A) Schematic representation of lithiation process of carbon-encapsulated silicon; (B) TEM images of time-dependent lithiation of Si nanoparticles in carbon shell. *(Reproduced from N. Liu, Z. Lu, J. Zhao, M.T. McDowell, H.-W. Lee, W. Zhao, Y. Cui, A pomegranate-inspired nanoscale design for large-volume-change lithium battery anodes, Nat. Nanotechnol. 9 (2014) 187–192 [41] with permission from Nature Publishing Group, Copyright © 2014).*

The further development of TEM techniques includes, for example, electrochemical cells with graphene windows, enabling researchers to detect single atoms and the atomic structure of single particles in situ, during electrochemical charge–discharge processes, under liquid electrolyte flow through the cell. Another recently introduced approach is differentially pumped environmental TEM, enabling atomic scale imaging with continuous gas flow in the sample area that, in combination with special sample holders, allows direct observation of the processes occurring in Li-air batteries or fuel cells.

7 IMPROVED INSTRUMENTATION AND INSPIRATIONS FOR NEW METHODS

7.1 New developments for standard techniques

Porosimetry, as mentioned in Section 3, is probably the most important method to determine pore sizes. However, ideally one would use a gas that mimics the electrolyte. The use of a variety of gases is a first step: the standard nitrogen is now complemented by argon, carbon dioxide, and hydrogen [17]. Differences in volume–pressure data of the various gases can be traced to the properties of the pores, which are otherwise very hard to determine. Porosimetry with liquid solvents is not common, with the exception of water and—for larger pores—mercury [17], both of which are of very limited use for energy storage devices.

Neutron depth profiling is a new spectroscopic method [3], which stands out for its in situ capacity. The spatial resolution in the beam direction can be better than 100 nm. The technique is based on the otherwise unwanted absorption of low-energy neutrons by ^{6}Li, which makes up some of the percentage of Li in the sample. The absorption is followed by fission to ^{4}He and ^{3}H. Both nuclei carry a high kinetic energy, which is progressively reduced when they have to travel through the sample.

NMR in situ and *NMR imaging* are very established standards for chemical analysis and for soft matter tomography, respectively. The problems of applying these techniques in situ have been overcome by ^{7}Li NMR imaging in model cells, but only on the mm scale [1]. Note that ^{7}Li is the main isotope.

SAXS is technically similar to XRD, but works at small angles, such that it is sensitive for distances of several nm. It requires either sophisticated tools or a synchrotron, but it can provide information on typical distances in materials with large lattice spacings, or materials with incorporated nanoparticles or pores. The in situ capability is very good [42], but technical problems, for example, the small angle between beam and surface, have apparently impeded its use for energy materials. *SANS* is the neutron analogue [43], which has huge technical requirements, similar to neutron diffraction (mentioned previously). The method is very well established especially for soft matter. Neither SAXS or SANS are new, but improving the sensitivity has lead to increasing usefulness for the electrodes, especially those based on light elements (lithium, carbon), and on small pores [43].

7.2 Inspirations from surface science techniques

The following methods were as yet not or only rarely used for the characterization of energy storage materials. They hint at new ways to reach ever higher resolution and/or specificity.

Scanning tunneling microscopy (STM) is the first scanning probe method developed for the nanoscale. It has a very successful and long history of in situ application on metal/electrolyte interfaces. As for AFM, the biggest obstacle for its use is that real materials are

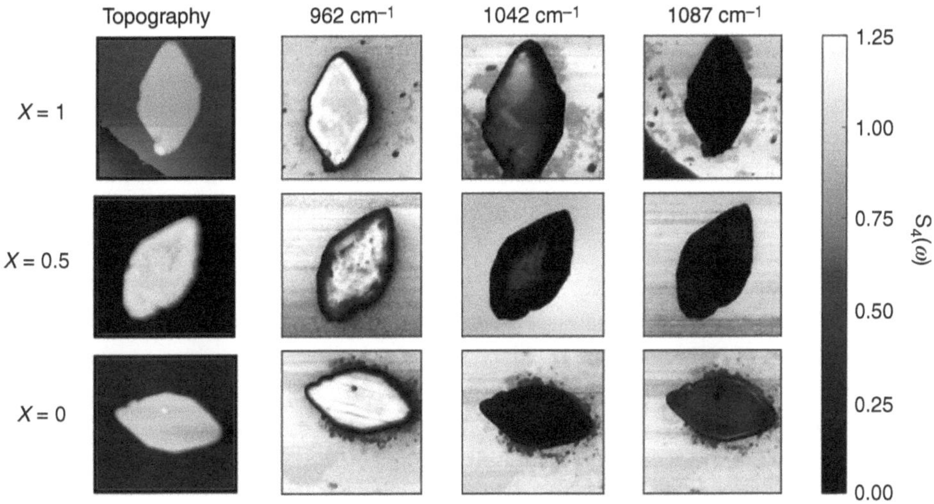

Figure 5.9 ***AFM topography and corresponding infrared near-field amplitude S4 (4th harmonic) images acquired in an AFM-FTIR-SNOM at 1087, 1042, and 962 cm^{-1} for $LiFePO_4$, $Li_{0.5}FePO_4$, and $FePO_4$ crystals, thus probing different vibrations.*** *(Reproduced from I.T. Lucas, A.S. McLeod, J.S. Syzdek, D.S. Middlemiss, C.P. Grey, D.N. Basov, R. Kostecki, IR near-field spectroscopy and imaging of single Li_xFePO_4 microcrystals, Nano Lett. 15 (2015) 1–7 [44] with permission from American Chemical Society, Copyright © 2015).*

insufficiently flat, so one would have to choose models as for AFM [24]. This makes sense only if lateral atomic resolution is required.

Based on AFM, various *scanning near field optical microscopes* (SNOM) have been designed. The common feature is a photon process with local intensity enhancement at the sharp probe tip. SNOMs thus are capable of local spectroscopy. They work at various wavelengths from infrared to the visible range, and include Raman spectroscopy. In addition to the AFM problems (given earlier), a SNOM requires sophisticated optics, which are often not compatible with in situ measurements. However, the high spatial resolution (down to the low nm range), and the wealth of spectroscopic information, suggest its use, see Fig. 5.9 for an example with IR.

Helium (and neon) ion microscopy is technically similar to SEM, but uses a beam of accelerated positive ions, created at an atomically sharp tip. The advantages include a spatial resolution that can be better than that of SEMs, low beam damage (when a neutralizing electron cloud is applied), and very high image depth (due to the point source). All could be favorable to image pores in unprecedented detail; however, the method requires a vacuum.

The *"nano-ESCA" XPS technique* [45] can be carried out on nanoscale areas on a surface when the emitted electrons are analyzed with a local probe. This method is still under development, but could be very promising to detect highly localized surface chemistries and defects, which are the essential ingredients of real nanoscale materials.

8 SUMMARY

Energy materials contain in most cases, even for well-established products, nanoscale structures. Quite often, a very good characterization can be achieved with standard macroscopic methods. However, constant improvements in design and analysis increasingly require specific nanoscale methods. Some are very well developed standards (electron microscopy), but new developments offer excellent chemical information and/or in situ capabilities, even up to an analysis of a complete battery product. The disadvantages are mainly a very high experimental effort, up to synchrotrons (for X-rays) and nuclear reactors (for neutrons). The holy grail is imaging nanomaterials with nanoscale structural and chemical resolution, thus visualizing and analyzing each single feature. As yet, electron microscopy is the only established method that can achieve this. X-ray microscopy and neutron methods, with their huge in situ capabilities, will almost certainly catch up, and SNOM might be the future.

GLOSSARY

AFM	Atomic force microscopy
BET	Brunauer–Emmett–Teller (isotherm for determining the internal area of a porous solid)
EBSD	Electron backscattered diffraction
ED	Electron diffraction
EDLC	Electrochemical double layer capacitor
EDS	Energy-dispersive X-ray spectroscopy
EELS	Electron energy loss spectroscopy
EM	Electron microscopy
ESCA	Electron spectroscopy for chemical analysis (XPS)
EXAFS	Extended X-ray absorption fine structure
(FT)IR	(Fourier transform) infrared
HR	High resolution
HR TEM	High-resolution TEM
LiB	Lithium ion battery
NMR	Nuclear magnetic resonance
SANS	Small angle neutron scattering
SAXS	Small angle X-ray scattering
SECM	Scanning electrochemical microscopy
SEI	Solid electrolyte interface
SEM	Scanning electron microscopy (or microscope)
SNOM	Scanning near-field optical microscopy
STM	Scanning tunneling microscopy
TEM	Transmission electron microscopy (microscope)
(TOF)SIMS	(Time of flight) secondary ion mass spectroscopy
UV–vis	Ultraviolet and visible
XANES	X-ray absorption near edge spectroscopy

XAS	X-ray absorption spectroscopy
XPS	X-ray photoelectron spectroscopy (ESCA)
XRD	X-ray diffraction
WDS	Wavelength dispersive X-ray spectroscopy

REFERENCES

[1] M.-K. Song, S. Park, F.M. Alamgir, J. Cho, M. Liu, Nanostructured electrodes for lithium-ion and lithium-air batteries: the latest developments, challenges, and perspectives, Mater. Sci. Eng. R Rep. 72 (2011) 203–252.

[2] P. Novák, D. Goers, L. Hardwick, M. Holzapfel, W. Scheifele, J. Ufheil, A. Würsig, Advanced in situ characterization methods applied to carbonaceous materials, J. Power Sources 146 (2005) 15–20.

[3] P.P.R.M.L. Harks, F.M. Mulder, P.H.L. Notten, In situ methods for Li-ion battery research: a review of recent developments, J. Power Sources 288 (2015) 92–105.

[4] C.P. Grey, N. Dupré, NMR studies of cathode materials for lithium-ion rechargeable batteries, Chem. Rev. 104 (2004) 4493–4512.

[5] A.I. Vogel, G.H. Jeffery, Vogel's Textbook of Quantitative Chemical Analysis, Longman Scientific & Technical, Harlow, UK, (1989).

[6] K. Urban, Probing the Nanoworld: Microscopies, Scattering and Spectroscopies of the Solid State, Forschungzentrum Jülich, Jülich, Germany, (2007).

[7] G.W.H. Höhne, W.F. Hemminger, H.-J. Flammersheim, Differential Scanning Calorimetry, Springer Science Mathplus Business Media, Berlin, Germany, (2003).

[8] A.J. Bard, L.R. Faulkner, Electrochemical Methods: Fundamentals and Applications, John Wiley & Sons, New York, NY, (1980).

[9] F.S. Gittleson, K.P.C. Yao, D.G. Kwabi, S.Y. Sayed, W.-H. Ryu, Y. Shao-Horn, A.D. Taylor, Raman spectroscopy in lithium–oxygen battery systems, ChemElectroChem 2 (2015) 1446–1457.

[10] J.C. Andrews, B.M. Weckhuysen, Hard X-ray spectroscopic nano-imaging of hierarchical functional materials at work, Chem. Phys. Chem. 14 (2013) 3655–3666.

[11] J.N. Weker, M.F. Toney, Emerging in situ and operando nanoscale X-ray imaging techniques for energy storage materials, Adv. Funct. Mater. 25 (2015) 1622–1637.

[12] Y.K. Chen-Wiegart, Z. Liu, K.T. Faber, S.A. Barnett, J. Wang, 3D analysis of a $LiCoO_2$–$Li(Ni_{1/3}Mn_{1/3}Co_{1/3})O_2$ Li-ion battery positive electrode using X-ray nano-tomography, Electrochem. Commun. 28 (2013) 127–130.

[13] J. Ayache, L. Beaunier, J. Boumendil, G. Ehret, D. Laub, Sample Preparation Handbook for Transmission Electron Microscopy, Springer, New York, NY, (2010).

[14] J. Li, C. Ma, M. Chi, C. Liang, N.J. Dudney, Solid electrolyte: the key for high-voltage lithium batteries, Adv. Energy Mater. 5 (2014) 1401408.

[15] L. Simonin, J.-F. Colin, V. Ranieri, E. Canévet, J.-F. Martin, C. Bourbon, C. Baehtz, P. Strobel, L. Daniel, S. Patoux, In situ investigations of a Li-rich Mn–Ni layered oxide for Li-ion batteries, J. Mater. Chem. 22 (2012) 11316.

[16] N. Sharma, V.K. Peterson, M.M. Elcombe, M. Avdeev, A.J. Studer, N. Blagojevic, R. Yusoff, N. Kamarulzaman, Structural changes in a commercial lithium-ion battery during electrochemical cycling: an in situ neutron diffraction study, J. Power Sources 195 (2010) 8258–8266.

[17] G.N. Okolo, R.C. Everson, H.W.J.P. Neomagus, M.J. Roberts, R. Sakurovs, Comparing the porosity and surface areas of coal as measured by gas adsorption, mercury intrusion, and SAXS techniques, Fuel 141 (2015) 293–304.

[18] P. Atkins, J. de Paula, Atkins' Physical Chemistry, OUP, Oxford, UK, (2010).

[19] M. Zarrabeitia, E. Castillo-Martınez, J.M.L.D. Amo, A. Eguıa-Barrio, M.Á. Muñoz-Márquez, T. Rojo, M. Casas-Cabanas, Identification of the critical synthesis parameters for enhanced cycling stability of Na-ion anode material $Na_2Ti_3O_7$, Acta Mater. 104 (2016) 125–130.

[20] N.M. Trease, L. Zhou, H.J. Chang, B.Y. Zhu, C.P. Grey, In situ NMR of lithium ion batteries: bulk susceptibility effects and practical considerations, Solid State Nucl. Magn. Reson. 42 (2012) 62–70.

[21] J.D.W. James, A. De Haseth, P.R. Griffiths, Fourier Transform Infrared Spectrometry, John Wiley & Sons, New York, NY, (2007).

[22] A.M. Bittner, M. Zhu, Y. Yang, H.F. Waibel, M. Konuma, U. Starke, C.J. Weber, Ageing of electrochemical double layer capacitors, J. Power Sources 203 (2012) 262–273.

[23] M. Bomio, P. Lavela, J.L. Tirado, ^{57}Fe Mössbauer spectroscopy and electron microscopy study of metal extraction from $CuFe_2O_4$ electrodes in lithium cells, Chem. Phys. Chem. 8 (2007) 1999–2007.

[24] S.-K. Jeong, M. Inaba, Y. Iriyama, T. Abe, Z. Ogumi, Surface film formation on a graphite negative electrode in lithium-ion batteries: AFM study on the effects of co-solvents in ethylene carbonate-based solutions, Electrochim. Acta 47 (2002) 1975–1982.

[25] R. Hiesgen, S. Sörgel, R. Costa, L. Carlé, I. Galm, N. Cañas, B. Pascucci, K.A. Friedrich, AFM as an analysis tool for high-capacity sulfur cathodes for Li–S batteries, Beilstein J. Nanotechnol. 4 (2013) 611–624.

[26] H. Bülter, F. Peters, J. Schwenzel, G. Wittstock, Spatiotemporal changes of the solid electrolyte interphase in lithium-ion batteries detected by scanning electrochemical microscopy, Angew. Chem. Int. Ed. 53 (2014) 10531–10535.

[27] Tokyo Instruments (Japan), Available from: http://www.tokyoinst.co.jp/en/products/detail/NF02/index.html

[28] J.-T. Li, J. Swiatowska, V. Maurice, A. Seyeux, L. Huang, S.-G. Sun, P. Marcus, XPS and ToF-SIMS study of electrode processes on Sn–Ni alloy anodes for Li-ion batteries, J. Phys. Chem. C 115 (2011) 7012–7018.

[29] C.B.C. David, B. Williams, Transmission Electron Microscopy, Springer, New York, NY, (2008).

[30] J.I. Goldstein, P.E. Gaithersburg, D.E. Newbury, D.C. Joy, C.E. Lyman, E. Lifshin, L. Sawyer, J.R. Michaev, Scanning Electron Microscopy and X-ray Microanalysis, Springer, US, (2003).

[31] G. Zheng, S.W. Lee, Z. Liang, H.-W. Lee, K. Yan, H. Yao, H. Wang, W. Li, S. Chu, Y. Cui, Interconnected hollow carbon nanospheres for stable lithium metal anodes, Nat. Nanotechnol. 9 (2014) 618–623.

[32] H. Liu, J. Wang, X. Zhang, D. Zhou, X. Qi, B. Qiu, J. Fang, R. Kloepsch, G. Schumacher, Z. Liu, J. Li, Morphological evolution of high-voltage spinel $LiNi_{0.5}Mn_{1.5}O_4$ cathode materials for lithium-ion batteries: the critical effects of surface orientations and particle size, ACS Appl. Mater. Interfaces 8 (2016) 4661–4675.

[33] D. Chen, S. Indris, M. Schulz, B. Gamer, R. Mönig, In situ scanning electron microscopy on lithium-ion battery electrodes using an ionic liquid, J. Power Sources 196 (2011) 6382–6387.

[34] P.R. Raimann, N.S. Hochgatterer, C. Korepp, K.C. Möller, M. Winter, H. Schröttner, F. Hofer, J.O. Besenhard, Monitoring dynamics of electrode reactions in Li-ion batteries by in situ ESEM, Ionics 12 (2006) 253–255.

[35] C.-M. Wang, W. Xu, J. Liu, J.-G. Zhang, L.V. Saraf, B.W. Arey, D. Choi, Z.-G. Yang, J. Xiao, S. Thevuthasan, D.R. Baer, In situ transmission electron microscopy observation of microstructure and phase evolution in an SnO_2 nanowire during lithium intercalation, Nano Lett. 11 (2011) 1874–1880.

[36] I. Diaz, A. Mayoral, TEM studies of zeolites and ordered mesoporous materials, Micron 42 (2011) 512–527.

[37] A. Mayoral, J. Coronas, C. Casado, C. Tellez, I. Dıaz, Atomic resolution analysis of microporous titanosilicate ETS-10 through aberration corrected STEM imaging, Chem. Cat. Chem. 5 (2013) 2595–2598.

[38] L. He, Y.B. Melnichenko, N.C. Gallego, C.I. Contescu, J. Guo, J. Bahadur, Investigation of morphology and hydrogen adsorption capacity of disordered carbons, Carbon 80 (2014) 82–90.

[39] J. Leschner, J. Biskupek, A. Chuvilin, U. Kaiser, Accessing the local three-dimensional structure of carbon materials sensitive to an electron beam, Carbon 48 (2010) 4042–4048.

[40] G. Maurin, C. Bousquet, F. Henn, P. Bernier, R. Almairac, B. Simon, Electrochemical intercalation of lithium into multiwall carbon nanotubes, Chem. Phys. Lett. 312 (1999) 14–18.

[41] N. Liu, Z. Lu, J. Zhao, M.T. McDowell, H.-W. Lee, W. Zhao, Y. Cui, A pomegranate-inspired nanoscale design for large-volume-change lithium battery anodes, Nat. Nanotechnol. 9 (2014) 187–192.

[42] G. Sandı, H. Joachin, R. Kizilel, S. Seifert, K.A. Carrado, In situ SAXS studies of the structural changes of polymer nanocomposites used in battery applications, Chem. Mater. 15 (2003) 838–843.

[43] H. Kurig, M. Russina, I. Tallo, M. Siebenbürger, T. Romann, E. Lust, The suitability of infinite slit-shaped pore model to describe the pores in highly porous carbon materials, Carbon 100 (2016) 617–624.
[44] I.T. Lucas, A.S. McLeod, J.S. Syzdek, D.S. Middlemiss, C.P. Grey, D.N. Basov, R. Kostecki, IR near-field spectroscopy and imaging of single Li_xFePO_4 microcrystals, Nano Lett. 15 (2015) 1–7.
[45] C. Wiemann, M. Patt, S. Cramm, M. Escher, M. Merkel, A. Gloskovskii, S. Thiess, W. Drube, C.M. Schneider, Probing buried layers by photoelectron spectromicroscopy with hard X-ray excitation, Appl. Phys. Lett. 100 (2012) 223106.

CHAPTER SIX

Electrochemical–Thermal Characterization and Thermal Modeling for Batteries

C. Ziebert, A. Melcher, B. Lei, W. Zhao, M. Rohde, H.J. Seifert
Karlsruhe Institute of Technology, Institute of Applied Materials—Applied Materials Physics, Eggenstein-Leopoldshafen, Germany

Contents

1 INTRODUCTION

Lithium-ion batteries (LIBs) have found a wide range of applications in the past three decades, like cell phones, tablet or laptop computers, and more recently their use in battery electric vehicles (BEV) or hybrid electric vehicles (HEV). The size of the batteries, as well as the capacity differ with the application. The use of LIBs in an BEV or HEV requires other batteries and battery configurations as in a cell phone. Therefore, physical phenomena occur in a wide range of time and length scales, from the atomic level up to the macroscopic level. Thermal behavior in particular is closely related to the context of the safety of battery systems.

Emerging Nanotechnologies in Rechargeable Energy Storage Systems

The thermal behavior of LIBs and their active materials depend on a large variety of internal and environmental physicochemical parameters. Additionally, the specific materials design and especially the use of nanoscale materials influences heat generation and heat dissipation during operation of the electrochemical cells. Calorimetry of the electrochemical cells, and their individual active materials is required to obtain quantitative thermal and thermodynamic data. The two most important experimental calorimetric devices, namely the isothermal heat conduction calorimeter (IHC) and the accelerating rate calorimeter (ARC), will be described in this chapter. It will be shown that an ARC can be used for in-operando thermal investigation of lithium-ion cells and battery packs during electrochemical cycling under isoperibolic and adiabatic environmental conditions, respectively. The results for cylindrical as well as prismatic-type pouch cell LIBs indicate a significant rise in surface temperatures under adiabatic conditions. Additionally, the careful calibration of the ARC allows the quantification of heat generation and dissipation in different operation modes. Moreover, it will be described how thermal runaway behavior and the related reaction mechanisms of the cells and materials can be investigated quantitatively.

One central point in the operation of LIBs is the aspect of cell safety, which corresponds to thermal stability. Several exothermic reactions can occur inside a cell during operation as its inner temperature is increasing. If the heat generation is larger than the dissipated heat to the surrounding space, this may cause heat to accumulate inside the cell and the chemical reactions will be accelerated, which leads to a further temperature increase, until a thermal runaway is reached. In these terms, a thermal runaway describes a rapid temperature increase in a very short time interval of the typical order of 100K/min and above [1].

Beyond the experiments, applying an approach based on mathematical modeling and the corresponding simulations of the battery physics promises a deeper theoretical understanding of the occurring phenomena inside the LIBs. Due to the geometric structure of a battery, their behavior is strongly affected by the interaction of the physics on varying length and time scales. The development of an appropriate electrochemical model starts with the work of Newman [2–4] based on the porous electrode theory [5] describing the diffusion dynamics and the charge transfer kinetics of the battery. This model is able to predict the electric response of a cell in an intercalation electrode system. It is adequate in small battery systems, where the temperature distribution can be assumed as constant. In larger battery cell systems the uniformity of the electrical potential along the current collectors gets lost and the temperature distribution must also be taken into account.

Therefore, the electrochemical model must be extended with thermal contributions and with contributions taking the different time and length scales into account. First thermal models for LIBs are given by Newman and Pals [6–8] using an inhomogeneous heat equation, where the electrochemical heat generation is used as heat source. This approach is applied to simple cells as well as for whole battery packs [9–13]. Parallel to this, the multidomain and multiscale character of the electrochemical model [14,15] leads to a new approach based on the mathematical theory of homogenization and the spatial averaging theorem [16]. Newer attempts can be found, for example, in Refs. [17–20]. The coupling between the multiscale, multidomain electrochemical and thermal model is introduced by Refs. [21–23] and is under development until today [24–28].

The main disadvantage of the models described earlier is that they cannot capture the phenomenon of a thermal runaway. A first attempt to overcome this disadvantage is given in [29], in which additional source terms based on an Arrhenius-type law are introduced. Spotnitz et al. [30,31] gave a first PDE modeling of a thermal runaway including reaction kinetics. In [32–34] the electrochemical-thermal model is extended with reaction kinetics based on an ODE formulation. Some newer results can be found in [35–37], where the PhD thesis of Tanaka [37] offers detailed insight into the modeling and simulation of a thermal runaway for different chemistries.

2 HEAT GENERATION IN LITHIUM-ION BATTERIES

Heat means a non-material-bound energy that can be transported over system boundaries. Since the heat Q is a process variable, it can only occur during state changes. Bernardi et al. [38] were the first to give an accurate description of the heat generation in Lithium-based electrochemical systems. Thereafter, for the heat flow $\frac{dQ}{dt} = \dot{Q}$ the following energy balance can be formulated in the cell:

$$\dot{Q} = \dot{Q}_{\text{rev}} + \dot{Q}_{\text{irrev}} + \dot{Q}_{\text{reac}} + \dot{Q}_{\text{mix}} \tag{6.1}$$

where the different terms are:

- Reaction entropy (Q_{rev}): Due to the storage and removal of lithium ions into the host lattice a reversible heat effect ($T\Delta S$) is generated. It has the same absolute value for charging and discharging and changes its sign with the current direction.
- Ohmic losses (Q_{irrev}): Both the transport of lithium ions through the electrolyte and the electrode material, and the conduction of the electrons in the electrodes and the current conductors are leading to losses. The resulting heat is always dissipated and does not change its sign even when reversing the processes.
- Heat of reaction by side reactions (Q_{reac}): Side reactions of the electrolyte with the active material (e.g., SEI formation or aging) may also be a source of heat. However, for the thermal behavior of cells studied under normal conditions, this term can be neglected.
- Enthalpy of mixing (Q_{mix}): During the cell operation, a gradient of the lithium concentration forms in the electrolyte, which relaxes back into an equilibrium state upon interruption of the current flow. This is accompanied by a change in entropy, which in turn results in a heat flow. However, this proportion is so small that it can be ignored. Thus the Ohmic losses and the change in entropy represent the most relevant sources for the heat flow.

2.1 Reversible and irreversible heat

2.1.1 Reversible heat

The reversible part of heat generation is directly related to the absolute temperature of a system and is connected to the entropy:

$$Q_{\text{rev}} = T \cdot \Delta S. \tag{6.2}$$

Together the entropy change and the enthalpy change ΔH describe the change in the Gibbs free energy ΔG during the charging and discharging of a lithium-ion cell:

$$\Delta G = \Delta H - T \cdot \Delta S. \tag{6.3}$$

When discharging the cell, the chemical energy is converted into electrical energy, according to

$$\Delta G = -n \cdot F \cdot \Delta U. \tag{6.4}$$

Here, n is the number of exchanged electrons, F is the Faraday constant, and ΔU is the potential difference between anode and cathode. From Eqs. (6.3) and (6.4) can be deduced that for $\Delta S \neq 0$ heat is generated during changes in the state of charge (SOC). This gives for the reversible heat flow:

$$\dot{Q}_{\text{rev}} = \frac{d}{dt}(T \cdot \Delta S) = I \cdot T \cdot \frac{dU_{\text{eq}}}{dT}, \tag{6.5}$$

with

$$\Delta S = n \cdot F \cdot \frac{dU_{\text{eq}}}{dT}. \tag{6.6}$$

where I is the electric current of the cell and U_{eq} is the equilibrium voltage, or open circuit voltage (OCV), of the cell. This means that for negative ΔS the cell cools under charge, unless this effect is not superimposed by the heat dissipation generated by the Ohmic losses.

2.1.2 Irreversible heat

The irreversible heat production is the sum of two contributions. The first part is the Gibbs free energy ΔG that describes the maximum amount of chemical energy that can be converted into electrical energy. The second part is the electrical work W_{el} that the cell can provide.

$$\Delta G = -n \cdot F \cdot U_{\text{eq}}, \tag{6.7}$$

$$W_{\text{el}} = n \cdot F \cdot U. \tag{6.8}$$

By summing the two terms and the formation of the first derivative with respect to time, the heat loss generated by Joule effects can be described as:

$$\dot{Q}_{\mathrm{irrev}} = \frac{d}{dt}(\Delta G + W_{\mathrm{el}}) = I(U - U_{\mathrm{eq}}). \tag{6.9}$$

2.2 Abuse leading to thermal runaway

During charging/discharging, the temperature of a cell is increased by heat generation in the inner parts until an equilibrium state between heat generation and heat dissipation to the environment is reached [39]. Therefore the temperature change of a cell depends on:

- the heat generation in the cell and
- the heat dissipation to the environment.

The corresponding equation is

$$mc_{\mathrm{p}}\frac{dT}{dt} = \dot{Q}_{\mathrm{g}} - \dot{Q}_{\mathrm{d}}, \tag{6.10}$$

with $\dot{Q}_{\mathrm{g}}$ in J/s as the generated heat per unit of time, $\dot{Q}_{\mathrm{d}}$ in J/s as the heat dissipated per unit of time, m as the mass of the cell in kg, and c_{p} in J/K as the heat capacity of the cell. The dissipated heat increases with increasing temperature difference between the cell and its environment. At a certain temperature difference ΔT_{bal} a steady state is reached, for which the following equation holds:

$$\dot{Q}_{\mathrm{g}} = \dot{Q}_{\mathrm{d}}. \tag{6.11}$$

However, if the heat generated cannot be dissipated fast enough, that is, the heat generation in the cell grows faster than the heat dissipation, the cell temperature rises continuously as shown in Fig. 6.1 [40]. As a result, a thermal runaway will occur. A thermal runaway occurs also when it is not possible to release heat at all, that is, under adiabatic conditions.

Thermal runaway will happen at relative high temperatures so that the battery is self-heating until it catches fire or even explodes. The heat generation is caused by some parasitic exothermic reactions and not by the regular charge/discharge reactions. There are usually three stages in a thermal runaway process [41]:

- *Low rate runaway*: Immediately after the LIB thermal abuse, at 80–90°C the solid electrolyte interphase (SEI layer) in anode begins to decompose, which will lead to a temperature increase.
- *Stable rate runaway*: Without the SEI layer and at higher temperatures, the anode and the electrolyte react exothermally at about 140°C. The cathode material releases oxygen.
- *High rate runaway*: Self-heating, oxygen evolving reactions happen, and the temperature increases rapidly.

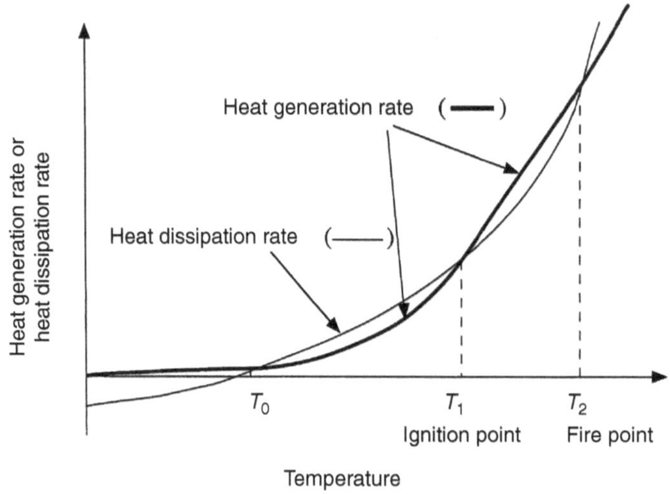

Figure 6.1 ***The balance between heat generation and dissipation rates which describes thermal runaway.*** *(Reprinted from S. Tobishima, J.-I. Yamaki, J. Power Sources 81–82 (1999) 882–886).*

3 ELECTROCHEMICAL-CALORIMETRIC MEASUREMENTS ON LITHIUM-ION BATTERIES

Quantitative measurements of the heat generation and dissipation in larger cells under regular conditions are necessary to understand the electrochemical and thermal processes in the cell and are important to design the thermal management system. According to the review article of Bandhauer et al. [42], the primary experimental equipment for this purpose are IHC and the accelerating rate calorimeter (ARC). The latter can be also used for safety tests where the cell is driven up to the thermal runaway because they are much more robust and especially designed to withstand even an explosion of the cell.

3.1 Isothermal heat conduction calorimetry

The IHC method uses a large heat sink (e.g., liquid or a metal heat sink) in close contact with the cell surface to keep the cell under isothermal conditions. Thus the battery remains at one temperature throughout operation. High accuracy thermopiles either placed near the surface of the cell or embedded inside the heat sink are then used to measure the heat rate. As in all calorimetric methods, special data processing or experimental procedures are usually required due to the long instrument time constants (i.e., time elapsed from heat generation to measurement [42]). In 1997, Saito et al. performed thermal studies of Sony US14500 cylindrical cells using a twin-type heat conduction calorimeter (Setaram C80-22) [43] and found a maximum heat generation rate of

0.03 W for a C/10 rate. Later, their comparison of 18650 cells from 4 different manufacturers at a 0.04 C rate [44] gave values between 0.03 and 0.04 W. Recently, they also compared the heat generation rates of self-assembled 18650 cells after long-time storage at different temperatures and state of charge [45]. The maximum observed heat generation rate for 1 C discharging was 0.085 W. Kobayashi et al. studied the heat generation rates of small prismatic NEC cells and Sony US18650 cells using a Calvet-type microcalorimeter (MMC5111-U) [46]. At a C/10 discharge rate, they reported maximum heat generation rates of 7 and 16 mW, respectively. Onda et al. used a thermostat-based setup to derive entropy changes [47] and a twin calorimeter to determine the specific heat capacity of Sony US18650G3 cells [48,49] and combined the results in an one-dimensional model of cell temperature distribution. At a discharge rate of 0.92 C, they measured a maximum heat generation rate of 1.4 W. From this literature review, it is clear that there exists a wide range of results for the heat generation rate, even for small cells of the same chemistry and form [50]. The previously mentioned articles all focused on the investigation of small cylindrical cells up to 18650 size or very small prismatic cells. According to Bandhauer et al., the influence of Joule heating, side reactions and of the nonuniform electrical distribution in the current collectors increases with increasing size of the cell [42]. Therefore only very recently, first reports that are dealing with calorimetric measurements of larger format pouch cells have been published [50–56]. An overview is given in Table 6.1. Xiao et al. [50] designed a new IHC calorimeter setup using two thermoelectric devices. They investigated 15.7 Ah C/$LiMn_2O_4$ pouch cells with respect to the reversible and irreversible heat generation. With these data they tried to express the heat generation in a new way by including enthalpy heating (depending on diffusion of lithium ions) and heat of mixing (caused by the change of the gradient of ion concentration). Chen et al. also developed a new IHC technique, which is shown schematically in Fig. 6.2A) and studied commercial 20 Ah C/$LiFePO_4$ pouch cells from A123 [51]. Their calorimeter is in the form of two identical slabs made of high-density polyethylene with known thermal properties that are attached to the back and front of the cell. The heat generated by the cell is conducted through the slabs to a constant temperature heat sink. Fig. 6.2B) shows as an example their measured effect of discharge rates on the heat generation rate of an A123 $LiFePO_4$ cell at an operating temperature of 20°C as a function of depth of discharge (varying from 0 to 1). An increase in battery heat generation throughout discharge at increased rates of discharge due to the higher irreversible heat generation is clearly visible. Reversible heat generation is a linear function of the current, whereas irreversible heat generation is a function of current squared; therefore increasing the discharge current causes a corresponding change in the proportion of the irreversible heat generation in the total heat generation of the battery. The total battery heat generation at 0.25°C is shown to be almost negligible when compared to the total amount of heat generated when the battery is discharged at 3 C at 20°C.

Table 6.1 Literature overview on calorimetric measurements of heat produced by large format pouch cells during charging and discharging at room temperature

	Cell type							Calorimeter					
			Dimensions										
			W	*L*	*T*								
Author	Manufacturer/ name	Capacity (Ah)	(mm)			Mass (g)	Materials (cathode/ anode)	Type	Mode	Max. heat generation at charge (W)	Max. heat gen. at discharge (W)	Total generated heat at charge (kJ)	Total gen. heat at discharge (kJ)
Xiao [50]	—	15.7	200	150	5.0	—	LMO/ carbon	IHC	Isothermal	4.1 C: 15	4.1 C: 75	—	—
										5.0 C: 22	5.0 C: 92	—	—
										5.7 C: 30	5.7 C: 100	—	—
										6.3 C: 30	6.3 C: 110	—	—
										7.0 C: 30	7.0 C: 122	—	—
Chen [51]	A123/AMP20 M1HD-A	20	227	160	7.2	496	LFP/ graphite	IHC	Isothermal	—	0.25 C: 0.87	—	—
										—	0.5 C: 3.32	—	—
										—	1.0 C: 4.92	—	—
										—	2.0 C:13.87	—	—
Abdul-Quadir [52]	Kokam/SLPB 100216216H	40	215	220	10.7	1100	NMC/ graphite	Adia-batic	Adiabatic	—	1.0 C: 11	—	—
										—	2.0 C: 33	—	—
Eddahech [53]	Kokam/SLPB 70205130P	12	220	132	6.5	327	NMC/ carbon	ARC	Isoperibolic	—	—	0.5 C: 0.332	0.5 C: 0.942
										—	—	1.0 C: 0.798	1.0 C: 1.897
										—	—	1.5 C: 1.485	1.5 C: 2.744
Nieto [54]	TENERGY/ 30123	10.5	59.5	157	9.8	210	LCO/ graphite	ARC	Adiabatic	—	1.0 C: 3.1	—	1.0 C: 5.5
										—	3.0 C: 14	—	3.0 C: 9.0
										—	5.0 C: 28	—	5.0 C: 11.5
Vertiz [55]	—	14	—	—	—	—	LFP/ graphite	ARC	—	—	0.5 C: 0.37	—	—
										—	1.0 C: 1.17	—	—
										—	3.0 C: 7.19	—	—
										—	5.0 C: 17.95	—	—

Schuster [56]	—	40	190	240	10.0	910	NMC/ graphite	ARC	Adiabatic	—	—	0.125 C:2.04	0.125 C:3.48
										—	—	0.25 C: 4.73	0.25 C: 4.47
										—	—	0.5 C: 7.88	0.5 C: 8.47
										—	—	1.0 C: 14.53	1.0 C: 14.86
									Isoperibolic	—	—	0.125 C:1.80	0.125 C:1.86
										—	—	0.25 C: 2.71	0.25 C: 4.45
										—	—	0.5 C: 7.53	0.5 C: 8.15
										—	—	1.0 C: 11.26	1.0 C: 13.40

Source: Reprinted from E. Schuster, C. Ziebert, A. Melcher, M. Rohde, H.J. Seifert, J. Power Sources 286 (2015) 580–589.

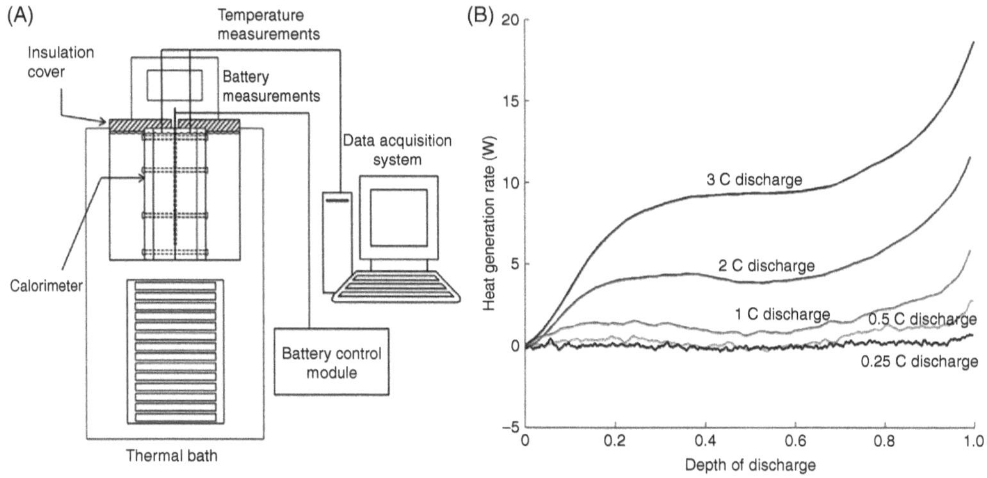

Figure 6.2 ***IHC.*** (A) Schematic of experimental setup, (B) effect of discharge rates on the heat generation rate of an A123 $LiFePO_4$ cell at an operating temperature of 20°C. *(Reprinted from K. Chen, G. Unsworth, X. Li, J. Power Sources 261 (2014) 28–37).*

3.2 Accelerating rate calorimetry

The ARC was first developed in the 1970s by Dow Chemical Company of Midland, Michigan, USA. It was designed to meet the potential needs for hazardous runaway reactions that may occur during the chemical processing, to detect and record the measurement values of time–temperature–pressure. Furthermore, an ARC provides a safe and adiabatic environment for operation and storage. In an adiabatic environment no heat exchange between the sample and the surroundings is possible, so that the heat energy of reactions can be monitored by the measured temperature change. In the ARC, the heat generation rate is determined by measuring the change in the cell surface temperature during charging and discharging. The principles of operation can be explained using the schematic setup of the EV-ARC calorimeter (made by Thermal Hazard Technology) in Fig. 6.3A. The calorimeter chamber with an inner diameter of 0.25 m and a height of 0.50 m has one heater and one thermocouple located in the lid and bottom, respectively, and two heaters and thermocouples (all type N) in the side wall. The cell under study is fixed on the lid of the calorimeter as shown in Fig. 6.3B. The calorimeter temperature is controlled by a main or so-called bomb thermocouple fixed onto the surface of the cell. The heaters adjust the required temperatures depending on the measurement conditions. These conditions can be either isoperibolic or adiabatic. Isoperibolic means that the calorimeter walls are maintained at a constant temperature by the integrated heaters, while measuring the variation of the surface temperature of the cells. In the adiabatic mode the heaters in the calorimeter chamber walls follow immediately any change of the bomb thermocouple, thus preventing transfer of heat from the cell to the walls. Up

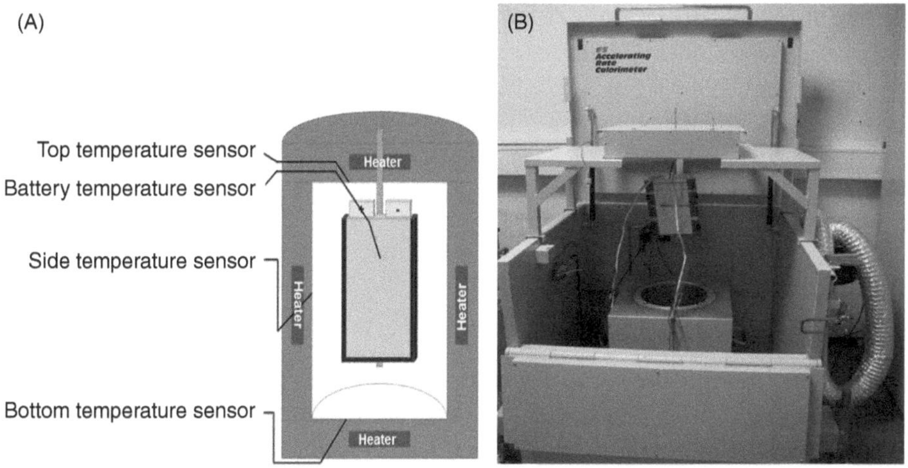

Figure 6.3 ***ARC.*** (A) Schematic of experimental setup and (B) photo of the EV-ARC system made by Thermal Hazard Technology.

to 24 additional thermocouples can be fixed onto the surface of the cell by heat sink paste and ceramic tape to measure the local temperature distribution. If the cell cools down, the temperature of the calorimeter remains constant since the calorimeter cannot cool down under these conditions.

Between 1998 and 2001, Selman's group investigated various commercial cylindrical 18650 cells with $LiCoO_2$ as cathode material in a commercially available ARC (ARC2000, Columbia Scientific Industries) [57–60]. At the maximum discharge rate of 1 C, they found a heat generation rate of 0.55 W [57]. They used the theoretical considerations of Bernardi et al. [38], the most commonly applied description of heat generation, which is based on the energy balance of a complete cell, to derive the heat dissipation rate from the thermal behavior. An adiabatic setup was also designed by Abdul-Quadir et al. [52] using Styrofoam material that completely covers the tested cell and is immersed in a water bath that is stirred. Using this setup they investigated commercial 40 Ah C/NMC pouch cells from Kokam up to a 2 C rate. Eddahech et al. [53], Nieto et al. [54], and Vertiz et al. [55] used a commercial ARC (EV-ARC, Thermal Hazard Technology) to study middle-sized pouch cells with capacities between 10.5 and 14 Ah. Schuster et al. investigated 40 Ah cells with the same equipment [56]. This was the first work in which heat data for large format pouch cells have been determined in the same calorimeter for isoperibolic and adiabatic modes, and in which these data have been compared to each other and to the data derived from measurements of the reversible and the irreversible heat. When looking again at Table 6.1 it becomes apparent that not much data about heat generation of large format pouch cells have been published so far.

3.2.1 Cycling under isoperibolic conditions

Cycling under isoperibolic conditions means that the measurement is conducted at constant environmental temperature T_{env}. This reflects the ideal conditions of a cell, freely suspended in space without any hindrance to dissipate the generated heat to the environment. In this case the heaters and thermocouples in the calorimeter chamber adjust a previously defined constant temperature and the cell temperature $T_s(t)$ varies according to the cycling:

$$T_{env} = \text{constant} \tag{6.12}$$

$$T_s(t) = T_{s_0} + \alpha \cdot t \tag{6.13}$$

with T_{env} as the calorimeter chamber temperature, $T_s(t)$ as the surface temperature of the cell, T_{s_0} as the initial value of the surface temperature, α as the heating rate, and t as the time [61]. Fig. 6.4A shows the voltage, current, and temperature profiles for isoperibolic cycling of an 18650 cell with $LiMn_2O_4$ cathode at T_{env} = 25°C and C/2 rate for 3 cycles. For the charging half cycles the temperature is continuously increasing. Then the temperature decreases to its starting value again during the following 1 h relaxation periods. For the discharge half cycles, a temperature increase at the beginning followed by a temperature decrease and then a temperature increase up to 5.5K close to the end of the discharge cycle can be seen. In Fig. 6.4B the difference between the surface temperature of the cell T_s and the temperature of the calorimeter chamber T_{env} is compared for different isoperibolic temperatures between 30 and 60°C for a 18650 cell with $LiFePO_4$ cathode and 2 C rate. In this range, the applied environmental temperature did not largely influence the cell thermal behavior. The maximum temperature increase, which is reached for the discharging half cycle (second peak), lies between 12 and 14K. After a complete electrochemical cycle the initial temperature is reached again.

3.2.2 Cycling under adiabatic conditions

An adiabatic environment means that there is no heat exchange with the environment of the cell, which is represented by a very high thermal resistance. This simulates the conditions of the worst case in a battery pack without cooling, in which many cells are closely packed together. Under the assumption that all cells are working similarly and in parallel, they generate at the same time a similar amount of heat. Thus, the environment of each cell has a similar temperature (quasiadiabatic state). This is realized in the calorimeter by the bomb thermocouple on the surface of the cell, which measures the temperature $T_s(t)$ and uses it as a parameter for controlling the temperature T_{env} of the calorimeter chamber via the heaters. The thermocouples of the calorimeter only supervise that the required temperature is reached.

$$T_{env} = T_s(t) = T_{env_0} + \alpha \cdot t \tag{6.14}$$

with T_{env_0} as the initial value of the calorimeter temperature [61].

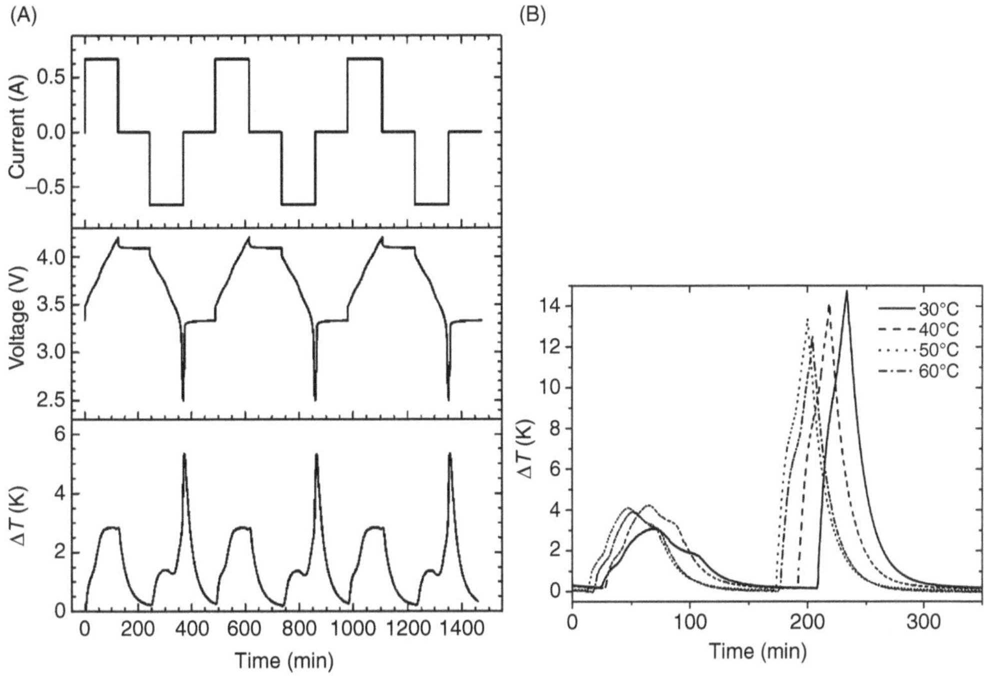

Figure 6.4 ***Isoperibolic cycling on 18650 cells.*** (A) Current profile (upper curve), voltage profile (middle curve), and temperature difference (lower curve) for a cell with $LiMn_2O_4$ cathode at $T_{env} = 25°C$ and C/2 rate. (B) Comparison of temperature differences for a cell with $LiFePO_4$ cathode at different temperatures T_{env} and 2 C rate.

Fig. 6.5 shows the voltage, current, and temperature profiles for adiabatic cycling of an 18650 cell with $LiMn_2O_4$ cathode starting at 25°C and C/2 rate for 3 cycles. It can be clearly seen that the cell temperature increases with every cycle by about 17K even at this relatively low C-rate. During the 1-h relaxation periods the cell cools down. Since active cooling is not possible with the ARC the temperature of the ARC remains constant. However, heat loss to the environment is possible via the cables.

3.2.3 Determination of heat data

In order to transfer the measured temperature data into the heat data, Eq. (6.10) has to be revisited. For adiabatic conditions $\dot{Q}_d = 0$, which gives:

$$\dot{Q}_g = m \cdot c_p \frac{dT}{dt} \tag{6.15}$$

and by integration follows:

$$Q_g = m \cdot c_p (T(t) - T(0)). \tag{6.16}$$

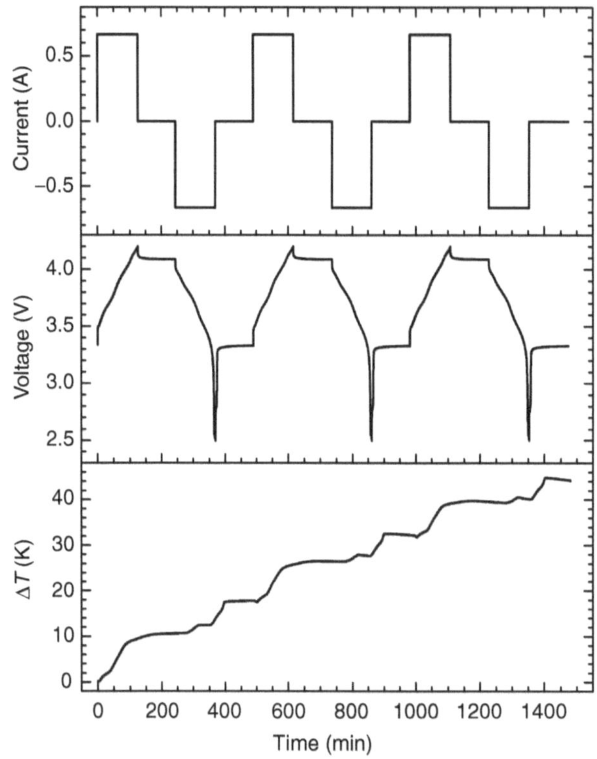

Figure 6.5 ***Adiabatic cycling on 18650 cell with $LiMn_2O_4$ cathode starting at 25°C and C/2 rate.*** Current profile (upper curve), voltage profile (middle curve), and temperature difference (lower curve).

Thus before the heat generation can be calculated from the acquired temperature curves under adiabatic conditions, the effective specific heat capacity has to be determined.

3.2.3.1 Effective specific heat capacity of a cell

The heat capacity of a substance is a physical property that indicates how much thermal energy has to be supplied to a material to increase its temperature by a certain amount. The necessary thermal energy for increasing the temperature increases with the amount of the substance. Therefore, the specific heat capacity c_p is considered, which refers to the mass m of a substance. There are different methods to determine the specific heat capacity of the cells. One possibility would be to determine the specific heat capacity of the individual components with subsequent mass-dependent integration across the entire cell. However, this method proves to be very expensive and impractical in the study of commercial cells. A more elegant method is to determine the effective specific heat capacity. For these measurements, the ARC combined with a power-controlled heater mat can be used. By applying a certain current I and voltage U to the mat over a defined time period $[t_1,t_2]$ the total supplied heat is:

$$\Delta Q = \int_{t_1}^{t_2} U \cdot I \, dt \tag{6.17}$$

A *sandwich* layout must be used, where the heater mat (area smaller than the cell surface area) is placed tightly between several cells, so that all heat generated by the mat is transferred to the cells by conduction. This is shown for two pouch cells in Fig. 6.6A. The maximum current and voltage applied to the heater mat are adjusted to achieve a heating rate of typically 0.02K/min. Under adiabatic conditions the temperature change of the two cells is measured in the ARC and the corresponding heat transferred from the mat to the cells is:

$$\Delta Q = c_{\mathrm{p}} \cdot m \cdot \Delta T_{\mathrm{adia}} \tag{6.18}$$

Thus when combining with Eq. (6.17), the effective specific heat capacity of the cell can be determined as a function of the temperature. Fig. 6.6B shows the measured effective specific heat of a pouch cell with 40 Ah capacity and with $Li(Ni_{0.33}Mn_{0.33}Co_{0.33})O_2$ (NMC) cathode. In the temperature range between 25 and 60°C the measured heat capacity for this cell increases from 1.07 to 1.19 J/gK.

Under isoperibolic conditions, however the heat dissipation rate can be described as:

$$\dot{Q}_{\mathrm{d}} = A \cdot h \cdot (T - T_{\mathrm{env}}) - \varepsilon\sigma(T^4 - T_{\mathrm{env}}^4) \tag{6.19}$$

where h is the heat transfer coefficient, A is the surface area of the cell, ε is the emissivity of the cell surface, and σ is the Stefan–Boltzmann constant. In Eq. (6.19), the first term

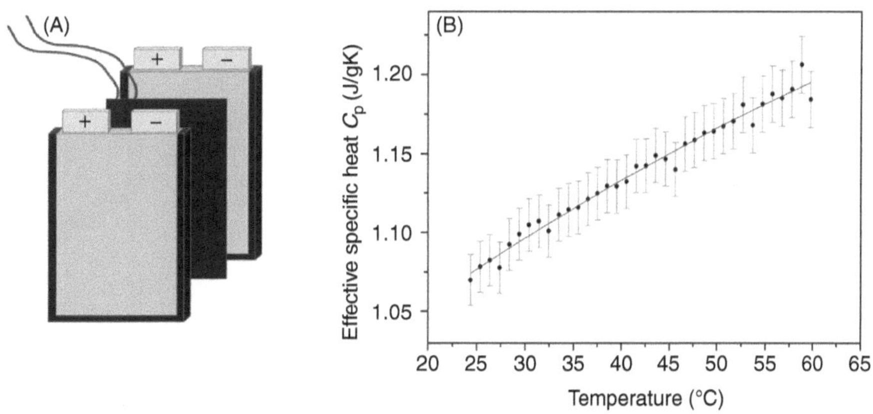

Figure 6.6 ***C_p measurements for pouch cells.*** (A) Sandwich setup and (B) effective specific heat of a pouch cell with $Li(Ni_{0.33}Mn_{0.33}Co_{0.33})O_2$ (NMC) cathode and 40 Ah capacity.

describes the heat transferred by convection and the second term the heat transferred by radiation. It has been shown for pouch cells that the radiation part can be neglected because the dissipation rate by radiation is less than 1% of the dissipation rate by convection [56]. On the contrast, Berndt showed that for lead acid batteries with a housing of styrene acrylonitrile resin (SAN) the heat loss by radiation is not negligible since the fraction of heat loss due to radiation was approximately 20% [39]. Thus, for the analysis of the measurements under isoperibolic conditions only one further factor is required, which is the heat transfer coefficient of the system and the cell under study.

3.2.3.2 Heat transfer coefficient

The most frequently used method for the determination of the heat transfer coefficient is the dummy method, which is described in Ref. [59]. Here, a dummy from a known aluminum alloy is used with the same geometry as the cell. This dummy cell is attached to the position of the cell in the calorimeter chamber and also has a thermocouple on the surface. The calorimeter is heated stepwise and the temperature behavior of the dummy cell is recorded during the adaptation of its temperature to the calorimeter temperature (Fig. 6.7A). In this case the following energy balance for a lumped heat transfer system under natural convective heating conditions can be used:

$$m \cdot c_{\mathrm{p}} \frac{dT}{dt} = A \cdot h \cdot (T - T_{\mathrm{env}}) \tag{6.20}$$

with the starting condition:

$$T = T_{\mathrm{env}} \quad \text{at} \quad t = 0. \tag{6.21}$$

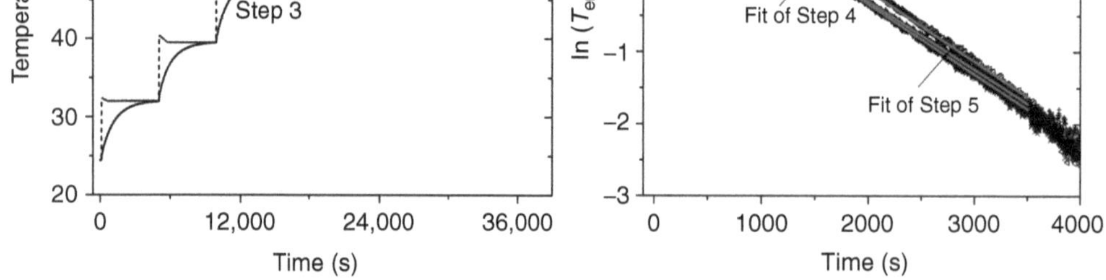

Figure 6.7 ***Determination of the heat transfer coefficient for a 18650 dummy cell made of AlMgSi$_{0.5}$ EN AW-6060 F22.*** (A) Measurement procedure and (B) linear fit for the determination of the slope.

The solution of the equation is:

$$\ln(T - T_{\text{env}}) = \frac{A \cdot h}{m \cdot c_{\text{p}}} t. \tag{6.22}$$

From the slope of the $\ln(T - T_{\text{env}})$ curve against the time t, the heat transfer coefficient h can be determined as shown in Fig. 6.7B. For example, for a 18650 dummy cell made of $AlMgSi_{0.5}$ EN AW-6060 F22, this results in $h = 10.9\ \text{W/m}^2\text{K}$ for the es-ARC calorimeter of THT.

In Fig. 6.8 the total generated heat data during discharge of a 40 Ah pouch cell with NMC cathode determined under adiabatic conditions Q_g^{adia} and under isoperibolic conditions Q_g^{iso} are compared. They show a generally good agreement. However, the heat data from the adiabatic measurements are always a bit higher. This might be due to the limitations of the experimental setup and the fact that the calorimeter was not able to actively cool during cell cooling, that is, in that time period, the temperature of the calorimeter remained constant.

3.2.4 Thermal runaway testing in an ARC

For thermal runaway testing in an ARC the *Heat–Wait–Seek* (HWS) method is usually applied. It starts in the *Heat mode* by heating up the cell in small temperature steps (e.g., 5°C, see Fig. 6.9). At the end of each temperature step the *Wait Mode* will be activated to reach thermal equilibrium over a certain time, which should be taken between 15 and 30 min. After reaching thermal equilibrium, the system enters *Seek Mode*, which seeks the temperature rate and ends with two possible modes—*Exotherm Mode* (under adiabatic conditions) or *Heat Mode*. The decision is related to the selected *Temperature Rate*

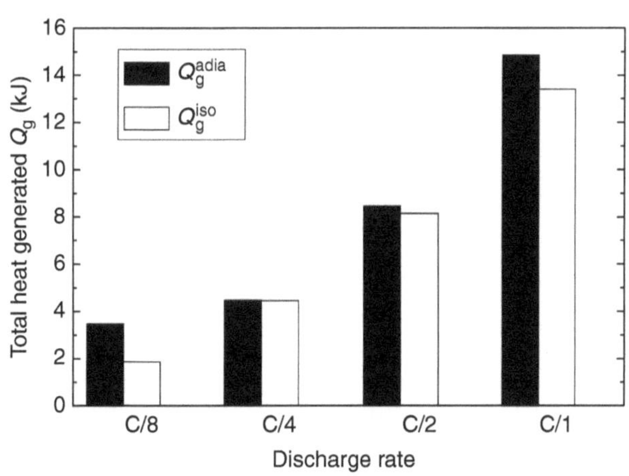

Figure 6.8 ***Comparison of total generated heat data during discharge of a 40-Ah pouch cell with NMC cathode determined under adiabatic conditions Q_g^{adia} and under isoperibolic conditions Q_g^{iso}.***

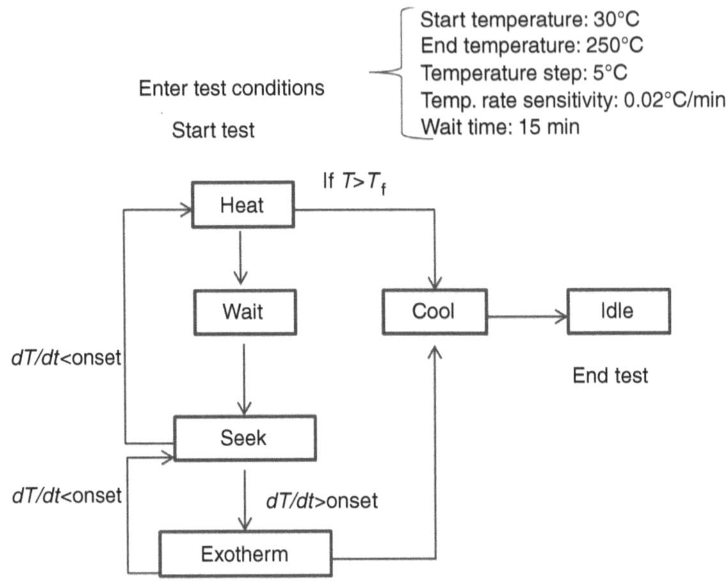

Figure 6.9 ***Processes of ARC Software Operations for HWS method.***

Sensitivity. If the measured temperature rate is larger than the selected *Temperature Rate Sensitivity* (e.g., 0.02°C/min), the system goes into *Exotherm Mode*. On the other hand, if the temperature rate is smaller, the system goes back into *Heat Mode*. If the temperature exceeds the *End temperature* value, the ARC switches off the heaters and starts to cool down, for example, by introducing pressurized air to the calorimeter chamber.

Fig. 6.10A shows a comparison of thermal runaway data for 18650 cells with different cathode materials. In the plot temperature versus time it can be clearly seen that the maximum temperature that is reached is the highest with 731°C for the NMC cathode while it reaches only 258°C for the LFP cathode. In the temperature rate versus temperature plot shown in Fig. 6.10B also the different stages of thermal runaway can be distinguished. The LFP cell does not leave the second stage, with temperature rates staying below 3°C/min. In contrast, the LMO cell reaches a maximum rate of 1285°C/min and the NMC cell even 7600°C/min.

4 THERMAL MODELING OF LITHIUM-ION BATTERIES

To model the thermal behavior of LIBs the main focus is the consideration of a single cell. It is assumed that the interior of the cell is separated from the environmental air due to the battery can, so the cell can be considered as a closed system. Under general working conditions, the LIBs are exposed to an electrical and/or a thermal load. As consequence, heat is generated inside the cell due to several electrochemical and chemical

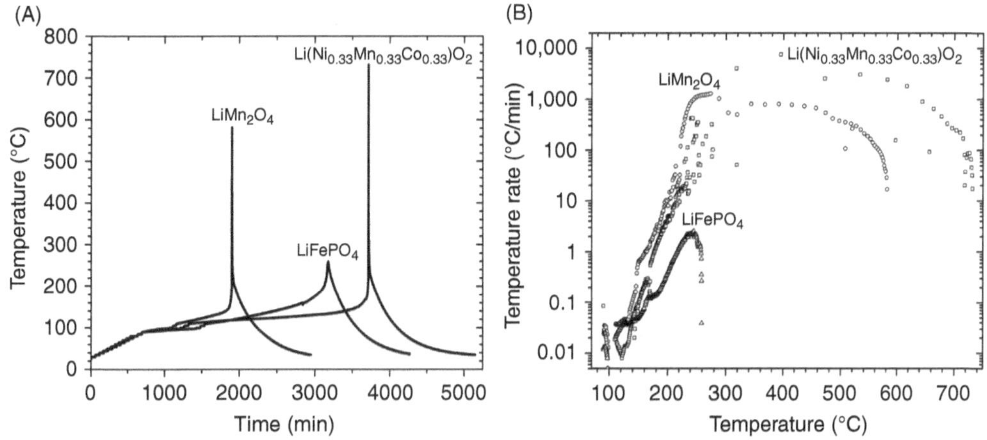

Figure 6.10 ***Comparison of thermal runaway data for 18650 cells with different cathode materials.*** **(A) Temperature versus time plot. (B) Temperature rate versus temperature plot.**

processes. If the heat generation inside the cell is smaller than the ability of the cell to dissipate heat to the environment, the cell is in a thermally stable state. If more heat is generated in the cell than can be dissipated to the environment the cell is in a thermally unstable state. The worst-case scenario in this second case is the occurrence of a thermal runaway. The critical aspect in the case of a thermal runaway is its relation to the energy conservation. With the help of the energy conservation, one is able to describe thermal characteristics like heat generation and heat dissipation. For our considerations we assume, that the cell geometry, respectively the cell volume are constant, that is, elastic processes and gas generation inside the cell are neglected. All physical parameters are constant with respect to time and space.

4.1 The energy conservation

If there is no heat convection inside the LIB, the general equation for the energy conservation can be derived from Fourier's law as initial boundary problem [62–64] with the parabolic differential equation:

$$\rho c_{\mathrm{p}} \frac{\partial T}{\partial t}(\mathbf{x},t) = \kappa \Delta T(\mathbf{x},t) + Q_{\mathrm{gen}}(\mathbf{x},t) \tag{6.23}$$

for the temperature $T:(\mathbf{x},t) \in \Omega \times \mathbb{R}^{+} \mapsto \mathbb{R}$ of the cell in K. The corresponding initial and boundary conditions are:

$$T(\mathbf{x},0) = T_0(\mathbf{x}), \forall \mathbf{x} \in \bar{\Omega}, \tag{6.24}$$

$$\mathbf{n} \cdot (\kappa \Delta T) = -h(T - T_{\mathrm{env}}) - \varepsilon\sigma(T^4 - T_{\mathrm{env}}^4), \forall \mathbf{x} \in \partial\Omega. \tag{6.25}$$

In this framework $\mathbf{x} \in \bar{\Omega} \subset \mathbb{R}^3$ is a spatial interior or surface point of the cell and $t \in \mathbb{R}^+ := \{t \in \mathbb{R}, t \geq 0\}$ is the time. T_{env} denotes the environmental temperature, T_0 is the initial temperature profile inside the cell at $t = 0$ s. It is assumed that the initial temperature at the boundary of the cell coincides with the environmental temperature at $t = 0$ s. Ω denotes the interior of the battery cell, $\partial\Omega$ represents the boundary of the cell and the closure $\bar{\Omega} = \Omega \cup \partial\Omega$ is the complete cell. Furthermore $\mathbf{n} \in \mathbb{R}^3$ is the outward pointing normal vector. h is the heat transfer coefficient, ε the emissivity, σ the Stefan-Boltzmann constant, that is, $\sigma = 5.670373 \cdot 10^{-8}\,\text{W/m}^2\text{K}^4$, ρ the density of the cell, the heat capacity is c_p and κ denotes the thermal conductivity.

The solution of Eq. (6.23) describes the temperature distribution inside the cell for times $t > 0$ and all spatial points $\mathbf{x} \in \Omega$. Eq. (6.24) represents the given initial distribution of the temperature in the LIB at $t = 0$ s. Finally Eq. (6.25) is the heat dissipation to the environment. The first term on the right-hand side of Eq. (6.25) is the heat dissipation due to convection, while the second term describes the heat dissipation due to radiation.

In the inhomogeneity Q_{gen} different heat sources are included. These contributions are coming from the heat generated by exothermic kinetic side reactions, Q_{exotherm} and from reversible and irreversible thermodynamic effects, which are represented by the electrochemical heat source, $Q_{\text{el-chem}}$:

$$Q_{\text{gen}}(\mathbf{x},t) = Q_{\text{exotherm}}(\mathbf{x},t) + Q_{\text{el-chem}}(\mathbf{x},t). \tag{6.26}$$

Thus, for Q_{exotherm} and $Q_{\text{el-chem}}$ in Eq. (6.26) an additional mathematical model must be formulated.

4.2 Identifying the electrochemical heat sources

If a LIB is exposed to an electric load, heat will be generated inside the cell due to the reversible and irreversible processes described in Section 2 by Eqs. (6.5) and (6.9). The critical variables in the electrochemical heat source are the equilibrium voltage U_{eq} and the derivation of the equilibrium voltage with respect to the temperature $\partial U_{\text{eq}}/\partial T$. A scheme of a LIB is shown in Fig. 6.11. In general a LIB consists of two current collectors, the positive and negative electrode and a separator between the two electrodes. The electrodes consist of a porous structure with active particles and pores filled with electrolyte. The domains of consideration are the cell domain, consisting of the electrode domains for cathode, anode, and separator. Due to their porous structure the electrode domain consists of the solid phase (particle domain) and a liquid phase filled with the electrolyte. All these variables have to be computed with a model based on the porous electrode model [5], which represents a multiscale multidomain approach (MSMD). Such an approach takes the physical and geometrical structure of the LIB on the different length scales and different geometrical domains into account. Charge transfer kinetics have to

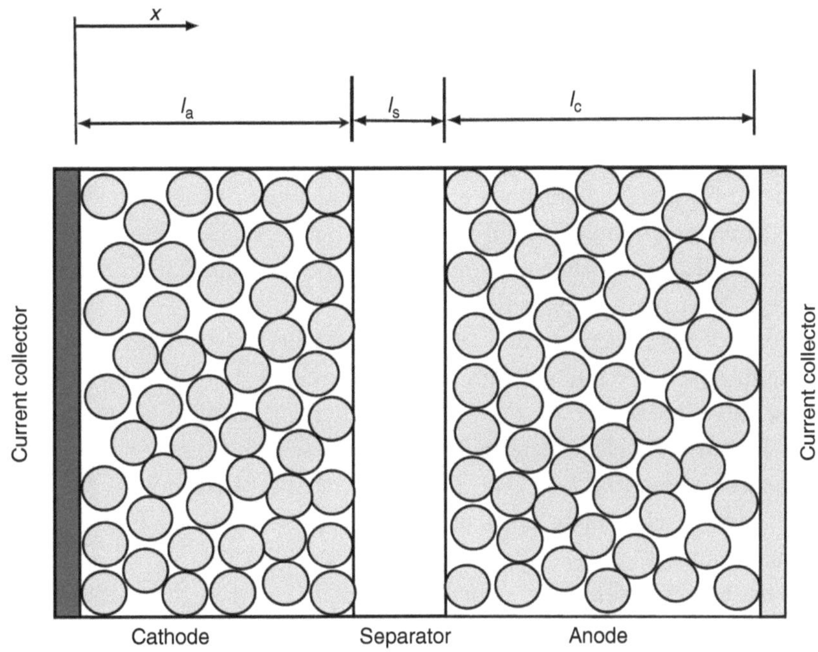

Figure 6.11 ***Scheme of a lithium-ion cell.***

be solved on the electrode–electrolyte interface. The transport of the Li-ions is modeled with a diffusion mechanism and the migration and diffusion of the Li-ions through the liquid electrolyte can be evaluated. The charge balances in the solid cathode and anode as well as the liquid electrolyte are also resolved in the corresponding matrices. To determine the electrochemical heat sources one has to apply the spatial averaging theorem [16] to the corresponding current densities and heat fluxes on each domain level. A short review can be found in Ref. [24], and a more detailed survey is given in [65].

4.3 Modeling the thermal runaway and exothermic heat sources

Exothermic reaction kinetics are closely related to thermal abuse mechanisms inside the LIB. Several exothermic chemical reactions can occur inside a battery as the temperature rises. This may generate heat, that accumulates inside the cell and accelerates the chemical reaction between the cell components, if the heat transfer to the surroundings is not sufficient [66], that is, if the heat generation rate exceeds the dissipation rate. External conditions for a temperature rise can be external heating, overcharging or -discharging, high current charging, nail penetration, external short, or others. In these cases a thermal runaway can occur in consequence with leak, smoke, gas venting, flames, and so forth, which leads to the destruction of the cells.

Several authors have given models to describe abuse behavior and thermal runaway. First attempts to describe the exothermic reaction kinetics are given by Ref. [29]. A simplified reaction-diffusion model is given in the works of Refs. [32–34]. The case of nail penetration is described in Ref. [67]. A catastrophe theoretic approach can be found in Ref. [68], which is based on a simplified ODE model. This model has been extended in Ref. [69]. To describe the thermal runaway of LIBs, one has to identify the main exothermic chemical reactions. Following Refs. [1,29,33,34,70], the general mechanism that leads to a thermal runaway can be described with respect to rising temperature in four main steps as follows:

1. SEI decomposition reaction: At $T > T_1$ the solid electrolyte interface (SEI) will decompose in an exothermic reaction $\Rightarrow$ heat source Q_{sei}.
2. Negative solvent reaction: At $T > T_2$ an exothermic reaction between the intercalated Li-ions and the electrolyte will start$\Rightarrow$ heat source Q_{ne}.
3. Positive solvent reaction: For $T > T_3$ an exothermic reaction between the positive material and the electrolyte takes place under the evolution of oxygen inside the cell$\Rightarrow$ heat source Q_{pe}.
4. Electrolyte decomposition: In a final exothermic reaction the electrolyte will decompose at $T > T_4 \Rightarrow$ heat source Q_{ele}.

For the exothermic heat source one assumes that the n independent exothermic reactions which occur in the LIB are governed by simple Arrhenius laws. Then the exothermic heat source is given as:

$$Q_{\text{exotherm}}(\mathbf{x},t) = \sum_{i=1}^{n} Q_i(\mathbf{x},t) \tag{6.27}$$

with

$$Q_i(\mathbf{x},t) = c_i(\mathbf{x},t) q_i A_i \exp\left(-\frac{E_{a,i}}{RT(\mathbf{x},t)}\right), \tag{6.28}$$

where $c_i(\mathbf{x},t)$ is the dimensionless concentration, q_i is the reaction enthalpy in J/g, A_i is the frequency factor in s^{-1}, $E_{a,i}$ is the activation energy in J/mol, and R is the universal gas constant, that is, $R = 8.314462$ J/molK. For further simplification it is assumed for the simulations, which will be shown in Section 5, that the concentration $c_i(\mathbf{x},t)$ is constant. In this so-called *constant fuel model* all time-spatial dynamics of the concentrations are neglected [64,71]. The exothermic heat sources defined in Eq. (6.28) are thus only dependent on the temperature T, that is, for $i \in \{sei, pe, ne, ele\}$

$$Q_i(\mathbf{x},t) = c_{i0} q_i A_i \exp\left(-\frac{E_{a,i}}{RT(\mathbf{x},t)}\right). \tag{6.29}$$

5 SIMULATIONS WITH COMSOL MULTIPHYSICS

The mathematical model described in Section 4 represents a simple MSMD, which has to be implemented into a software. This will be demonstrated in the following with the software *COMSOL Multiphysics* as an example. This software provides a so-called *Battery and Fuel Cell Module.* This module includes some predefined templates for the modeling of batteries and fuel cells and therefore it can also be applied to the modeling of LIBs. To implement the model from the last section in *COMSOL Multiphysics*, the user has to define the porous electrode model, the solid or the constant fuel model, respectively, and the thermal model, their coupling, the geometry of the cell, the simulation parameters and the discretization in time and space.

For the simulations presented in the next sections, a cylindrical 18650 cell with $LiCoO_2$ chemistry is chosen for the sake of simplicity. From the material database in *COMSOL Multiphysics* a Li_xC_6 anode, a Li_xCoO_2 cathode and as electrolyte 1:1 EC:DEC with a $LiPF_6$ salt is used. The internal spirally wound geometry is not resolved. From the reaction-diffusion system of the exothermic reactions inside the LIB, only the additional heat sources are used to extend the heat transport equation with additional heat sources (constant fuel model).

In a cylindrical geometry, equations for the potentials at the current collector and the temperature field have to be formulated in cylindrical coordinates. For convenience an axial-symmetric model in the r–z-plane is considered:

$$-\frac{1}{r}\varepsilon_{\pm}\sigma_{\pm}^{\text{eff}}\frac{\partial}{\partial r}\left(r\frac{\partial\phi_{\pm}}{\partial r}\right)-\varepsilon_{\pm}\sigma_{\pm}^{\text{eff}}\frac{\partial^2\phi_{\pm}}{\partial z^2}=J_{\pm}, \tag{6.30}$$

$$\frac{\partial(\rho_{\text{batt}}c_{\text{p,batt}}T)}{\partial t}-\left(\frac{1}{r}\frac{\partial}{\partial r}\left(\kappa_{\text{T,rad,batt}}r\frac{\partial T}{\partial r}\right)+\kappa_{\text{T,z,batt}}\frac{\partial^2 T}{\partial z^2}\right)=Q_{\text{gen}}. \tag{6.31}$$

Here $J_{\pm}$ and Q_{gen} denote the current density and the heat source, respectively. The initial and boundary conditions of the temperature field T are given in Eqs. (6.24) and (6.25). The effective thermal conductivity (radial, angular direction) $\kappa_{\text{T,\{rad,z\},batt}}$, density ρ_{batt}, and heat capacity $c_{\text{p,batt}}$ are given as follows [32], where $\mathcal{J}=\{pos., neg., cc_{pos}, cc_{neg}, sep\}$ denotes the index set to the corresponding parameter in cathode, anode, separator, and current collector:

$$\kappa_{\text{T,z,batt}}=\frac{\sum_{i\in\mathcal{J}}\kappa_{T,i}L_i}{L_{\text{batt}}},\kappa_{\text{T,rad,batt}}=\frac{L_{\text{batt}}}{\sum_{i\in\mathcal{J}}\frac{L_i}{\kappa_{T,i}}}, \tag{6.32}$$

$$\rho_{\text{batt}}=\sum_{i\in\mathcal{J}}\frac{\rho_i L_i}{L_{\text{batt}}},c_{\text{p,batt}}=\sum_{i\in\mathcal{J}}\frac{c_{p,i}L_i}{L_{\text{batt}}}. \tag{6.33}$$

The main physical parameters of the simulations can be found in Tables 6.2 and 6.3. The additional simulation parameters of the exothermic reactions are taken from Refs. [32,33]. In the simulations the radius of the particles in the positive and negative electrode $r_{p,\{pos,neg\}}$ is chosen as parameter, that is, $r_{p,\{pos,neg\}} \in \{10,100,1000,10000\}$ nm with $r_{p,pos} = r_{p,neg}$. It is the purpose of the following simulation case study to show the impact of the particle radii on the temporal evolution of the overall mean cell temperature

$$\bar{T}(t) = \frac{1}{\Omega}\int_{\Omega} T(\mathbf{x},t)\, d\Omega. \tag{6.34}$$

Therefore two kinds of simulations will be compared. First adiabatic simulations, where the convective heat transfer coefficient is $h = 0\ \mathrm{W/m^2K}$. In these simulations a thermal runaway occurs. We distinguish between a so-called *oven test or oven trial*, where the cell will be heated from exterior due to the increase of the environmental temperature, and the case where the LIB will be heated from interior due to an applied current load. In the second case, isoperibolic simulations are considered with a convective heat transfer coefficient of $h = 20\ \mathrm{W/m^2K}$.

In the oven test, the increase of the environmental temperature is linear with a constant heating rate of $dT_{env}/dt = 5 \cdot 10^{-2}$ K/s. If a current load is applied, the current load is chosen as a rectangular periodic current profile. The amplitude is chosen as $I_{app} = 40\ \mathrm{A/m^2}$ and $I_{app} = 80\ \mathrm{A/m^2}$ and the period is $t_{period} = 500$ s, where the LIB will be charged/discharged with a pulse length of $t_{pulse} = 250$ s each. The maximal time span in all simulation cases is $t_{sim} = 1.5 \cdot 10^4$ s.

The simulations were performed in *COMSOL Multiphysics* Version 5.2. For the time integration a backward differentiation formula (BDF) integration scheme is chosen with a minimum order of 1 and a maximum order of 5 using a variable step size with a maximum time step of 1 s and an absolute tolerance of 0.001. Since the spatial discretization in *COMSOL Multiphysics* is based on the *Finite Element Method* (FEM) an adaptive spatial discretization in the three models of the particle domain, the electrode and the cell domain is used. The model of the particle domain is solved automatically in the *Battery and Fuel Cell Module*. Therefore only a spatial discretization for the electrode domain and the cell domain is needed. The electrode domain is one-dimensional and the cell domain is modeled due to symmetry reasons in the angular direction of a cylindrical cell as two dimensional. In the electrode domain, the maximum element size in the discretization is chosen as 1 μm. And in total, the discretization contains 168 elements. For the one-dimensional finite element discretization in the electrode domain, quadratic basis functions were chosen. In the cell domain the spatial discretization is performed in the *r–z*-plane using 2266 triangular elements with the element size in the interval $[3.9 \cdot 10^{-4}, 8.45 \cdot 10^{-2}]$ m and quadratic basis functions.

Table 6.2 Simulation parameters

Parameter	Value	Parameter	Value
Initial neg. solid phase concentration	$c_{s,neg,0} = 7{,}917\frac{mol}{m^3}$	Initial pos. solid phase concentration	$c_{s,pos,0} = 16{,}002\frac{mol}{m^3}$
Initial temperature	$T_0 \in [273.15{,}373.15]$ K	Initial electrolyte salt concentration	$c_{l,0} = 2{,}000\frac{mol}{m^3}$
Neg. solid phase Li-diffusivity	$D_{s,neg} = 3.9 \cdot 10^{-14}\frac{m^2}{s}$	Pos. solid phase Li-diffusivity	$D_{s,pos} = 1 \cdot 10^{-13}\frac{m^2}{s}$
Neg. solid phase vol.-fraction	$\varepsilon_{s,neg} = 0.471$	Pos. solid phase vol.-fraction	$\varepsilon_{s,pos} = 0.297$
Neg. electrolyte vol.-fraction	$\varepsilon_{e,neg} = 0.357$	Pos. electrolyte vol.-fraction	$\varepsilon_{e,pos} = 0.444$
Neg. max. solid phase concentration	$c_{s,max,neg} = 26{,}390\frac{mol}{m^3}$	Pos. max. solid phase concentration	$c_{s,max,pos} = 22{,}860\frac{mol}{m^3}$
Neg. electrode thermal conductivity	$K_{T,neg} = 1.04\frac{W}{mK}$	Pos. electrode thermal conductivity	$K_{T,pos} = 1.58\frac{W}{mK}$
Neg. current collector thermal cond.	$K_{T,cc,neg} = 298.15\frac{W}{mK}$	Pos. current collector thermal cond.	$K_{T,cc,pos} = 170\frac{W}{mK}$
Neg. electrode density	$\rho_{neg} = 1{,}347 \cdot 33\frac{kg}{m^3}$	Pos. electrode density	$\rho_{pos} = 2{,}328 \cdot 5\frac{kg}{m^3}$
Neg. current collector density	$\rho_{neg,cc} = 8{,}933\frac{kg}{m^3}$	Pos. current collector density	$\rho_{pos,cc} = 2{,}770\frac{kg}{m^3}$
Neg. reaction rate coefficient	$k_{neg} = 2 \cdot 10^{-11}\frac{m}{s}$	Pos. reaction rate coefficient	$k_{pos} = 2 \cdot 10^{-11}\frac{m}{s}$
Neg. electrode heat capacity	$c_{p,neg} = 1{,}437.4\frac{J}{kgK}$	Pos. electrode heat capacity	$c_{p,pos} = 1{,}269 \cdot 21\frac{J}{kgK}$
Neg. current collector	$c_{p,cc,neg} = 385\frac{J}{kgK}$	Pos. current collector heat capacity	$c_{p,cc,pos} = 875\frac{J}{kgK}$
Separator density	$\rho_{sep} = 1{,}008.98\frac{kg}{m^3}$	Separator collector thermal cond.	$K_{T,sep} = 0.344\frac{W}{mK}$
Separator heat capacity	$c_{p,sep} = 1{,}978.16\frac{J}{kgK}$		
Battery radius	$r_{batt} = 9$ mm	Battery height	$h_{batt} = 65$ mm
Mandrel radius	$r_{mand} = 2$ mm		
Cell thickness	$L_{batt} = 157\ \mu m$	Thickness battery canister	$d_{can} = 0.25$ mm
Thickness neg. current collector	$L_{neg,cc} = 7\ \mu m$	Thickness pos. current collector	$L_{pos,cc} = 10\ \mu m$
Length pos./neg. electrode	$L_{pos/neg} = 55\ \mu m$	Length separator	$L_{sep} = 30\ \mu m$

Table 6.3 Exothermic simulation parameters

$x \in \{sei, pe, ne, e\}$, $y \in \{c, p, e\}$	Reaction heat, H_x (J/kg)	Frequency factor, A_x (s^{-1})	Activation energy, E_A (J/mol)	Volume content, W_y (kg/m³)
SEI reaction	$2.57 \cdot 10^5$	$1.667 \cdot 10^{15}$	$1.3508 \cdot 10^5$	$1.39 \cdot 10^3$
Neg. solvent reaction	$1.714 \cdot 10^6$	$2.5 \cdot 10^{13}$	$1.3508 \cdot 10^5$	$1.39 \cdot 10^3$
Pos. solvent reaction	$3.14 \cdot 10^5$	$6.667 \cdot 10^{13}$	$1.396 \cdot 10^5$	$1.3 \cdot 10^3$
Electrolyte decomp.	$1.55 \cdot 10^5$	$5.14 \cdot 10^{25}$	$2.74 \cdot 10^5$	$5 \cdot 10^2$

5.1 Adiabatic simulations up to a thermal runaway

To simulate an oven trial, an additional boundary condition is implemented in the model, which describes the increasing environmental temperature during heating by a surface heat source Q_{surf}, with $Q_{surf} = 1809\ W/m^2$. This value is equivalent to a constant heating rate of 5K/min. The radii of the particles at positive and negative electrode are chosen as mentioned earlier. In all four cases the simulation aborts at a time $t_{TR} \approx 888$ s with a mean cell temperature $\overline{T}_{cell} > 900K$. Therefore we can conclude that a thermal runaway in the LIB occurred and that there is no influence of the size of the particles in the anode and cathode on the thermal runaway time t_{TR}. If the periodic current profile is applied the thermal runaway time decreases with increasing particle radius. The results are shown in Table 6.4 and Figs. 6.12 and 6.13.

As seen in Fig. 6.13 the time evolution of the overall mean cell temperature $\overline{T}_{mean}$ can be split into three regions. A first zone, where $\overline{T}_{mean}$ shows a linear or almost linear rise in temperature, an intermediate zone, where the heating rate rises in a nonlinear fashion, and the Thermal Runaway zone, where the temperature increase goes exponential with the temperature. For the first two current profiles one can see that for $r_p = 10$ nm, the heating rate is the lowest, while for $r_p = 10{,}000$ nm, the heating rate is the highest. For the particle radius $r_p = 100$ nm, the corresponding temperature curve is very close to the temperature curve of $r_p = 10$ nm in the first zone and starts to diverge in the second zone. For the particle radius $r_p = 1000$ nm the temperature curve diverge from the beginning from the curves of the radii $r_p = 10$ nm. In total the heating rate is correlated both to the amplitude of the applied current profile and to the radius of the particle. The higher the amplitude and the larger the particle radius, the higher is the heating rate in the first two zones. In the Thermal Runaway zone the heating rates are almost identical and very high for all particle radii.

Table 6.4 Thermal runaway time t_{TR} (s) as function of the particle radius r_p (nm) for the oven trial (a) and the electrical current load with amplitude $I_{app} = 40\ A/m^2$ (b) and $I_{app} = 80\ A/m^2$ (c)

t_{TR} (s)

r_p (nm)	10	100	1,000	10,000
(a) Oven trial	886.86	888.08	888.27	888.29
(b) $I_{app} = 40\ A/m^2$	5,438.5	5,395.9	4,996.9	3,322
(c) $I_{app} = 80\ A/m^2$	1,871.4	1,847.8	1,724.4	1,242.6

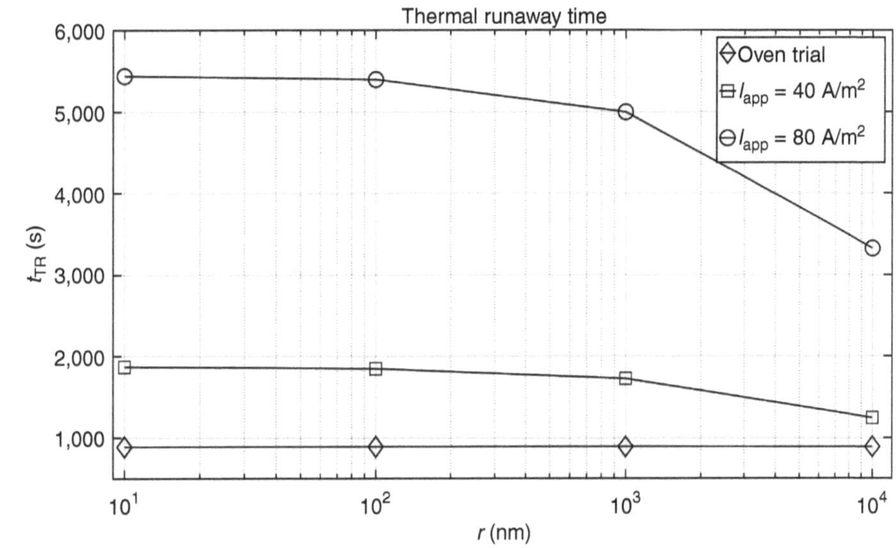

Figure 6.12 ***Thermal runaway time t_{TR} as function of the particle radius r_p.***

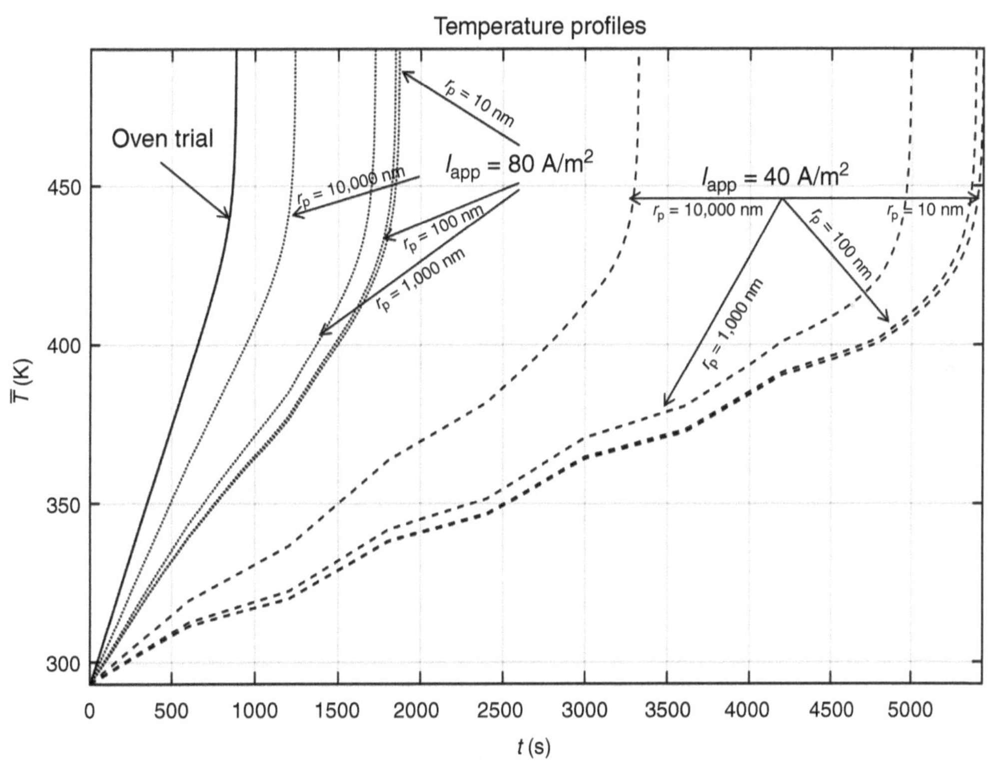

Figure 6.13 ***Time evolution of mean cell temperature during oven trial and electric load.***

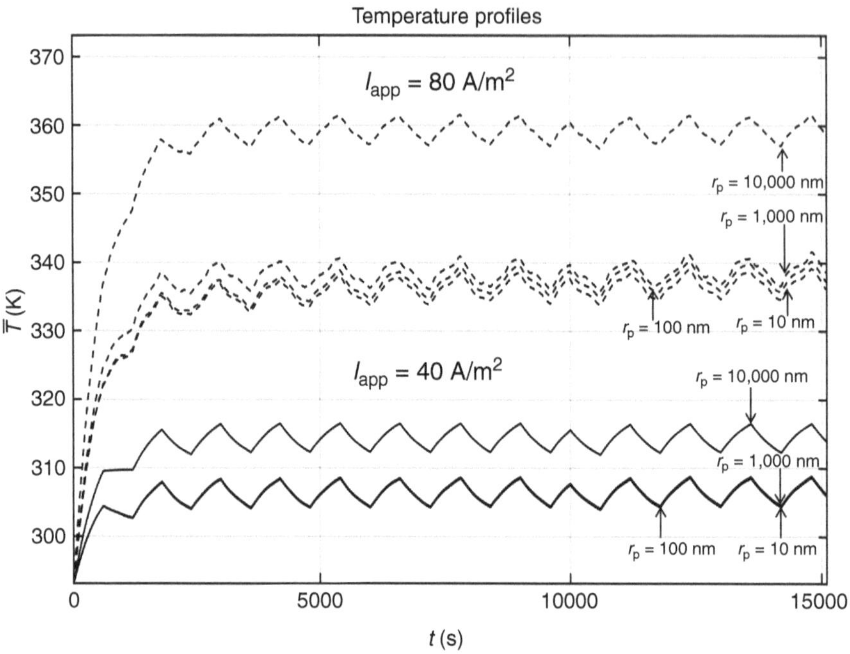

Figure 6.14 ***Time evolution of mean cell temperature during electric load where no thermal runaway occurs.***

5.2 Isoperibolic simulations of cell cycling

In this section, the case where no thermal runaway occurs is considered. The temporal evolution of the spatial averaged mean cell temperature $\overline{T}(t)$ is shown in Fig. 6.14 for the applied current profiles $I_{app} = 40\ \mathrm{A/m^2}$ and $I_{app} = 80\ \mathrm{A/m^2}$, with respect to the particle radius $r_p \in \{10,100,1000,10000\}$ nm. From Fig. 6.14 the temporal evolution can be divided into two parts. For $t \lesssim 2,500\,\mathrm{s}$ the LIB shows a transient behavior and for $2,500\,\mathrm{s} \lesssim t \lesssim 15,000\,\mathrm{s}$ a *quasiperiodic* behavior (we define here *quasiperiodic behavior* in the following sense: the length L of one time period of the signal is defined as $L := \overline{L} \pm \Delta L$, where ΔL is a component of unpredictability due to errors in the physical model, the numerical computation, and so forth). To consider the quasiperiodic part of Fig. 6.14 we write the overall mean cell temperature $\overline{T}(t)$ in the following sense:

$$\overline{T}(t) = \overline{T}_{\mathrm{mean}} \pm \Delta\overline{T}(t), \tag{6.35}$$

where

$$\overline{T}_{\mathrm{mean}} = \frac{1}{t_2 - t_1}\int_{t_1}^{t_2} \overline{T}(t)\, dt \tag{6.36}$$

Table 6.5 Temporal mean cell temperature $\overline{T}_{mean}$ and its deviation from the initial temperature T_0 for the electrical current load with amplitude $I_{app} = 40\ A/m^2$ and $I_{app} = 80\ A/m^2$

	(a) $\overline{T}_{mean}$ (K)				(b) $\overline{T}_{mean} - T_0$ (K)			
r_p (nm)	10	100	1,000	10,000	10	100	1,000	10,000
$I_{app} = 40\ A/m^2$	306.37	306.38	306.56	314.39	13.22	13.23	13.41	21.24
$I_{app} = 80\ A/m^2$	336.39	337.44	338.62	359.30	43.24	44.29	45.47	66.15
	(c) $\Delta\overline{T}_{mean}(I_{app})$(K)				(d) $\Delta\overline{T}_{mean}(r_p)$(K)			
r_p (nm)	10	100	1,000	10,000	10	100	1,000	10,000
	30.02	31.06	30.05	44.91	—	0.01	0.18	7.83
	—	—	—	—	—	1.05	1.18	20.68

is the time-independent mean value of $\overline{T}(t)$ and $\Delta\overline{T}(t)$ the time-dependent deviation from $\overline{T}_{mean}$. In Table 6.5(a) $\overline{T}_{mean}$ is listed. For both applied current profiles the time-independent mean values are very close together for $r_p \in \{10,100,1000\}$ nm, while for $r_p = 10000$ nm, $\overline{T}_{mean}$ shows a drastic increase. In Table 6.5(b) the deviation of $\overline{T}_{mean}$ from the initial temperature $T_0 = 293.15$K is shown, which gives the overall temperature increase due to the electric profile with respect to the radius of the particles. One can see that for $r_p \in \{10,100,1000\}$ nm, this increase is of the same order, while for $r_p = 10000$ nm this increase is more significant. To compare this difference in Table 6.5(d) the difference of the mean time-independent temperature with respect to two different particle radii is defined as

$$\Delta\overline{T}(r_p) = \overline{T}(r_{i,p}) - \overline{T}(r_{p,j}), \quad (i,j) \in \{10,100,1000,10000\}\text{nm} \tag{6.37}$$

for a fixed current profile. In an analog way the difference $\Delta\overline{T}(I_{app})$ for a fixed particle radius with respect to the applied current profile is defined and can be seen in Table 6.5(c). Again the difference in the two current profiles is comparable and approximate 30K for $r_p \in \{10,100,1000\}$ nm with a difference of 1K with respect to the particle radius, while for $r_p = 10000$ nm the difference is 45K for the two current profiles, and with respect to the particle radii it is 15K.

In summary one can see that the simulations predict for the two current profiles a similar thermal behavior in the overall mean cell temperature $\overline{T}(t)$ for $r_p \in \{10,100,1000\}$ nm, while for $r_p = 10000$ nm for both current profiles a larger heating in the time evolution was found.

6 CONCLUSIONS

It has been shown in this chapter that both calorimetry and modeling of electrochemical cells are required to obtain quantitative thermal and thermodynamic data for regular use and abuse conditions. Moreover, it has been demonstrated that an accelerating rate calorimeter is a versatile and powerful instrument for in-operando thermal investigation of lithium-ion cells during electrochemical cycling under isoperibolic and adiabatic environmental conditions as well as for safety tests. The observed significant rise in the cell surface temperatures under adiabatic conditions emphasizes the need for a well-adapted thermal management system in practical applications of LIBs. The simulation case study revealed that the use of nanoscale materials influences heat generation and heat dissipation during operation of the electrochemical cells and thus has to be considered in order to develop safer cells. This first attempt also makes clear that further experimental work and further simulations will be needed to study the influence of nanoscale materials on the electrochemical and thermal properties of lithium-ion cells more thoroughly. For the simulation of thermal runaway the coupled electrochemical thermal model based on porous electrode theory of Newman had to be extended with a simple combustion model coming from reaction kinetics, including various types of heat sources based on an Arrhenius law. It has been explained how this model was realized using the commercial finite element package COMSOL Multiphysics. Of course the quality of such simulations could be further improved by, for example, using the solid fuel instead of the constant fuel model and by using more parameters that have been determined by experiments. Especially two aspects are important. First the detailed microstructure or nanostructure of the electrodes has not been taken into account. This could be done by using experimental methods, such as X-ray tomography or focused ion beam in order to generate more realistic 3D-models in the particle and the electrode domains. In addition the separator is considered, as in most studies, as a continuous medium with porosity that can be described by the Bruggeman equation. For improved thermal runaway simulation the temperature-dependent parameters affecting separator performance, such as porosity degree or pore size should be taken into account as well as venting effects. Still, many efforts will be needed to make fully seamless multiscale simulations from atomistic to cell scale possible.

GLOSSARY

(a)	Abbreviations
ARC	Accelerating rate calorimeter
EV	Electric vehicle
FEM	Finite element modeling
HEV	Hybrid electric vehicle
HWS	Heat–wait–seek
IHC	Isothermal heat conduction calorimeter
LCO	$LiCoO_2$
LFP	$LiFePO_4$
LIB	Lithium-ion battery
LMO	$LiMn_2O_4$

NMC	$Li(Ni_{0.33}Mn_{0.33}C_{0.33})O_2$
OCV	Open circuit voltage
ODE	Ordinary differential equation
PDE	Partial differential equation
SAN	Styrene acrylonitrile resin
SEI	Solid electrolyte interface
SOC	State of charge
(b)	Coordinate systems
(r, θ, z)	Cylinder coordinates
(x_1, x_2, x_3)	Cartesian coordinates
(c)	Differential operators
d/dt	Time derivative operator
$D = \partial^2/\partial x_1^2 + \partial^2/\partial x_2^2 + \partial^2/\partial x_3^2$	Laplacian operator
$\nabla = (\partial/\partial x_1, \partial/\partial x_2, \partial/\partial x_3)^T$	Nabla operator
(d)	Physical constants
$F = 96{,}485{,}336$ C/mol	Faraday constant
$R = 8.314462$ J/molK	Universal gas constant
$\sigma = 5.670373 \cdot 10^{-8}$ W/m^{2}K^4	Stefan–Boltzmann constant
(e)	Independent variables
$\mathbf{x}$	Spatial interior or surface point
t	Time (s)
(f)	Geometric notations
$\mathbf{n}$	Outwards pointing normal vector
$\mathbb{R}^+$	Nonnegative real numbers
$\mathbb{R}^3$	Three dimensional Euclidean space
Ω	The interior of the LIB
$\bar{\Omega}$	The complete LIB
$\partial\Omega$	Surface of the LIB
(g)	Physical variables
H	Enthalpy (J)
S	Entropy (J/K)
ΔG	Gibbs free energy (J)
ΔH	Enthalpy change (J)
ΔU	Potential difference cathode/anode (V)
U_{eq}	Equilibrium voltage, open circuit voltage (V)
U	Voltage (V)
I	Electric current (A)
W_{el}	Electrical work (J)
$T(\mathbf{x},t)$	Temperature (K) at time t
$\bar{T}(t)$	Spatial averaged overall cell temperature (K) at time t
$T_s(t)$	Cell surface temperature (K) at time t
$c_i(\mathbf{x},t)$	Dimensionless concentration [1] in point $\mathbf{x}$ at time t for $i \in \{sei, ne, pe, ele\}$
$J_\pm$	Current density (A/m^2)
$\phi_\pm$	Electrical potential (V)
$\dot{Q}$	Total heat flow (J/s)
$\dot{Q}_g$	Generated heat (J/s)
$\dot{Q}_g$	Dissipated heat (J/s)
$Q_{el\text{-}chem}$	Electrochemical heat source (W/m^3)
Q_{ele}	Exothermic heat source (W/m^3) from the decomposition of the electrolyte

$Q_{exotherm}$	Heat source (W/m^3) from exothermic side reactions
Q_{gen}	Total heat source (W/m^3)
Q_{irrev}	Thermodynamic heat source (W/m^3) from irreversible processes
Q_{mix}	Heat from enthalpy mixing (W/m^3)
Q_{ne}	Exothermic heat source (W/m^3) from the lithium electrolyte reaction at the anode
Q_{pe}	Exothermic heat source (W/m^3) from the lithium electrolyte reaction at the anode
Q_{reac}	Heat from side reactions (W/m^3)
Q_{rev}	Thermodynamic heat source (W/m^3) from reversible processes
Q_{sei}	Exothermic heat source (W/m^3) from the SEI decomposition
(h)	Initial and boundary values
$T_0(\mathbf{x})$	Initial temperature distribution/field (K) at time $t = 0$ s
$T_{s_0}(\mathbf{x})$	Initial surface temperature of the cell (K) at time $t = 0$ s
T_{env}	Environmental temperature (K) or the calorimeter chamber temperature
(i)	Physical parameters
α	Heating rate (K/s)
A_i	Frequency factor (s^{-1}) for $i \in \{sei, ne, pe, ele\}$
c_p	Heat capacity of the LIB (J/kgK)
$c_{p,batt}$	Effective heat capacity of the LIB (J/kgK)
$E_{a,i}$	Activation energy (J/mol) for $i \in \{sei, ne, pe, ele\}$
h	Convective heat transfer coefficient (W/m^2K)
ε	Emissivity in [1]
$\varepsilon_\pm$	Electric permittivity (F/m)
κ	Thermal conductivity of the LIB (W/mK)
$\kappa_{T,\{rad,z\},batt}$	Effective thermal conductivity of the LIB (W/mK) in radial and z direction
m	Mass of the cell (kg)
n	Number of exchanged electrons [1]
q_i	Reaction enthalpy (J/g) for $i \in \{sei, ne, pe, ele\}$
ρ	Density (kg/m^3)
ρ_{batt}	Effective density (kg/m^3)
$\sigma_\pm$	Electric conductivity (S/m)

REFERENCES

[1] D.P. Abraham, E.P. Roth, R. Kostecki, K. McCarthy, S. MacLaren, D.H. Doughty, Diagnostic examination of thermally abused high-power lithium-ion cells, J. Power Sources 161 (2006) 648–657.

[2] M. Doyle, T.F. Fuller, J. Newman, Modeling galvanostatic charge and discharge of the lithium/polymer/insertion cell, J. Electrochem. Soc. 140 (6) (1993) 1526–1533.

[3] M. Doyle, T.F. Fuller, J. Newman, simulation and optimization of the dual lithium ion insertion cell, J. Electrochem. Soc. 141 (1) (1994) 1–10.

[4] M. Doyle, J. Newman, Comparison of the modeling predictions with experimental data from plastic lithium ion cells, J. Electrochem. Soc. 143 (6) (1996) 1890–1903.

[5] J. Newman, W. Tiedemann, Porous-electrode theory with battery applications, AIChE J. 21 (1) (1975) 25–41.

[6] J. Newman, W. Tiedemann, Temperature rise in a battery module with constant heat generation, J. Electrochem. Soc. 142 (4) (1995) 1054–1057.

[7] C.R. Pals, J. Newman, Thermal modeling of lithium/polymer battery. I. Discharge behavior of a single cell, J. Electrochem. Soc. 142 (10) (1995) 3274–3281.

[8] C.R. Pals, J. Newman, Thermal modeling of lithium/polymer battery. II. Temperature profiles in a cell stack, J. Electrochem. Soc. 142 (10) (1995) 3282–3288.

[9] S. Al Hallaj, H. Maleki, J.S. Hong, J.R. Selman, Thermal modeling and design considerations of lithium-ion batteries, J. Power Sources 83 (1999) 1–8.
[10] S.C. Chen, C.C. Wan, Y.Y. Wang, Thermal analysis of lithium-ion batteries, J. Power Sources 140 (2005) 111–124.
[11] Y. Inui, Y. Kobayashi, Y. Watanabe, Y. Watase, Y. Kitamura, Simulation of temperature distribution in cylindrical and prismatic lithium ion secondary batteries, Energy Convers. Manag. 48 (2007) 2103–2109.
[12] C. Zhu, X. Li, L. Song, L. Xiang, Development of the theoretically based thermal model for lithium ion battery pack, J. Power Sources 223 (2013) 155–164.
[13] J. Hu, L. Tao, L. Jing, Temperature field analysis and thermal dissipation structure optimization of lithium-ion battery pack in PEVs, Res. J. Appl. Sci. 7 (1) (2014) 62–71.
[14] C.Y. Wang, W.B. Gu, B.Y. Liaw, Micro-macroscopic coupled modeling of batteries and fuel cells. I. Model development, J. Electrochem. Soc. 145 (10) (1998) 3407–3417.
[15] W.B. Gu, C.Y. Wang, B.Y. Liaw, Micro-macroscopic coupled modeling of batteries and fuel cells. II. Application to nickel cadmium and nickel-metal hydride cells, J. Electrochem. Soc. 145 (10) (1998) 3418–3427.
[16] F.A. Howes, S. Whitaker, The spatial averaging theorem revisited, Chem. Eng. Sci. 40 (8) (1985) 1387–1392.
[17] A. Latz, J. Zausch, O. Iliev, Modeling of species and charge transport in Li-ion batteries based on non-equilibrium thermodynamics, in: I. Dimov, S. Dimova, N. Kolkovska (Eds.), NMA, 2010, LNCS 6046, Springer, Berlin, Heidelberg, 2011, pp. 329–337.
[18] G. Richardson, G. Denault, C.P. Please, Multiscale modeling and analysis of lithium-ion battery charge and discharge, J. Eng. Math. 72 (2012) 41–72.
[19] M. Landstorfer, T. Jacob, Mathematical modeling of intercalation batteries at the cell level and beyond, Chem. Soc. Rev. 42 (2013) 3234–3252.
[20] A. Latz, J. Zausch, Multiscale modeling of lithium ion batteries: thermal aspects, Beilstein J. Nanotechnol. 6 (2015) 987–1007.
[21] W.B. Gu, C.Y. Wang, Thermal-electrochemical modeling of battery systems, J. Electrochem. Soc. 147 (8) (2000) 2910–2922.
[22] C.Y. Wang, V. Srinivasan, Computational battery dynamics (CBD)-electrochemical/thermal coupled modeling and multi-scale modeling, J. Power Sources 110 (2002) 364–376.
[23] K. Kumaresan, G. Sikha, R.E. White, Thermal model for lithium-ion cell, J. Electrochem. Soc. 155 (2) (2008) A164–A171.
[24] L. Cai, R.E. White, Mathematical modeling of lithium ion battery with thermal effects in COMSOL Inc. Multiphysics (MP) software, J. Power Sources 196 (2011) 5985–5989.
[25] G.H. Kim, K. Smith, K.J. Lee, S. Santhanagopalan, A. Pesaran, Multi-domain modeling of lithium-ion batteries encompassing multi-phyiscs in varied length scales, J. Electrochem. Soc. 158 (8) (2011) A955–A969.
[26] M. Guo, G.H. Kim, R.E. White, A three-dimensional multi-physics model for a Li-ion battery, J. Power Sources 240 (2013) 80–94.
[27] K.J. Lee, K. Smith, A. Pesaran, G.H. Kim, Three dimensional thermal-, electrical-, and electrochemical-coupled model for cylindrical wound large format lithium-ion batteries, J. Power Sources 241 (2013) 20–32.
[28] A. Latz, J. Zausch, Thermal-electrochemical lithium-ion battery simulations on microstructure and porous electrode scale, ECS Trans. 69 (1) (2015) 75–81.
[29] T.D. Hatchard, D.D. MacNeil, A. Basu, J.R. Dahn, Thermal model of cylindrical and prismatic lithium-ion cells, J. Electrochem. Soc. 148 (7) (2001) A755–A761.
[30] R. Spotnitz, J. Franklin, Abuse behavior of high-power lithium-ion cells, J. Power Sources 113 (2003) 81–100.
[31] R.M. Spotnitz, J. Weaver, G. Yeduvaka, D.H. Doughty, E.P. Roth, Simulation of abuse tolerance of lithium-ion battery packs, J. Power Sources 163 (2007) 1080–1086.
[32] G.H. Kim, A. Pesaran, R. Spotnitz, A three-dimensional thermal abuse model for lithium-ion cells, J. Power Sources 170 (2007) 476–489.
[33] P. Peng, Y. Sun, F. Jiang, Thermal analyses of $LiCoO_2$ lithium-ion battery during oven tests, Heat Mass Transf. 50 (2014) 1405–1416.
[34] P. Peng, Y. Sun, F. Jiang, Numerical simulations and thermal behavior analysis for oven thermal abusing of $LiCoO_2$ lithium-ion battery, CIESC J. 65 (2) (2014) 647–657.

[35] C.F. Lopez, J.A. Jeervarajan, P.P. Mukherjee, Characterization of lithium-ion battery thermal abuse behavior using experimental and computational analysis, J. Electrochem. Soc. 162 (10) (2015) A2163–A2173.
[36] M. Chen, Q. Sun, K. Wu, B. Liu, P. Peng, Q. Wang, A thermal runaway simulation on a lithium titanate battery and battery module, Energies 8 (2015) 490–500.
[37] N. Tanaka, Modeling and simulation of thermo-electrochemistry of thermal runaway in lithium-ion batteries, PhD Thesis, University of Stuttgart, 2015. Available from: http://elib.uni-stuttgart.de/opus/volltexte/2015/10112/
[38] D. Bernardi, E. Pawlikowski, J. Newman, A general energy balance for battery systems, J. Electrochem. Soc. 132 (1985) 5–12.
[39] D. Berndt, Maintenance-Free Batteries, Research Studies Press Ltd., Baldock, Hertfordshire, England, (2003).
[40] S. Tobishima, J.-I. Yamaki, A consideration of lithium cell safety, J. Power Sources 81–82 (1999) 882–886.
[41] K. Malgorzata, Lithium-Ion Battery Materials and Engineering: Current Topics and Problems from the Manufacturing Perspective, Springer, London, (2014).
[42] T.M. Bandhauer, S. Garimella, T.F. Fuller, A critical review of thermal issues in lithium-ion batteries, J. Electrochem. Soc. 158 (2011) R1–R25.
[43] Y. Saito, K. Kanari, K. Takano, Thermal studies of a lithium-ion battery, J. Power Sources 68 (1997) 451–454.
[44] Y. Saito, K. Takano, K. Kanari, A. Negishi, K. Nozaki, K. Kato, Comparative studies of thermal behaviors of various lithium-ion cells, J. Power Sources 97–98 (2001) 688–692.
[45] Y. Saito, M. Shikano, H. Kobayashi, Heat generation behavior during charging and discharging of lithium-ion batteries after long-time storage, J. Power Sources 244 (2013) 294–299.
[46] Y. Kobayashi, N. Kihira, K. Takei, H. Miyashiro, K. Kumai, N. Terada, R. Ishikawa, Electrochemical and calorimetric approach to spinel lithium manganese oxide, J. Power Sources 81 (1999) 463–466.
[47] K. Onda, H. Kameyama, T. Hanamoto, K. Ito, Experimental study on heat generation behavior of small lithium-ion secondary batteries, J. Electrochem. Soc. 150 (2003) A285–A291.
[48] K. Onda, T. Ohshima, M. Nakayama, K. Fukuda, T. Araki, Thermal behavior of small lithium-ion battery during rapid charge and discharge cycles, J. Power Sources 158 (2006) 535–542.
[49] T. Ohshima, M. Nakayama, K. Fukuda, T. Araki, K. Onda, Thermal behavior of small lithium-ion battery during rapid charge and discharge cycles, Elec. Eng. Japan 157 (2006) 1521–1528.
[50] M. Xiao, S.-Y. Choe, Theoretical and experimental analysis of heat generations of a pouch type $LiMn_2O_4$/carbon high power Li-polymer battery, J. Power Sources 241 (2013) 46–55.
[51] K. Chen, G. Unsworth, X. Li, Measurements of heat generation in prismatic Li-ion batteries, J. Power Sources 261 (2014) 28–37.
[52] Y. Abdul-Quadir, T. Laurila, J. Karppinen, K. Jalkanen, K. Vuorilehto, L. Skogström, M. Paulasto-Kröckel, Heat generation in high power prismatic Li-ion battery cell with $LiMnNiCoO_2$ cathode material, Int. J. Energy Res. 38 (2014) 1424–1437.
[53] A. Eddahech, O. Briat, J.M. Vinassa, Thermal characterization of a high-power lithium-ion battery: potentiometric and calorimetric measurement of entropy changes, Energy 61 (2014) 432–439.
[54] N. Nieto, L. Diaz, J. Gastelurutia, I. Alava, F. Blanco, J.C. Ramos, A. Rivas, Thermal modeling of large format lithium-ion cells, J. Electrochem. Soc. 160 (2013) A212–A217.
[55] G. Vertiz, M. Oyarbide, H. Macicior, O. Miguel, I. Cantero, P. Fernandez de Arroibae, I. Ulacia, Thermal characterization of large size lithium-ion pouch cell based on 1d electro-thermal model, J. Power Sources 272 (2014) 476–484.
[56] E. Schuster, C. Ziebert, A. Melcher, M. Rohde, H.J. Seifert, Thermal behavior and electrochemical heat generation in a commercial 40 Ah lithium ion pouch cell, J. Power Sources 286 (2015) 580–589.
[57] J.S. Hong, H. Maleki, S. Hallaj, L. Reddey, J.R. Selman, Electrochemical-calorimetric studies of lithium-ion cells, J. Electrochem. Soc. 145 (1998) 1489–1501.
[58] S. Hallaj, R. Venkatachalapathy, J. Prakash, J.R. Selman, Entropy changes due to structural transformation in the graphite anode and phase change of the $LiCoO_2$ cathode, J. Electrochem. Soc. 147 (2000) 2432–2436.
[59] S. Hallaj, J. Prakash, J.R. Selman, Characterization of commercial Li-ion batteries using electrochemical-calorimetric measurements, J. Power Sources 87 (2000) 186–194.

[60] J. Selman, S. Al Hallaj, I. Uchiba, Y. Hirano, Cooperative research on safety fundamentals of lithium batteries, J. Power Sources 97–98 (2001) 726–732.
[61] S.M. Sarge, G.W.H. Höhne, W. Hemminger, Calorimetry, Wiley-VCH, Weinheim, Germany, (2014).
[62] J. Bebernes, D. Eberly, Mathematical Problems from Combustion Theory, Applied Mathematical Sciences, vol. 83, Springer, New York, NY, 1989.
[63] V. Volpert, Elliptic Partial Differential Equations, Volume 2: Reactions-Diffusion Equations, Birkhäuser, Basel 2014.
[64] D.A. Frank-Kamenetskii, Diffusion and Heat Transfer in Chemical Kinetics, Plenum Press, New York, NY, (1969).
[65] A. Melcher, C. Ziebert, M. Rohde, B. Lei, H.J. Seifert, Modeling and simulation of the thermal runaway behavior of cylindrical Li-ion cells—computing of critical parameters, Energies 9 (2016) 292.
[66] G. Guo, B. Long, B. Cheng, S. Zhou, B. Cao, Three-dimensional thermal finite element modeling of lithium-ion battery in thermal abuse application, J. Power Sources 195 (2010) 2393–2398.
[67] K.-C. Chiu, C.-H. Lin, S.-F. Yeh, Y.-H. Lin, K.-C. Chen, An electrochemical modeling of lithium-ion battery nail penetration, J. Power Sources 251 (2014) 254–263.
[68] Q. Wang, P. Ping, J. Sun, Catastrophe analysis of cylindrical lithium ion battery, Nonlinear Dyn. 61 (2010) 763–772.
[69] Q. Wang, P. Ping, X. Zhao, G. Chu, J. Sun, C. Chen, Thermal runaway caused fire and explosion of lithium ion battery, J. Power Sources 208 (2012) 210–224.
[70] D. Lisbona, T. Snee, A review of hazards associated with lithium and lithium-ion batteries, Process Saf. Environ. Prot. 89 (2011) 434–442.
[71] N.N. Semenoff, Zur Theorie des Verbrennungsprozesses, Z. Phys. 48 (1928) 571.

CHAPTER SEVEN

Life Cycle Assessment of Nanotechnology in Batteries for Electric Vehicles

L. Oliveira, M. Messagie, S. Rangaraju, M. Hernandez, J. Sanfelix, J. Van Mierlo
Mobility, Logistic and Automotive Technology Research Center (MOBI), Vrije Universiteit Brussel, Brussels, Belgium

Contents

1 INTRODUCTION

1.1 Problem setting and environmental concerns related to nanotechnology

Engineered nanomaterials (ENMs) and the development of innovative nanoscale technology is a growing field that promises plenty of opportunities for improvement and optimization of our current industrial practices [1–3].

The growing use of nanomaterials and nanoproducts entails responsible development, ensuring that innovation does not interfere with safe practices for the environment and human well-being. As is often the case with novel technology, uncertainties on the potential environmental, health, and safety concerns related to nanotechnology have arisen [4,5]. It has been demonstrated, however, that nanotechnology applications can bring numerous benefits, for example, to the field of medicine; in electronics as

Emerging Nanotechnologies in Rechargeable Energy Storage Systems

semiconductors, batteries, and energy efficient systems; in communications and information systems; among many others, including providing benefits to the environment [1]. Given the rapid expansion of nanotechnology applications, environmental scientists have taken on the task to understand and clearly define all potential environmental concerns and benefits associated to nanomaterials and nanotechnology. Life cycle assessment (LCA) methodology has an important role in achieving this.

In general, LCAs address the potential environmental and human health impacts for any product system during its entire life cycle, including extraction of raw material, manufacture, use, end-of-life (EOL) treatment, recycling, and final disposal. LCA is backed by international standards, namely ISO 14040 and 14044 [6,7], numerous papers, and case studies. LCA accounts for impact categories, such as climate change (CC), toxicity of water and air bodies, ozone depletion, and particulate matter generation, among others. As exemplified in Fig. 7.1, all inputs and outputs corresponding to each life cycle stage are the subject of a quantitative and qualitative analysis, by means of characterization, aggregation, and interpretation of results, to define their potential environmental impacts.

So far, LCA is seen as an appropriate tool for comparing new technologies to already existing technologies [8]. Such is the case of nanoproducts, made from incorporated nanomaterials (materials with at least one dimension of 1–100 nm), and nanoparticles (particles with all three dimensions within 1–100 nm) [9]. LCAs are able to show opportunities for preventing pollution and reducing materials consumption through the use of nanoproduction, use and disposal, as seen in Fig. 7.1. On the other hand, for nanoscale materials and technology, there are certain limits to the application of LCA, in particular with respect to the assessment of toxicity of ENPs and large-scale production impacts [1].

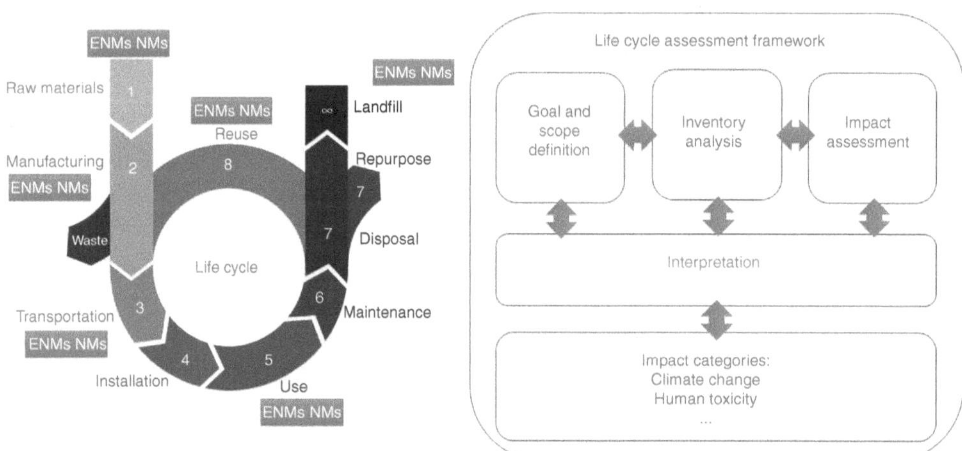

Figure 7.1 ***Life cycle assessment (LCA) methodology and the environmental impacts of nanotechnology.*** *ENMs*, Engineered nanomaterials; *NMs*, nanomaterials.

Several studies in the LCA field have focused on nanomaterials and nanoproducts, as well as the drawbacks and limitations for the application of this methodology [10–13]. The following sections give a state of the art overview of nanotechnology and their potential environmental impact as identified by LCA analyses.

In addition, a LCA case study is performed later in this chapter, where the application of Si nanowire (SiNW) structures to battery elements, specifically anodes, is analyzed. The comparison is made between nickel manganese cobalt oxide (NMC) battery with a graphite anode and a similar one, doped with SiNWs at the anode side. Results demonstrate that batteries behave, environmentally wise, differently throughout the manufacturing and recycling stages. Very interesting trade-offs are discussed further ahead in the chapter.

ENMs are designed to provide a wide variety of advantageous functions in different fields of applications. These advantageous functions are the result of intrinsic and extrinsic attributes of the nanoscale materials, which differentiates them from their bulk material counterparts [14]. The intrinsic attributes of the ENMs, such as chemical composition, particle size distribution, surface area, and morphology, as well as extrinsic attributes, for example, the manufacturing, transport, and storage of the ENMs, is what allows for optimized performance in terms of strength, resistivity, reactivity, and electrical behavior, among others [15]. As identified by Dhawan and Sharma [14], the size of nanoparticles is perhaps the most influential property causing a deviation in the behavior of nanomaterial from that of the bulk; as particle size decreases a greater proportion of atoms is located at the surface. Hence, nanoparticles have a greater surface area when compared to larger particles, which, combined with quantum effects, can for example affect their reactivity and strength [4].

These unique properties of nanomaterials are often associated with positive expectations in areas, such as optimization of materials and energy usage [5]. On top of functionality advantages, nanomaterials are expected to bring positive impacts to the environmental performance of the products by means of weight reduction, greater energy efficiency, or reduced emission of waste and detrimental gases. As per Ref. [15], examples of improved environmental performance by nanoproducts use are carbon nanotubes (CNTs) and aerogels that can be used in nanocomposites in automotive applications to reduce weight and therefore increase the fuel efficiency; thus, reducing emissions of gases and particles to the atmosphere. Another example is nanomaterials in the catalytic conversion technologies, reducing unburned hydrocarbons, particulate matter, and other emissions in vehicle engines.

The different chemical compositions and manufacturing processes associated with nanotechnology, have been commonly related to negative environmental impacts, particularly in the stages of raw materials mining and processing, and also in the very energy intensive processes in the manufacturing of the components, as per Refs. [8,9]. Unlike many conventional manufacturing techniques, nanomanufacturing requires unique facilities and process designs, potentially originating additional or at least different types of environmental impact comparing to conventional manufacturing. Throughout the production, use, and disposal of ENMs, the release of nanoparticles to water, soil, and air bodies could also contribute to environmental impact, which is a topic currently addressed by environmental scientists [5].

There are ongoing efforts in the application of LCA to the field of nanotechnology; however, in spite of the necessity for LCA assessments directed to ENMs, so far there is still a lack of comprehensive assessments focused on this topic [16–18]. As described by Miseljic and Olsen [17], only a small fraction of the studies cover some aspects of the specific life cycle of nanomaterials.

A large part of the studies performed so far mainly focuses on the manufacturing stage of the nanotechnology. Moreover, the majority of them limit their focus to the issues of energy and materials consumption, not considering nanospecific fate, transport, and toxicity issues [16]. Data on production of ENMs are scarce, for which most of the studies rely on generic data, and in the cases when primary data are used, normally this one is not openly disclosed due to confidentiality agreements with the industry [17].

The life cycle stages of use and disposal of the ENMs is often not considered in the assessment, resulting in an availability of mainly cradle-to-gate studies. This situation is generally attributed to the lack of data and knowledge to represent these life cycle stages of the products. In the case of the disposal of the ENMs, it is still difficult to define whether these materials will eventually be recycled, landfilled, or disposed off as waste for incineration. There is also a lack of studies addressing the release of nanoparticles to the environment [17]. For the moment, only one study, by Walser et al. [19], deals with potential toxicity impacts in the releases of nanoparticles, particularly nanosilver. When making reference to emissions related to the manufacturing stage, generally these refer to better understood emissions, such as carbon dioxide, acidifying pollutants, biological oxygen demand, and nitrogen dioxide, among others. Environmental pathways of nanoparticles are shown in Fig. 7.2.

The limited coverage of LCA studies in the field of ENMs can be associated with two main challenges inherent to the LCA methodology: first, the lack of reliable and appropriate inventory data for the materials used in the products, manufacturing processes of nanospecific technology, as well as data to represent use and EOL scenarios [1]. And second, the availability of characterization factors for ecotoxicological impacts of ENMs, and other physicochemical characteristics of ENPs, which are not adequately understood in relation to environmental behavior and effects [17]. Fig. 7.2 shows the potential pathways of release and fate of ENMs and particles. There is a need for further development of impact assessment methods dedicated to nanospecific technology. However, this one is constrained by the availability of nanospecific databases and laboratory tests.

So far, as reviewed by Gavankar et al. [16], several tools and techniques have been used to quantitatively assess the fate, transport, or toxicity of chemicals and bulk materials,

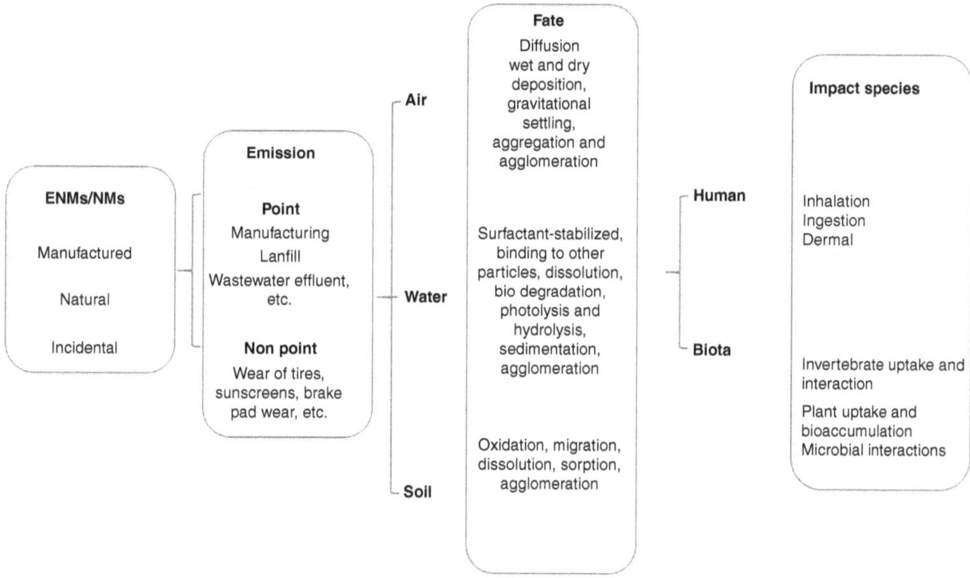

Figure 7.2 ***Environmental pathways for the release of nanomaterials to the environment [11].*** *SiNW,* Silicon nanowires.

including USEtox [20], CalTOX [21], and QSAR [22]. However, their usefulness in the field of nanomaterials is still questionable, as they do not take into consideration the intrinsic and extrinsic properties of nanomaterials [16,17].

In a cradle-to-gate comparison between products with ENMs and without, results of the studies show that nanoproducts tend to demand more energy and have an inferior environmental profile compared to traditional materials, as identified by [23]. Khanna et al. [24] quantified the energy requirements for CNFs and found this energy demand to be 13–50 times higher than the production of primary aluminum for the same mass basis. Similar results, in terms of high-energy consumption, were found for CNFs over steel and polypropylene. Meyer et al. [25] compared socks with and without nanomaterials, and found that although the use of nanocomponents is small, the effects of manufacturing can be significant depending on the manufacturing techniques, and the scale of production at regional and global levels. In addition, Miseljic and Olsen [17] found that 1 g of Ag or TiO_2 poses a higher ecotoxicity impact, up to c. 3.5 orders of magnitude, than the production of 1 g polymer (PP, PE, and PET average) or 1 Wh of grid mix electricity from Scandinavia.

On the other hand, Lloyd and Lave [26] showed that the use stage for ENM products is more environmentally friendly than that for their conventional counterpart products, by assessing clay–polypropylene ENMs in car body panels, instead of steel or aluminum in light-duty vehicles. The composite design used in the analysis showed potential benefits in reducing energy use and environment discharges; however impacts, such as potential ENP release were not considered. Similarly, Khanna and Bakshi [10] showed that the use stage for ENM products, used in body panels in automobiles, is better than for conventional materials, as an improved functionality is achieved.

In general, it seems that the quantity of use of these ENMs and resulting benefits will decide their energy and environmental impacts.

Christensen [8] identified important parameters to be considered when performing an LCA study focused on nanotechnology:

1. Include scarce materials, such as indium, cerium, lithium, and determine up to which point they can be recycled.
2. Include energy consumption during manufacturing of the nanomaterials, as it has been identified previously as considerable.
3. Include toxic materials that are generated during the production of nanomaterials, throughout the life cycle, or those used for the production.

1.2 Life cycle assessments and battery nanotechnology

In the field of traction Li-ion batteries, the use of nanomaterials is seen as promising given their convenient mechanical and electrical properties [27,28]. Much of the research efforts is oriented to the use of nanomaterials in battery anodes, using novel materials to increase the overall energy density of the battery [28,29]. So far, it has been difficult to assess with precision whether the use of ENMs in batteries is justified from an environmental point of view. The technical advantages of nanomaterials in batteries, such as high-charge densities, high-current capability, good surface area per mass, fast charge/discharge time cycles, and so forth, are clearly beneficial for the overall functioning of the battery. However, as recent LCA studies have revealed, there are concerns related to the impact contributions of the battery nanomaterials during production [30]. As with other nanoscale processes in different applications, high-material and energy intensity and the use of toxic chemicals are the main environmental concerns raised in environmental assessments [1,25,30].

Wender and Seager [31] used LCA to quantify the energy trade-offs associated to nanoenabled Li-ion batteries. The assessment is based on laboratory-scale data in the manufacturing of single wall carbon nanotubes (SWCNT) through laser vaporization production method. The authors reported on the range of materials and energy inputs required per kilowatt hour of battery energy storage capacity, as well as waste outputs of SWCNT manufacturing. Results of this study show that the energy demand of freestanding SWCNT anode manufacturing is 2 orders of magnitude larger than the sum of all other Li-ion battery–manufacturing processes [31]. Results are expected to be similar for other nanomanufacturing processes. Similarly, Amarakoon et al. [30] made an environmental analysis in the application of SWCNTs in Li-ion batteries for vehicles. The report identified the energy consumption of SWCNT anodes to be significantly greater than for conventional graphite anodes, as currently produced in laboratory scale. Both studies referred to the possibility, importance, and challenges of reducing the energy required in the production of SWCNTs in future commercial scales.

More recently, Li et al. [28] performed a study focused on SiNW used in Li-ion battery anodes. The study covered the complete life cycle of this new battery, including characterization and reporting of nanowastes and nanoparticles emissions from the SiNW synthesis. For the majority of environmental impacts considered, including global warming, human toxicity (HT), and photochemical oxidation potential, the battery use contributes to more than half of the impact. The manufacturing of the SiNW anode contributes to 15, 10, and 17% of total life cycle impacts, for the previously mentioned impact categories, respectively. This impact is mainly attributed to the large amounts of embedded energy and toxic chemicals. In a comparison with conventional graphite anode–based batteries, SiNW anode batteries show larger environmental impacts between 6% and 43% for the impact categories considered. The largest difference is found in the category of HT, and it is attributed to the use of toxic chemicals (HF and HNO_3) in the synthesis of the SiNW materials. The authors conclude that the difference in impact generated by the two battery types is moderate, considering lab-scale inventory data uncertainties and the potential reduction of environmental impact in the scale-up of production levels.

As stated by Seager et al. [27], the total contribution to the environmental impact of nanotechnology in batteries will be the result of a balance between impact during production, and benefits delivered during their use in the electric vehicle (EV) applications, as well as the reach of recyclability in the sector. These benefits will increase proportionally in the lifetime of the battery cells and the energy inputs and pollution outputs.

Three main challenges remain in the application of LCA for nanotechnology for batteries:

1. the lack of inventory data at both the laboratory scale and commercial scale,
2. the development of nanospecific methodologies for impact assessment, and
3. the further inclusion of use and EOL cycle stages in the studies performed.

1.3 Life cycle assessment methodology

The process-based LCA (P-LCA) [32] methodology is a tool to evaluate the environmental burdens associated with a product, process, or activity. The method is mostly standardized and according to ISO 14040 [33]. P-LCA can assist in:

- identifying opportunities to improve the environmental performance of products at various points in their life cycle;
- informing decision makers in industry, government, or nongovernment organizations (e.g., for the purpose of strategic planning, priority setting, and product or process design or redesign);

- the selection of relevant indicators of environmental performance, including measurement techniques; and
- marketing [e.g., implementing an ecolabeling scheme, making an environmental claim, or producing an environmental product declaration (EPD)].

1.3.1 Goal and scope definition

This phase documents the description of the purpose and scope of the study, the definition of the functional unit, the description of the system boundaries, and the methodological approach implemented.

1.3.1.1 Goal and scope

A LCA study can be performed for different purposes, for example, comparing the environmental performance of a product with a competitor that has the same function, or determining the life cycle stages that contribute more to certain impact categories. Based on this information it is possible to propose environmental improvements that lead to a reduction in the overall damages. Thus, the aim of the study has to be defined with precision. Furthermore, the target audience of the study has to be specified, as well as if the results will be used in a comparative assertion to be disclosed to the public (e.g., EPD).

1.3.1.2 Functional unit

The functional unit defines what is being studied in coherence with the goal and scope, and describes the function of the system. One of the aims of defining a functional unit is to have a reference point to normalize the data collection and the results obtained in the impact assessment, as well as to have a reference flow that links all the processes involved in the analysis. Therefore, the functional unit has to be clearly defined and it has to be measurable. Another purpose of the functional unit is to facilitate the comparison of the results when different systems are analyzed in the study.

1.3.1.3 System boundaries

The system boundaries are understood as the definition of what is included within the studied system and what remains outside. The system boundaries determine which unitary processes should be included in the LCA. These boundaries have to be consistent with the goal and scope as previously defined. The criteria implemented to establish the system boundaries have to be described and justified in this phase of the study.

1.3.2 Inventory analysis

The inventory phase comprises the process of compiling the data to quantify the input and output flows (mass; energy; emissions to air, water, and soil; and coproducts) during the entire life cycle of the studied system. The collected inventory has to be in accordance with the defined functional unit and the system boundaries.

The data collection for the inventory concerns two types of data: primary and secondary. Primary data refers to data collected with direct measurements or calculated by the LCA practitioner, normally used in the main stages of the life cycle of the analyzed system. Secondary data is the data obtained from databases, are more generic, and are used for the background processes.

The inventory phase is one of the largest phases in a LCA, as collecting data is a laborious task and it requires a large investment with regard to time and resources. The data quality requirements, previously defined, will be considered during the preparation of the inventory.

1.3.3 Impact assessment

The impact assessment is the third of the four phases of a LCA study. The purpose of the impact assessment is to evaluate the results of the life cycle inventory (LCI) analysis performed in the previous phase, to understand the environmental implications during the entire life cycle; hence, the term *life cycle impact assessment* (LCIA) [34]. In general, this phase implies the association of inventory data with specific environmental impacts, and the quantification of those impacts. The detail level, the impact methods selected, and the methodology to calculate them depend on the goal and scope of the study.

Two types of methods are distinguished to classify and characterize the impacts: midpoint, or problem-oriented approach, and endpoint, or damage-oriented approach. The term midpoint expresses the fact that this point is located somewhere on an intermediate position between the LCI results and the damage (endpoint) on the impact pathway. Midpoint methods quantify the environmental impact of different impact categories, such as global warming potential, acidification, and so forth. These results are subjected to lower modeling requirements than endpoint methods, and therefore they have lower uncertainty. The endpoint methods analyze the last environmental effect, thereby identifying and analyzing the damage of those impacts on human beings and nature. On one hand, endpoint methods are more difficult to estimate, but on the other hand, they provide results in a more understandable way for any type of audience.

1.3.4 Interpretation

In this LCA phase, the LCIA results are combined with an evaluation and analysis of the impact, consistent with the defined goal and scope. The interpretation of the results can be implemented as recommendations for decision making. Sensibility analysis can be performed to analyze the fluctuation in the results due to a variation in relevant processes of the studied system. The consistency of the results is of high importance and it can be evaluated by experts in the area or compared with similar studies.

2 CASE STUDY: USE OF NANOMATERIALS IN LI-ION BATTERY ANODES

2.1 Goal and scope of the analysis

2.1.1 Goal

The analysis focuses on the comparison of two different types of NMC-based Li-ion batteries, both with carbon-based cathode materials and one with selected nanomaterials implemented in the anode. The intended use of the analysis is to highlight the environmental burdens and benefits of using nanomaterials in anodes for Li-ion batteries based on NMC cathodes. The target audience for such results is OEMs that would like to use NMC-based batteries in their applications, and would like to estimate the benefits and damages of their choice in different battery compositions and the potential role of nanomaterials use.

The functional unit for the analysis is 1 kWh of energy stored and delivered (over the course of the vehicle lifetime, 200,000 km over the course of 10 years). This choice will reflect more impacts of the manufacturing of the battery system. If a different functional unit were to be chosen, such as impacts per kilometer driven, the functionality of the battery itself could be seen as a promoter of electrical mobility (while being a fundamental part of an EV). This problem is very common in almost every study that addresses LCAs of battery systems for EV applications. For example, on one hand, using the kilowatt hour problem of multifunctionality from a first life and consequential second one can be easily tackled (i.e., first life traction battery, second life, and stationary application). This approach leaves the vehicle itself out of the picture, its efficiency, and other important factors. In addition, the comparison with other types of vehicles is not very suitable, as there are not many means of relating both. On the other hand, using kilometers driven, solves some of the earlier issues, but only up to the EOL stage of the product. The multifunctionality of the different battery systems would not be fully exploited, as this would not be reflected in the results. The choice for kilowatt hour as the preferred functional unit serves the purpose and the scope of the analysis by comparing mostly manufacturing and EOL (recycling) scenarios for the two battery systems.

2.1.2 Scope

The system boundaries comprehend the resource extraction, manufacturing, assembly, distribution, usage, and disposal/EOL stages. There is no specific geotemporal scope for the analysis. Data quality is directly related to the accuracy of the life cycle inventories. Uncertainty analysis is not part of the study itself.

2.2 Life cycle inventory of Si nanowire-based batteries and conventional graphite anode-based batteries

2.2.1 Battery characterization

One of the biggest advantages of using nanoscale anode materials is the reduction of the total weight of a battery pack, alongside with the increased capacity per mass unit.

Table 7.1 Battery characterization parameters [28]

Type	Capacity (kWh)	Weight (kg)	Topology	Charge discharge efficiency (%)	Cycle life (80% SOH) (cycles)	Total lifetime energy delivered (kWh)	Energy density (Wh/kg)
NMC cathode (graphite-based anode)	43.2	360	36 Modules of 12 cells each	90+	~3.000	116.640	140
NMC cathode (SiNW anode)	43.2	120	12 Modules of 12 cells each	95+	~1.000	41.040	360

NMC, Nickel manganese cobalt oxide.

Table 7.1 illustrates the main characteristics from the battery systems at pack level (integration in EV), which are relevant to the environmental life cycle analysis.

Within the vehicle application, an average energy consumption of 0.2 kWh/km is considered (measured empirically throughout a multiyear study of a fleet of EVs) over the course of 200,000 km (normal driving distance over a 5- to 10-year span) [35,36]. Both batteries do, at least, match the requirements; such is the case of SiNW–NMC battery that exceeds in total available energy at vehicle EOL in the case of the conventional NMC. The advantage highlighted is both the material and weight reduction by the use of the nanomaterial anode over the conventional.

2.2.2 Manufacturing stage

The manufacturing material requirements for both battery systems are shown in Tables 7.2 and 7.3. The subprocesses that lead to the main components are not displayed. The amount of materials is presented in total weight/mass and total percentage. The conventional NMC inventory is sourced from Majeau-Bettez et al. [29], which is widely recognized as a reference work in the field. The SiNW production and battery manufacturing process is sourced from Li et al. [28].

2.2.3 Use stage

The use stage of the battery pack within an EV setting is dependent on two factors: EV efficiency and performance of the electricity mix used to charge the battery system. The assumption that EV efficiency is considered for the analysis. The electricity mix selected was the one modeled by Rangaraju et al. [37,38] and refers to the electricity supply scenario in Belgium for the year 2014. The shares of each production unit can be seen in Table 7.4.

Table 7.2 Life cycle inventory (LCI) of conventional graphite anode/NMC (1:1:1) cathode Li-ion battery pack [29]

Component	Subassembly	Mass (kg)	Percent mass (%)
Anode (94:3:3)	Artificial graphite	45.00	12.50
	Carbon black	1.44	0.40
	Copper foil	19.08	5.30
Cathode (92:4:4)	Aluminum foil	17.28	4.80
	Carbon black	2.16	0.60
	NMC (1:1:1)	48.60	13.50
Binder	SBR/CMC	1.44	0.40
	PVDF	2.16	0.60
Separator	Celgard polyethylene	7.92	2.20
Casing	Poly–aluminum–poly	44.28	12.30
Electrolyte	Lithium hexaflourophosphate ($LiPF_6$, 1 M)	4.32	1.20
	Ethyl carbonate and ethyl methyl carbonate	37.80	10.50
BMS	Copper wires	3.60	1.00
	Stainless steel	2.88	0.80
	Printed circuit board	0.72	0.20
Pack housing	Polyethylene terephthalate	61.20	17.00
PCS	Stainless steel	1.80	0.50
	Aluminum	58.32	16.20
	Total	360	100

BMS, Battery-management system; CMC, carboxymethylcellulose; PCS, passive cooling system; PVDF, polyvinylidene fluoride; SBR, styrene-butadiene rubber.

2.2.4 End of life

The EOL scenario was treated as a hydrometallurgical process where the battery materials are either recovered toward a secondary use or disposed to inert landfill after energetic recovery. This process is also known as the Umicore process [39]. The nanomaterials present in the anode are not yet designed for recovery after service life. The collection rates for the batteries and recovery of the materials were updated to reflect real world scenarios. These values are embedded in the life cycle model and their scores included in the results [40].

3 LIFE CYCLE IMPACT ASSESSMENT

The results of the LCIA include the LCI of all the environmentally significant activities, such as the materials input, energy consumptions, and environmental discharges. They are converted into communicable environmental impact indicators, for example, CC. First, the LCI data have been modeled using the widely used Ecoinvent database [41], in SimaPro v8.3, a dedicated life cycle modeling software. The "Recipe2009" [41]

Table 7.3 LCI of the SiNW anode/NMC (1:1:1) cathode Li-ion battery pack [28]

Component	Subassembly	Mass (kg)	Percent mass (%)
Anode (7:2:1)	SiNW	10.80	9.00
	Carbon black	3.12	2.60
	Copper foil (12 μm)	9.36	7.80
Cathode (92:4:4)	Aluminum foil	10.26	8.55
	Carbon black	0.54	0.45
	NMC (1:1:1)	11.88	9.90
Binder	SBR + CMC binder (10–20 μm, 1:1)	1.56	1.30
Separator	PE (20 μm, 1.2 g/cm^3)	2.64	2.20
Casing	Poly–aluminum–poly (200 μm)	14.76	12.30
Electrolyte	Lithium hexaflourophosphate ($LiPF_6$, 1 M)	12.24	10.20
BMS	Copper wires	1.20	1.00
	Stainless steel	0.96	0.80
	Printed circuit board	0.24	0.20
Pack housing	Polyethylene terephthalate	20.40	17.00
PCS	Stainless steel	0.60	0.50
	Aluminum	19.44	16.20
	Total	120.00	100.00

Table 7.4 Belgium electricity mix for the year of 2014 [37]

Production unit	Coal	Fuel	Gas	Nuclear	Hydro	Wind	Solar	Wood	BFG	Imports
Production share	3%	0%	21%	40%	2%	3%	4%	1%	4%	21%

impact assessment method has been used for the analysis. With this selected method, it is possible to assess 18 different environmental indicators, such as CC, fossil depletion, acidification, and eutrophication. However, in this study we assessed only three important impact categories, namely (1) CC, which is the most commonly communicated indicator in form of CO_2 equivalent emissions; (2) cumulative energy demand (CED); and (3) HT.

In the model, several assumptions are made to allow a simplification of the calculation process. As per energy densities, it is assumed that an amount of 2.5 kg of SiNW–NMC battery and 8.9 kg of conventional NMC battery is needed to provide 1 kWh of energy (the selected functional unit). The model reflects the functionality of the battery itself in providing energy (kWh) over the vehicle lifetime and not kilometers driven; it is assumed that is the function of a vehicle and not a battery system directly. The rest of the power train is not considered in the model in this study. It would be expected that in a model that uses vehicle kilometers as a functional unit, the advantages of an extra number of cycles of a conventional NMC battery would clash with the reduced weight over the vehicle lifetime of the nanomaterial NMC anode battery. Nevertheless, this is not the case, as energy delivery is the main point under analysis. A second life model for the battery systems is also not addressed in the analysis as it lies outside the defined scope.

3.1 Climate change

Climate change (also known as global warming) is caused by the emissions of greenhouse gases (GHG) from the anthropogenic activities. Human activities, mainly the burning of fossil fuels in energy production and industrial processes, deforestation, and so forth have emitted a large amount of GHG, including carbon dioxide. GHG act like a blanket, absorbing infrared radiation energy from earth and maintaining the global temperature. This phenomenon is commonly known as the greenhouse effect. However, the excessive emissions of these gases leads to overheating of Earth and, thereby, causing severe damage to the ecosystem and human health. In the "Recipe" method, all the GHG substances [e.g., methane (CH_4) and dinitrogen monoxide (N_2O)] are characterized to the equivalent of CO_2 and it is expressed in kilogram CO_2 equivalent (CO_{2eq}).

Fig. 7.3 compares the climate change potential of the two different battery technologies analyzed in this study. It is visible that the SiNW battery has a higher CO_{2eq} value compared to the conventional NMC battery. The slight difference is due mainly to the production and the recycling stages of the batteries. The $CO_{2eq.}$ emissions in the use stages of the batteries correspond to the emissions associated with the electricity used for charging the batteries (i.e., the Belgian average electricity in this case). Use stage comparison compares the efficiencies of the two battery chemistries. As the difference between the efficiencies of the NMC/SiNW (97%) and NMC/graphite battery (90%)

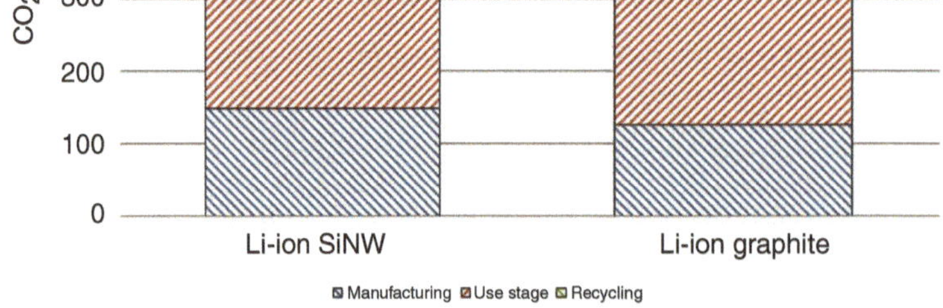

Figure 7.3 ***Comparison of life cycle climate change (CC) potentials of NMC/SiNW and NMC/graphite battery.***

is not large, there is not a huge difference in the magnitudes of the use stage emission of the two batteries. Moreover, in both chemistries, the use stage contributes the largest share of the CO_{2eq} emissions. This is because the manufacturing and recycling processes are one-time processes in the life cycle, whereas the charging electricity is needed throughout the life cycle.

In the NMC/graphite battery, the CO_2 emissions in the production phase arise mainly from the materials and manufacturing of the carbon-based positive electrodes, followed by negative electrodes, cell container, and the battery-management systems (BMS). Also, the net electricity used in the manufacturing and assembly processes of the unit contributes to a significant portion of emissions. In the case of NMC/SiNW battery, the production of the SiNW anode is the main contributing process for CO_{2eq} emissions in the production stage.

During the recycling stage of both batteries, there is little CO_{2eq} benefit from the material recovery, mainly of aluminum and copper. Nevertheless, the main driver for the recycling of NMC batteries is the cobalt and energy recovery through a pyrometallurgical method (highly energetic). It is assumed that the recovered materials replace virgin materials, and thereby avoid the emissions associated in the life cycle of virgin materials. However, in this case, the $CO_{2eq.}$ benefit of the material recycling is not greater than the emissions during the recycling processes and thus, a positive score for this impact category is achieved.

3.2 Cumulative energy demand

This indicator quantifies the entire demand of primary energy in all the unit processes throughout the life cycle stages of the battery technologies. It considers different forms of energy sources, for example, fossil fuels, nuclear, nonrenewable biomass, and renewable sources, and it is expressed in megajoules (MJ). However, in this study, we consider only the nonrenewable forms of energy demand in the life cycle.

Fig. 7.4 shows the energy demand scores of the two batteries. The manufacturing and the recycling of the NMC/SiNW battery is more energy intensive than the conventional NMC/graphite battery. The production of SiNW for the anode is responsible for a large fraction of the energy demand in the manufacturing of the NMC/SiNW battery. The use stage energy demands correspond to the primary energy (e.g., coal, nuclear fuel, oil, and natural gas) used in the electricity production. In general, in both batteries, the electricity consumptions (and thereby the energy used in the upstream of electricity production during the production and recycling stages) make up the majority of the energy demand.

3.3 Human toxicity

This indicator quantifies the potential of the analyzed system to cause human health effects due to the emissions of toxic substances in the life cycle. Recipe method takes

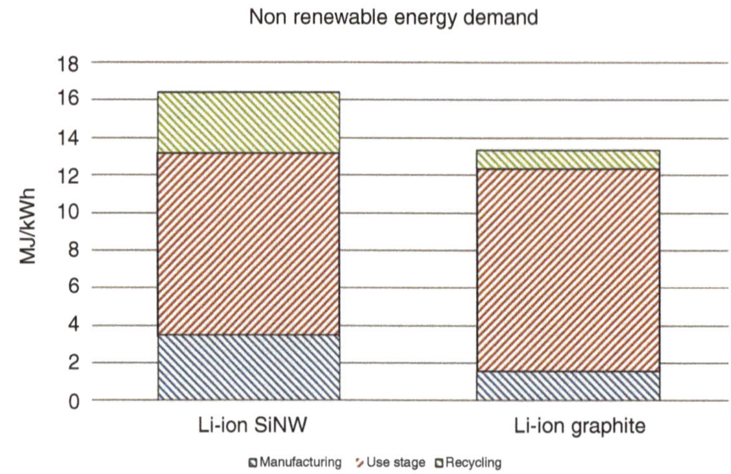

Figure 7.4 ***Comparison of life cycle cumulative energy demand (CED) of NMC/SiNW and NMC/graphite battery.***

into account the environmental persistence and accumulation of toxic substances in the food chain, together with the level of toxicity of those chemicals. Dichlorobenzene (1,4-DB), which is a well-known pesticide, is used as a reference substance for normalizing the toxicity of the other substances contributing in this impact category [41].

Fig. 7.5 shows the HT potentials of the two batteries. Unlike in the CC and energy demand indicators, the NMC/SiNW battery has a better score for HT, compared to the

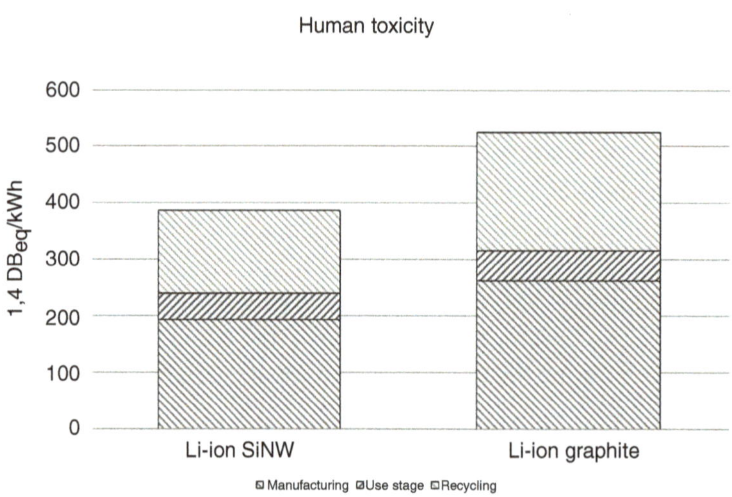

Figure 7.5 ***Comparison of life cycle human toxicity (HT) potentials of NMC/SiNW and NMC/graphite battery.*** 1,4 DB_{eq}, 1,4-Dichlorobenzene equivalent.

NMC/graphite battery. One of the reasons can be the fact that the LCI, available for the emissions and effects of the toxic substances from the manufacturing stage of the NMC/SiNW battery, is very limited due to the lack of studies on environmental damages of nanomaterials. At the present moment there is no impact assessment method that fully characterizes the ecotoxicity levels of nanomaterials.

The recycling stage of this battery (as well as the NMC) is modeled as per the general industrial practice, in which only the standard set of materials (e.g., cobalt, aluminum, copper, and iron) is recovered. The nanomaterials present in the cells are not designed or purposed for any kind of recycling process.

The contribution analysis in the manufacturing stage of NMC/SiNW battery reveals that the toxic substances are discharged mainly in the mining of raw materials used in the production of the anode and cathode materials, and the specific metals used in the printed wiring boards of the BMS. For instance, the mining of gold, which is used in the printed wiring boards, involves a large amount of sulfidic tailings (it contains a large amount of toxic substances) in the mining fields. Comparable to the NMC/SiNW, the toxic emission in the manufacturing stage of the NMC/graphite battery originates mainly from the BMS, followed by the positive and negative electrodes. In the recycling stages of both batteries, a significant amount of human toxic emissions are avoided by recycling, mainly aluminum, copper, and iron. However, the benefits are not sufficient to negate the emissions in the recycling processes. The use stage emissions occur mainly in the mining of fuels, such as nuclear and coal in the electricity production chain.

4 DISCUSSION AND CONCLUSIONS

Both batteries fulfill their respective functions as providers of energy for electrical vehicle mobility. Although, very different outputs from the assessment can be seen from the different impact categories.

The inclusion of SiNWs in the anode does provide more capacity at the expense of life cycle. Another evident benefit is the outstanding weight reduction that favors its application as an EV battery. The total cycles that the analyzed NMC/SiNW battery was designed to endure sit closely to the 1000 mark (full charge/discharge cycles), which with the rated capacity, seems adequate for the 200,000 km over 10-year span benchmark.

The conventional NMC/graphite battery benchmarked against the NMC/SiNW system also proved to be more than sufficient as an EV-energy provider. This battery system, with the increased number of cycles is completely adequate (under normal conditions) not only for the EV operation, but also for a robust second life application as backup or primer for renewable energy systems (i.e., residential PV) and regular residential backup. The number of cycles exhausted from the battery after 200,000 km is short above 1,400, leaving an interesting margin for an application that precedes its EOL [35,36].

Analyzing the impact assessment results, specifically climate change, the scores are comparable, being the recycling stage the process where the differences are more noticeable when comparing to other product stages. Also concerning climate change, there are no obvious benefits in EOL procedures as these are mostly energy intensive and dependent. This benefit would be observed in other impact categories from "ReciPe2009," such as metal and fossil depletion. A similar trend is observed in the CED impact category as the energy intensiveness of recycling and production processes. Although the battery weight of the SiNW is one-third of the conventional NMC, apparently the material requirements are reduced as well. Unfortunately, the production processes of the nanomaterials require an outstanding amount of material input per output unit. The energetic requirements for this stage are also significantly high when compared with the benchmark NMC battery.

Although these caveats are significant, it cannot be overlooked that the weight reduction impact on the EV use stage can be of paramount importance. This weight/size reduction enables design opportunities that can potentate the usage of the EV. Even a higher capacity battery system, enabling more driven kilometers is a possibility (albeit the extra cost for the buyer). The application of SiNW materials at an anode level also enables the production of "tailor-made" batteries.

Finally, the use stage is totally dependent on the performance of the electricity mix. For this analysis, the Belgian electricity mix for 2014 was used as a benchmark. This mix is dominated by nuclear electricity production, as well as by natural gas. In a mix with a significant better environmental performance, less impact would be observed throughout all the impact categories. Even going a step further, both batteries would benefit equally from a 100% renewable energy mix. With minimal energy conversion impacts, the use stage could be considered mostly clean.

Nanomaterials pose excellent opportunities for battery systems and EV applications. The introduction of new materials, such as graphene nanosheets and nanospheres is a step toward the right direction. Environmentally, more information related to the manufacturing and disposal of the materials should be disclosed to correctly account for the eventual damages to the ecosystems.

GLOSSARY

BMS Battery-management system
CC Climate change
CED Cumulative energy demand
CMC Carboxymethyl cellulose
CNF Carbon nanofiber
CNT Carbon nanotube
$CO_{2eq.}$ Carbon dioxide equivalent
1,4-DB_{eq} 1,4-Dichlorobenzene equivalent

ENM	Engineered nanomaterial
ENP	Engineered nanoparticle
EOL	End of life
EV	Electric vehicle
FU	Functional unit
HT	Human toxicity
LCA	Life cycle assessment
LCI	Life cycle inventory
LCIA	Life cycle impact assessment
MJ	Megajoule
NMC	Nickel manganese cobalt oxide
PCS	Passive cooling system
PE	Polyethylene
PET	Polyethylene terephthalate
P-LCA	Process-based LCA
PP	Polypropylene
PVDF	Polyvinylidene fluoride
SBR	Styrene–butadiene rubber
SiNW	Silicon nanowire
SWCNT	Single wall carbon nanotube

REFERENCES

[1] M.A. Curran, P. Frankl, R. Heijungs, A. Köhler, S.I. Olsen, Nanotechnology and life cycle assessment, in: A Systems Approach to Nanotechnology and the Environment: Synthesis of Results Obtained at a Workshop, Washington, DC, October 2–3, 2006.

[2] S. Panero, B. Scrosati, M. Wachtler, F. Croce, Nanotechnology for the progress of lithium batteries R&D, J. Power Sources 129 (2004) 90–95.

[3] J. Loeffler, Overview on promising nanomaterials for industrial applications, Nanomaterial Roadmap 2012, Sixth Framework Programme, NanoRoad SME European Project, Karlsruhe, 2005.

[4] The Royal Society and The Royal Academy of Engineering, Nanoscience and nanotechnologies: opportunities and uncertainties, London, 2004. Available from: https://royalsociety.org/topics-policy/publications/2004/nanoscience-nanotechnologies/

[5] C. Bauer, J. Buchgeister, R. Hischier, W.R. Poganietz, L. Schebek, J. Warsen, Towards a framework for life cycle thinking in the assessment of nanotechnology, J. Clean. Prod. 16 (2008) 910–926.

[6] I.S. ISO 14040—Environmental management—Life cycle assessment—Principles and frameworks (2006).

[7] I. S. ISO 14044 International standard assessment—Requirements and guidelines (2006).

[8] F.M. Christensen, Reflections from an OECD workshop on environmental benefits of nanotechnology, Int. J. Life Cycle Assess. 15 (2010) 137–138.

[9] ISO/TS 27687:2008, Nanotechnologies—Terminology and definitions for nano-objects—Nanoparticle, nanofibre, and nanoplate (2008).

[10] V. Khanna, B.R. Bakshi, Carbon nanofiber polymer composites: evaluation of life cycle energy use, Environ. Sci. Technol. 43 (2009) 2078–2084.

[11] G.F. Grubb, B.R. Bakshi, Appreciating the role of thermodynamics in LCA improvement analysis via an application to titanium dioxide nanoparticles, Environ. Sci. Technol. 45 (2011) 3054–3061.

[12] M.L. Healy, L.J. Dahlben, J.A. Isaacs, Environmental assessment of single-walled carbon nanotube processes, J. Ind. Ecol. 12 (2008) 376–393.

[13] H. Greijer, L. Karlson, S.-E. Lindquist, A. Hagfeldt, Environmental aspects of electricity generation from a nanocrystalline dye sensitized solar cell system, Renew. Energy 23 (2001) 27–39.

[14] A. Dhawan, V. Sharma, Toxicity assessment of nanomaterials: methods and challenges, Anal. Bioanal. Chem. 398 (2010) 589–605.

[15] D. Lazarevic, G. Finnveden, Life Cycle Aspects of Nanomaterials, KTH, Royal Institute of Technology, Stockholm, Sweden, (2013).
[16] S. Gavankar, S. Suh, A.F. Keller, Life cycle assessment at nanoscale: review and recommendations, Int. J. Life Cycle Assess. 17 (2012) 295–303.
[17] M. Miseljic, S.I. Olsen, Life-cycle assessment of engineered nanomaterials: a literature review of assessment status, J. Nanoparticle Res. 16 (2014) 2427.
[18] R. Hischier, Framework for LCI modelling of releases of manufactured nanomaterials along their life cycle, Int. J. Life Cycle Assess. 19 (4) (2014) 838–849.
[19] T. Walser, E. Demou, D.J. Lang, S. Hellweg, Prospective environmental life cycle assessment of nanosilver T-shirts, Environ. Sci. Technol 45 (2011) 4570–4578.
[20] R.K. Rosenbaum, T.M. Bachmann, L.S. Gold, M.A.J. Huijbregts, O. Jolliet, R. Juraske, et al. USEtox—the UNEP-SETAC toxicity model: recommended characterisation factors for human toxicity and freshwater ecotoxicity in life cycle impact assessment, Int. J. Life Cycle Assess. 13 (2008) 532–546.
[21] T.E. McKone, K.G. Enoch, CalTOX, A multimedia total exposure model spreadsheet user's guide (2002).
[22] T. Puzyn, D. Leszczynska, J. Leszczynski, Toward the development of "nano-QSARs": advances and challenges, Small 5 (2009) 2494–2509.
[23] D. Kushnir, B.A. Sandén, Energy requirements of carbon nanoparticle production, J. Ind. Ecol. 12 (2008) 360–375.
[24] V. Khanna, B.R. Bakshi, L.J. Lee, Carbon nanofiber production, J. Ind. Ecol. 12 (2008) 394–410.
[25] D.E. Meyer, M.A. Curran, M.A. Gonzalez, An examination of silver nanoparticles in socks using screening-level life cycle assessment, J. Nanopart. Res. 13 (2011) 147–156.
[26] S.M. Lloyd, L.B. Lave, Life cycle economic and environmental implications of using nanocomposites in automobiles, Environ. Sci. Technol. 37 (2003) 3458–3466.
[27] T.P. Seager, R.P. Raffaelle, B.J. Landi, Sources of variability and uncertainty in LCA of single wall carbon nanotubes for Li-ion batteries in electric vehicles, in: 2008 IEEE International Symposium on Electronics and the Environment, IEEE, San Francisco, CA, USA, May 19–22, 2008, pp.1–5.
[28] B. Li, X. Gao, J. Li, C. Yuan, Life cycle environmental impact of high-capacity lithium ion battery with silicon nanowires anode for electric vehicles, Environ. Sci. Technol. 48 (2014) 3047–3055.
[29] G. Majeau-Bettez, T.R. Hawkins, A.H. Stromman, Life cycle environmental assessment of lithium-ion and nickel metal hydride batteries for plug-in hybrid and battery electric vehicles, Env. Sci Technol. 45 (2011) 4548–4554.
[30] S. Amarakoon, J. Smith, B. Segal, Application of life-cycle assessment to nanoscale technology: lithium-ion batteries for electric vehicles, 2013, pp. 1–119. Available from: http://www.epa.gov/dfe
[31] B.A. Wender, T.P. Seager, Towards prospective life cycle assessment: single-wall carbon nanotubes for lithium-ion batteries, in: 2011 IEEE International Symposium on Sustainable Systems and Technology, IEEE, Chicago, IL, USA, May 16–18, 2011, pp. 1–4.
[32] B.P. Weidema, T. Ekvall, R. Heijungs, Guidelines for application of deepened and broadened LCA, Deliverable D18 of Work Package, 5 (2009) 17. CALCAS Project.
[33] J. Ling-Chin, O. Heidrich, A.P. Roskilly, Life cycle assessment (LCA)—from analyzing methodology development to introducing an LCA framework for marine photovoltaic (PV) systems, Renew. Sustain. Energy Rev. 59 (2016) 352–378.
[34] N. Hooftman, L. Oliveira, M. Messagie, T. Coosemans, J. Van Mierlo, Environmental analysis of petrol, diesel and electric passenger cars in a Belgian urban setting, Energies 9 (2016) 84.
[35] J. Sanfélix, M. Messagie, N. Omar, J. Van Mierlo, V. Hennige, Environmental performance of advanced hybrid energy storage systems for electric vehicle applications, Appl. Energy 137 (2015) 925–930.
[36] L. Oliveira, M. Messagie, S. Rangaraju, J. Sanfelix, M. Hernandez Rivas, J. Van Mierlo, Key issues of lithium-ion batteries—from resource depletion to environmental performance indicators, J. Clean. Prod. 108 (2015) 354–362.
[37] S. Rangaraju, L. De Vroey, M. Messagie, J. Mertens, J. Van Mierlo, Impacts of electricity mix, charging profile, and driving behavior on the emissions performance of battery electric vehicles: a Belgian case study, Appl. Energy 148 (2015) 496–505.

[38] M. Messagie, J. Mertens, L. Oliveira, S. Rangaraju, J. Sanfelix, T. Coosemans, et al. The hourly life cycle carbon footprint of electricity generation in Belgium, bringing a temporal resolution in life cycle assessment, Appl. Energy 134 (2014) 469–476.
[39] X. Zeng, J. Li, N. Singh, Recycling of spent lithium-ion battery: a critical review, Crit. Rev. Env. Sci. Tec. 44 (2014) 1129–1165.
[40] D.A. Notter, M. Gauch, R. Widmer, P. Wäger, A. Stamp, R. Zah, et al. Contribution of Li-ion batteries to the environmental impact of electric vehicles, Environ. Sci. Technol. 44 (2010) 6550–6556.
[41] M. Goedkoop, R. Heijungs, M. Huijbregts, A. De Schryver, J. Struijs, R. Van Zelm, ReCiPe Report, first ed., 2009, pp. 1–132. Available from: http://publication/uuid/F2E6AB20-5106-408B-AE9A-7005578437C6

CHAPTER EIGHT

Safety of Rechargeable Energy Storage Systems with a focus on Li-ion Technology

A. Pfrang, A. Kriston, V. Ruiz, N. Lebedeva, F. di Persio
European Commission, Joint Research Centre (JRC), Directorate for Energy, Transport and Climate, Energy Storage Unit, Petten, The Netherlands

Contents

1 INTRODUCTION

Nanomaterials provide a large surface-area-to-volume ratio for an energy conversion processes to occur, potentially reducing the diffusion length [1] and hence improving kinetics, rate capability [2], and thermal performance [3] and enabling new electrode reactions [4,5]. Regarding mechanical integrity, nanostructured materials

Emerging Nanotechnologies in Rechargeable Energy Storage Systems

can accommodate large-volume expansion/contraction changes, preserving the integrity of the electrode, leading to a stable cycle performance [5]. At the same time, nanomaterials provide increased surface area for solid–electrolyte interface (SEI) formation, which in some cases may lead to poor coulombic efficiency (higher lithium consumption and irreversible capacity) and reduced Li^+ storage sites (due to a potentially lower degree of crystallinity) [1]. On the negative side, a high-electrolyte/electrode surface area leads to an increase in nondesirable side reactions. This may be detrimental for thermal stability and cycle life [6]. Also, nanostructuring might decrease the volumetric energy density of active materials due to poor particle packing [4]. Overall, for each material a trade-off of properties—performance, cost, environmental impact, and safety—needs to be evaluated to find an optimum compromise for a specific application.

In general, rechargeable energy storage systems (RESS) exhibit a progressive capacity fade until the remaining capacity is too low for the specific application and the RESS thereby reaches its end of life. Under certain circumstances though, safety-relevant events can occur during operation or storage. For the introduction of new RESS technologies—and especially for technologies using nanotechnology—it is crucial not only to understand related safety risks, but also to consider how a new technology may be received by the media and the public (e.g., vehicle fires are relatively common for conventional vehicles [7], but the media coverage for an electric vehicle (EV) fire is much higher [8]). In the following, the most relevant RESS hazards, predominant failure scenarios, common risk mitigation techniques and safety tests—including those relevant for nanomaterials—are presented.

2 HAZARDS

Any potential source of harm is considered a hazard and in the following section, hazards arising from RESS are discussed. It should be noted that these hazards do not necessarily apply—at least not to the same extent—to all types of RESS [e.g., batteries, supercapacitors (SCs)]; however, nanostructuring of materials can have an influence on certain properties. Hazards of a mechanical/physical, electrical, and chemical nature are described.

While the introduction of Li-ion batteries has led to great improvement in energy density, the use of flammable organic electrolyte introduced new safety issues when compared to technologies employing aqueous electrolytes (e.g., NiMH, NiCd, and lead acid). In parallel, there has been an increase in the use of SCs containing organic electrolytes in a range of applications, due to their combined high power, high energy, and excellent durability. These organic-based technologies require a modified approach toward cell and system design, storage, handling, and operation.

2.1 Mechanical/physical hazards

Mechanical/physical hazards include the structural damage caused by fire or explosions that can eventually lead to human injuries (e.g., for manufacturing workers, battery users, or disposal personnel) or equipment damage [9].

2.1.1 Fire

The potential RESS fire hazard has been brought to public attention by several battery-related safety events that made worldwide news [8,10]. When considering new technologies, it seems reasonable to compare their safety level with that of an existing technology used for the same application to reach a realistic safety assessment, for example, the safety of an electric car should be compared to the safety of a combustion engine car. For internal combustion engine cars, on average, 31 highway vehicle fires were reported per hour leading to death of 1 person a day in the United States [11], compared to 3 fires reported from around 20,000 Tesla cars all over the world [12]. It is important to understand this hazard and to mitigate related risks. Protection is provided in most situations and RESS fires rarely occur when strict quality control is guaranteed during production, specified operating conditions are respected, and appropriate risk mitigation measures are taken.

RESS fires can be a consequence of abuse, for example, upon exposure to an external fire, to high temperatures, or because of inadequate electrical charging (see also Section 3). Under these circumstances, substantial amounts of potentially flammable gas mixtures (Section 2.3) can be emitted from a RESS (e.g., during cell venting). These gases, when the lower flammability limit is reached, can be ignited either by sparks from high-voltage RESS, by self-ignition during thermal runaway, or by external ignition [13,14]. Flammable gases can also be released during the formation process after cell production. Manufacturing faults might lead to safety relevant events during this process [15], and consequently specific safety measures are implemented in formation plants.

Overall Li-ion batteries are more susceptible to fire hazards than other electrochemical power sources, such as SCs [16]. For this reason the focus of this section will be on this technology.

The energy chemically stored within a Li-ion battery can be estimated by the heat of combustion of its components and amounts up to approximately 10 times the energy electrically stored [13,15,17,18]. Therefore, even a fully discharged Li-ion battery cell contains significant chemical energy with the corresponding potential for heat release. The overall heat of combustion of a Li-ion battery can be estimated to several kJ/g of battery mass or around 100 kJ/Ah of battery capacity [15,17,19,20]. The relevant combustible components of a battery will be described further, but it should be noted that combustion of the electrolyte provides an important contribution to the overall heat release. The estimated heat release during combustion and gas release during thermal runaway from Li-ion batteries can be found in Table 8.1 and information about the temperature ranges where specific exothermic reactions typically occur at a significant rate is discussed in Section 3.1.

Table 8.1 Estimated heat release during combustion and gas release during thermal runaway from Li-ion batteries

Property	Order of magnitude estimate	References
Chemically stored energy/electrically stored energy	Ratio approx. 10:1	[15,17,21]
Heat of combustion/mass	Several kJ/g	[17,19,20]
Heat of combustion/capacity	Few 100 kJ/Ah	[15,17,19]
Volume of gas during thermal runaway (at normal pressure and room temperature)/capacity	Few L/Ah	[13,17,21,22]

The electrolyte used in batteries and SCs is generally composed of one or several organic solvents (e.g., ethyl carbonate and acetonitrile), which contain solvated ions provided by salts (e.g., $LiPF_6$, $TEABF_4$). The electrolyte typically represents 8–16 wt.% of the overall cell weight [17,19]. The organic solvents are often highly volatile and flammable and their decomposition can lead to a significant temperature increase. The details of the decomposition depend on many parameters, such as electrolyte salt concentration or electrolyte solvent mixtures [13]. Flame retardants are typically added for flammability reduction in commercial Li-ion batteries, but not in SCs [23]. The flash points of common pure electrolyte solvents are below 150°C [15,22] and around 30°C for common electrolyte solvent blends [24]. For further relevant physical properties of electrolyte solvents see [25,26].

Another important contribution to the heat generated during a fire, originates from the separator—often made from polyethylene (PE), polypropylene (PP), or a combination thereof—which is easily combustible and can contribute, depending on composition, around 45 kJ/g of separator.

Thermal stability of cathode materials in Li-ion batteries has been intensively studied, often with a focus on assessing the onset temperature of self-sustained exothermic reactions [27,28] (Section 4.1).

Also anode materials, typically consisting mainly of graphite or some other form of carbon, can contribute to heat release as their thermal stability is reduced by interaction with intercalated lithium. While the heat release is typically smaller than that originating from reactions on the cathode, reactions at the anode start at lower temperatures [22].

In the case of double-layer SCs, both electrodes are fabricated with carbon-based materials (e.g., activated carbons). The stability toward elevated temperatures is significantly higher than in the case of battery electrodes. For example, SCs are able to withstand at least 1,000 h of operation at temperatures ranging from 70 to 90°C [29], while self-heating can already occur in this temperature range for Li-ion batteries (Section 3.1).

Different materials, typically hard metal or hard plastic, are used for RESS casings. In case of fire, the heat released from the plastic casing can constitute a relevant fraction of the overall heat release. Therefore, the casing also has to be considered for the assessment of fire safety [15].

Generally, Li-ion battery fires are difficult to extinguish, but intervention tactics should rather be to first evacuate the area to avoid exposure of persons to fire, heat, and potentially toxic gases, followed by containment of the event by trained personnel only (or even simply letting the battery burnout), which can be complemented by active firefighting [30]. Lithium is intercalated into cathode and anode materials and no or very little free lithium metal is present in Li-ion cells. Therefore proven strategies for suppressing metal fires are not appropriate for Li-ion battery fires [15].

If a cell is on fire within a multicell assembly, heat is introduced into neighboring cells, which can cause thermal runaway (Section 3.1) of those cells, which is denoted as cell-to-cell propagation. Consequently, overall heat generation and severity of the event can potentially exceed that of a single cell. To avoid such propagation, removal of heat, for example, adding large amounts of water to the battery for cooling is advisable [30,31]. After successful extinction of a fire, ignition can reoccur [15], particularly for Li-ion batteries, and appropriate measures (e.g., observation and continued cooling) should be taken.

2.1.2 Explosion

The internal pressure of a cell increases when gases are generated internally without pressure release. This occurs, for example, when heated electrolyte evaporates or decomposes [13] and can lead to bulging of cells. If the pressure increases further, the cell may undergo events from a relatively benign rupture to violent detonation [9]. While pouch cells can swell already when operating at 60°C [32], hard cases will only bulge after sufficient heating. The seals of pouch cells open at fairly low pressure, while cells with hard casing typically feature dedicated vents that are activated at a predefined pressure (Section 4.2). The content of cylindrical cells (e.g., a complete jelly roll) can be ejected depending on the cell's design. In the case of internal obstructions to gas release after pressure buildup, for example, by deformation of electrode layers, internal pressure can reach a sufficient level for material ejection [15,33]. The fuel/air mixture resulting from gas emission typically contains flammable electrolyte vapor, which can lead to explosion upon ignition [13]. Explosion typically involves release of accumulated gases and electrolyte and scattering of RESS components. The severity of the explosion depends on the energy involved and the rate at which that energy is released [9].

Emission of gas in the order of a few liters per ampere hour capacity was observed [13,22] and estimated during thermal runaway [17]. The volume of generated gas can give an indication of the pressure buildup potentially leading to explosion. Roth and Orendorff found that around 1.2 L/Ah of gas was emitted from 18,650 cells with 5 different cathode materials—lithium iron phosphate ($LiFePO_4$ or LFP), lithium cobalt oxide ($LiCoO_2$ or LCO), lithium manganese oxide ($LiMn_2O_4$ or LMO), lithium nickel manganese cobalt oxide (NMC), and lithium nickel cobalt aluminum oxide (NCA)—during thermal runaway, an amount that cannot be contained by standard cell enclosures [13].

In most cells, an explosion—rather than lower pressure gas release—will occur only if the integrated safety features (e.g., vent valves) fail, which is typically observed only under extreme conditions. However, in the event that it occurs, exposed individuals may suffer physical injuries and equipment can be damaged (either by the deflagration itself or by the corrosive nature of the expelled materials). After an explosion, thorough ventilation of the area and cooling of the battery are advisable.

2.2 Electrical hazards

Electrical hazards can be grouped into direct and indirect hazards. Electric shock and arcing are direct electrical hazards that can cause immediate injuries or damage upon direct contact with the RESS. Indirect hazards, such as fire and toxic gas release, are the consequences of overheating caused by, for example, Joule heating during electrical abusive conditions (e.g., short circuit and overcharge).

The hazard level of electric shock depends on the overall voltage of the RESS system. For automotive applications a 300–600 V battery system is common, which may cause death or very serious injuries upon touching live parts [34]. Voltage levels lower than 60 V (e.g., 48 V nominal voltage systems) are not considered to be dangerous, and regulations and standards do not require the application of safety measures.

Arcing can damage connectors and fuses (both internal and external), which may destabilize protection devices and/or compromise the functionality of the battery system [35]. Arc energy mainly depends on the current and as cells with lower impedance generate higher current, they are susceptible to a more serious arcing.

Joule heating may cause overheating of the auxiliary components and/or the cells, which may generate faulty operation and/or thermal runaway (Section 3.1).

The hazard level of electric shock in RESS is not expected to change when nanostructured materials are used instead of the same conventional material. The standard potential of reaction, until the same chemical compounds involved, are the same, therefore the measurable voltage level of a nanomaterial-based electrode does not differ considerably from conventional ones. On the other hand nanoscale materials have substantially higher surface area and therefore can produce higher currents [36], which in turn can create a more severe arcing than cells based on conventional battery materials. Furthermore cells with higher current capabilities may decrease the needed tripping time for the activation of protective devices (Section 4.2), therefore nanostructure-based cells with properly dimensioned protection devices are not expected to exhibit a decreased electrical safety performance.

2.3 Chemical hazards

Many of the materials used in contemporary Li-ion battery cells are hazardous and may be toxic, flammable, and/or corrosive.

Active cathode materials contain transition metal ions, such as Co, Ni, Mn, and Fe cations. While materials containing Co and Ni cations are suspected of causing cancer in humans and are considered skin-sensitizing [37,38], materials containing Mn and Fe cations are regarded as being less hazardous [39,40] and therefore safer alternatives to Co- and Ni-containing active cathode materials.

Among active anode materials, such as graphite, activated charcoal, and some types of mesoporous carbons, are considered nonhazardous, while other types of mesoporous carbons are suspected carcinogens [41]. Lithium titanate (LTO) can cause irritation of skin, eyes, and respiratory ways and is harmful if swallowed [42]. Silicon dioxide particles, although chemically inert, can cause damage to lungs through prolonged or repeated exposure if inhaled [43]. Contemporary electrolytes for Li-ion batteries are known to be flammable, toxic, and/or corrosive depending on the exact composition of the electrolyte mixture. As mentioned in Section 2.1.1, reports on incidents with Li-ion batteries catching fire have made the public aware of their flammability hazard and have triggered massive research on the mechanisms initiating such events and the ways to make operation, storage, transportation and recycling of Li-ion batteries safer [13,44,45]. Chemical toxicity hazards related to exposure to the battery components—such as electrolyte and its decomposition products—are, however, less widely recognized. Although threshold values for gas concentration of electrolyte components are mentioned in a number of standards, such as SAE J2464:2009 [46], SAE J2929:2013 [47], and UL 2580:2013 [48], corresponding release of liquid electrolyte from Li-ion battery cells has not been quantitatively assessed for applications other than space and military ([49] and references therein) until recently [26]. Properties, hazard identification, and inhalation toxicity of various aprotic solvents commonly used in Li-ion battery electrolytes, as well as that of hydrogen fluoride (HF) as a decomposition product of the widely used $LiPF_6$ salt are reviewed in a recent study [26]. Calculations show that at room temperature a small-electrolyte release can result in the formation of a potentially toxic atmosphere. For most contemporary electrolyte components this corresponds to a release of less than c. 250 mL in a volume occupied by a medium-size car with a clearance of 1 m, that is, c. 62 m^3 [26]. For some toxic and highly volatile solvents, for example, 1,2-dimethoxyethane (DME), 2-methyl-tetrahydrofuran (2-Me-THF), 1,3-dioxolane (1,3-DL), and diethyl carbonate (DEC), the volume of the solvent required to evaporate to create a potentially harmful atmosphere at room temperature is below 15 mL [26].

Under abuse conditions side reactions may take place in Li-ion battery cells at an accelerated rate and lead to a buildup of gaseous products from thermal and/or electrocatalytic decomposition of cell components (primarily of electrolyte). Also, generation of gas occurs during the cell formation process and in cells with manufacturing defects, whose presence can lead to, for example, internal short circuit (ISC). The excessive internal cell pressure is then relieved, normally in a manner intended by design, that is, through a vent or a seal [9]. This process is called venting. Extensive research has shown that gases generated in and vented from Li-ion batteries typically include carbon dioxide (CO_2), carbon monoxide (CO), hydrogen (H_2), oxygen (O_2), light C_1–C_5 hydrocarbons, for example, methane and ethane, and fluorine-containing compounds, such as hydrogen fluoride (HF) and fluoroorganics, such as ethyl fluoride [13,44,50–53]. It is obvious that many of these gas mixture constituents are flammable, toxic, and corrosive. However, as mentioned in [44], "...cell vent gas toxicity could be more strongly dependent upon minor gas components, such as fluorinated compounds, for example, HF, COF_2, and F_2." Upon cell venting, solid particles of active cathode and anode materials, as well as aerosol of electrolyte may be released. Hence, exposure to a complex mixture of solid, liquid, and gaseous cell components and products of their decomposition needs to be considered when evaluating the toxic hazard associated with venting of Li-ion battery cells. SCs with acetonitrile-based electrolyte were found to emit hydrogen cyanide and cyanogen and this has led Japan to forbid the use of acetonitrile in SCs [54].

With increasing interest in application of nanomaterials in Li-ion batteries [1,5,6,55,56], specific hazards posed by this class of materials deserve thorough consideration. Although existing experimental data suggest that exposure to nanomaterials can be dangerous, a question of whether nanomaterials may be more toxic than their bulk equivalents is still being debated [57]. Reasons for concern are, however, justified, because enhanced toxicity of nanomaterials may be predicted on the basis of size-related characteristics. Nanomaterials were shown to be able to reach intracellular environment, neural axons, and cross biological barriers, such as blood–brain barrier, that is, to penetrate into biological systems normally interdicted to bulk materials [57–59]. Due to an increased surface-to-volume ratio and hence higher proportion of reactive surface atoms, nanomaterials can catalyze the formation of reactive oxygen species in biological tissues known to cause oxidative stress and cellular damage [57–59]. Similar to bulk materials, parameters, such as the shape of nanoparticles, their surface charge, agglomeration degree, and chemical composition play an important role in defining their toxicity [57,60]. Furthermore, some nanomaterials, for example, carbon nanotubes and titanium dioxide, have been shown to exhibit a high biopersistence, thereby increasing the duration of exposure to these toxic agents. In short, nanomaterials have the toxic properties of large particles, in addition to those acquired by virtue of their small size [57].

When assessing the severity of a chemical hazard, it is important to evaluate the setting in which exposure to a certain chemical agent occurs. Exposure to Li-ion battery materials can occur in a number of scenarios that may be occupational or accidental. Workers may be exposed to materials at different stages of the supply chain where materials are being used, handled, or processed. These not only include manufacturing, storage, and transport of materials, electrodes, and cells, but also dismantling and recycling of Li-ion batteries and maintenance and cleaning of the relevant equipment.

Existing EU legislation on worker protection also applies to nanomaterials, although it does not refer explicitly to these materials [59]. Of particular relevance are the Framework Directive 89/391/EEC, the Chemical Agent Directive 98/24/EC, Carcinogen and Mutagen Directive 2004/37/EC, as well as the legislation on chemicals: Registration, Evaluation, Authorisation, and Restriction of Chemicals (REACH) and Classification, Labelling, and Packaging (CLP) [59,61]. Guidance and tools together with information on good practice on management of nanomaterials is available at the websites of the European Agency for Safety and Health at Work (EU-OSHA) [59], The European Chemicals Agency [61], The National Institute for Occupational Safety and Health (NIOSH) [62], the US Environmental Protection Agency (EPA) [63], and The Organisation for Economic Co-operation and Development (OECD) [64].

Accidental exposure to chemical materials differs from occupational exposure, in that it occurs unexpectedly, can affect not only professionals but also the general population, and the duration of exposure is normally shorter than a working day. For accidental exposure, the occupational exposure limit values referred to in the EU legislation mentioned earlier are less appropriate and a different set of threshold values, the so-called acute exposure threshold values, are defined for accidental release of and exposure to chemical materials [65,66].

3 FAILURE SCENARIOS

This section describes a nonexhaustive list of examples of frequently observed cell failures under foreseen abuse conditions. Modification of RESS that was not intended by the RESS manufacturer represents the unforeseen abuse, which is not discussed in detail in the following scenarios. Such modification might be carried out by amateurs (e.g., modification of RESS software or other works related to tuning of EVs to increase power, works on electrically powered drones or toys) who are not trained on RESS safety and consequently are not fully aware of the related risks. Even though legal responsibility for such safety events might lie with the actor, measures—such as protection of RESS software against modification or provision of information or training—should be taken to avoid such safety events.

While no specific abuse test results are available for nanostructured RESS, the possible effect of using nanosized materials is discussed here based on the phenomena presented in previous Chapters 1–3 and 6.

3.1 Overheating

Thermal runaway describes the rapid self-heating of a cell driven by exothermic reactions of cell materials whereby the stored chemical energy is released. While thermal runaway is a possible consequence of many failures (e.g., short circuit, overcharge, and mechanical deformation), this phenomenon will be discussed within the context of overheating. The possible causes of overheating include excessive Joule heating (e.g., due to an aging-related increase of internal resistance), failure of the battery's cooling system, or exposure to high-external temperature (e.g., in a fire), among others.

In all situations potentially leading to thermal runaway, the critical question is if the initiating fault or situation causes enough heating to lead to a self-supporting exothermic reaction or, in other words, if the rate of heat generation exceeds the rate of heat loss. Self-heating does not always result in thermal runaway (as rapid temperature increase does not follow in all cases). An adiabatic environment—as implemented in abuse testing in an accelerated rate calorimeter (ARC)—can be considered the worst case scenario.

Li-ion batteries are more likely to experience thermal runway than SCs. The latter technology does not release sufficient thermal energy during, for example, fast charging or discharging. Their internal resistance is an order of magnitude lower than that of batteries, which results in lower Joule heating [16]. In addition, self-heating reactions do not occur in SCs; they operate at lower voltages, which implies a lower electrolyte decomposition, and they can work safely for long periods at both high and low temperatures (e.g., from −40 to +65°C).

Safety of Li-ion technology can be improved, not only by reducing the probability of an initiating fault and the related heat generation, but also by improving the heat dissipation. The probability of reaching thermal runaway, and also the time taken until it fully develops, depends not only on rates of heat generation and loss, but also on many other parameters, such as operation strategy (e.g., voltage, current or power limits) and implementation of protective devices (Section 4.2).

The precise temperature for a certain type of reaction to occur at significant rate depends on the cell chemistry. In the following, an overview of typical reactions occurring with increasing temperature is given for Li-ion batteries.

The upper temperature limit for storage and operation of Li-ion batteries is typically below 60°C [15]. The reason for this limit is not only increased degradation at higher temperatures, but also proximity to cell self-heating temperatures (e.g., 70–90°C [15]). The onset temperature of self-heating has been observed to depend on the state of charge (SOC) where lower SOC corresponds to higher onset temperature [45,67]. The protective SEI layer between anode and electrolyte will start to exothermically decompose significantly in the temperature range of 75–100°C [15,22,45,68]. At higher temperatures (>120°C) exothermic reaction between intercalated lithium and electrolyte can occur [68].

If the temperature increases further, the separator—whose main function is to prevent direct electrical contact between anode and cathode—can fail. Commonly applied shutdown separators soften at temperatures in the range from 130 to 140°C, which leads to a closure of pores and a shutdown of current. At temperatures significantly above

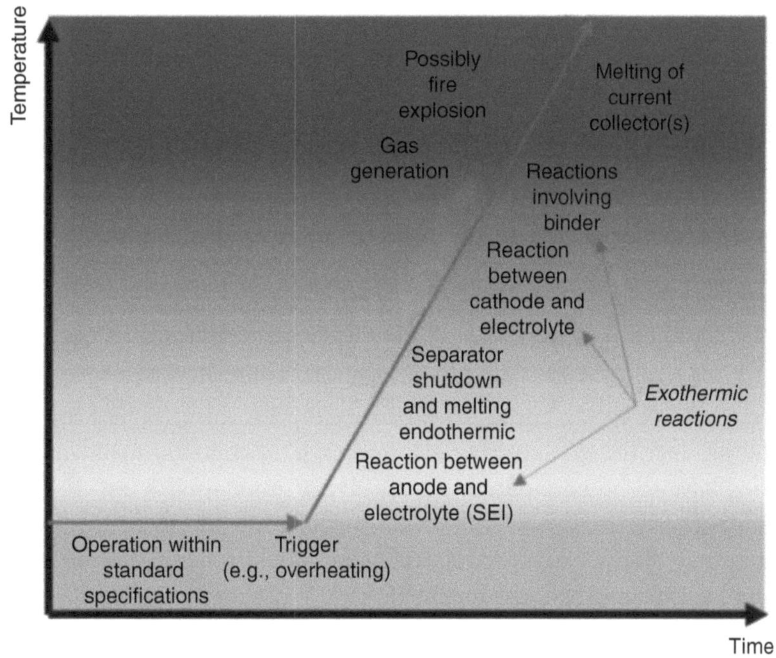

Figure 8.1 ***Typical evolution of thermal runaway in a Li-ion cell.*** The scales are given in arbitrary units and the reactions shown separately here may occur also simultaneously. *SEI*, solid–electrolyte interface.

150°C, typical separators will melt—the separator will "break down"—and thereby allow contact between anode and cathode [15,20,69], while ceramic separators will remain intact up to higher temperatures (for further details see Section 4.1.3).

Cathode materials react significantly with electrolyte exothermically, often in a temperature range from 150 to 350°C, depending on the combination of materials [22,27,28], producing oxygen. This oxygen might react further with the electrolyte solvents. Fig. 8.1 shows the typical evolution of thermal runaway in a Li-ion cell.

Also fluorinated binders, such as polyvinilidene fluoride (PVDF), have been found to react exothermically with lithiated carbon at significant rate at temperatures above 200°C [68]. High-rate gas generation is typically observed during the "thermal runaway peak," commonly in a temperature range from 250 to 350°C [13] (Sections 2.1.2 and 2.3).

Cell temperature during thermal runaway can exceed 600°C [15]. The melting point of aluminum, 660°C, can be reached in that case, however the melting point of copper (1085°C) is not. Nevertheless, liquid aluminum can form alloys with copper that exhibit melting points as low as 550°C [70]. Consequently, damage to both aluminum and copper current collectors can be expected.

It should be noted that overheating can lead to cell venting and also to fire and explosion (Section 2.1).

For cases where nanotechnology enables the use of completely different chemistry, cell behavior under overheating conditions might deviate significantly from the behavior sketched earlier, but no experimental evidence is currently available for the assessment of thermal behavior.

3.2 Mechanical deformation

Mechanical deformation can result from drop, crush, or mechanical shock (i.e., rapid acceleration or deceleration) or comprise penetration, but the probability of the occurrence of such events depends strongly on the RESS application. While a severe crush of a battery is highly improbable for batteries in consumer electronics devices, a crush of an automotive battery may reasonably occur in a vehicle crash. On the other hand, a drop is less probable for an automotive battery, but can be expected to occur several times during product life in consumer electronics. Consequently mechanical deformation scenarios have to be considered specifically for each application (for related tests, see Section 5.2).

A probable outcome of severe mechanical deformation is the creation of an ISC, which then results in local heating and potentially in thermal runaway [15]. However mechanical damage may not always lead to an immediate safety event, mild mechanical damage can also lead to hazardous situations that develop later in the battery life. This especially applies to cells with soft casing like pouch cells. Such delayed failure (e.g., developing over many cycles) might be due to electrode or separator degradation, which in turn can lead to Li plating or to the formation of a hole in the separator. Scrapping, quarantining, or specific monitoring of cells with suspected latent damage is recommended [15].

The severity of a safety event immediately following mechanical deformation is directly linked to the rate of heat generation and dissipation. Whereas low-impedance internal shorting might lead to low-cell heating and benign failure, high-impedance shorting could lead to significant heat generation and—assuming insufficient heat removal—rapid temperature increase and thermal runaway. Therefore cells that are mechanically damaged at the electrode edges—which can typically lead to higher impedance shorting—are usually more prone to thermal runaway than cells damaged perpendicular to the electrode surface [15]. It has also been demonstrated that the location of the ISC in a cell has a significant effect on severity of the safety event [71].

Mechanical deformation can not only lead to ISC, which is potentially followed by cell thermal runaway, venting, fire, and explosion (Section 2.1), but also directly to rupture of the RESS enclosure and leakage/release of battery materials (Section 2.3).

3.3 External short circuit

External short circuit is one of the reasons of cell overheating that may trigger self-heating and potentially thermal runaway. Failure of the external devices [e.g., battery-management system (BMS), charger, and cabling], loss of isolation, or misuse may create

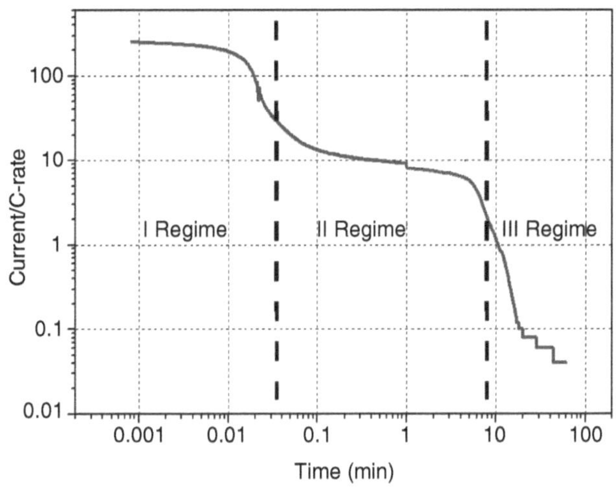

Figure 8.2 ***Current decay of short-circuit current during an isothermal short-circuit test of a 4.5 mAh NCA-Gr coin cell (by EL-CELL) battery.*** The test was performed at 25°C with a MACCOR cycler by applying 0 V at the terminals.

a direct connection of the positive and negative terminals of the RESS. The magnitude of the current created in the short circuit depends on the RESS internal resistance, its voltage, and on the external circuit resistance. An example of short-circuit behavior [72] of a Li-ion battery cell can be divided into three regimes as illustrated in Fig. 8.2. In the first regime (I), the initial short circuit current is 2–3 orders of magnitude higher than the rated battery current. In this regime, the resulting current is governed by the ohmic resistance only [73], because the time constants of other transport processes are in the range of few 100 ms [74], therefore they are not the rate-limiting steps in this time scale. The current is supplied by both faradaic processes (i.e., discharge) and capacitive processes. As the highest current regime ends quickly, the average temperature of the RESS does not reach the onset temperature of thermal runaway. Later, the current drops significantly but the cell's temperature increases continuously [35,73,75] and may reach the onset temperature of self-heating or thermal runaway. However, local hot spots may form earlier, which may trigger unwanted events.

Based on Okazaki et al. model [76] the initial short-circuit current can be calculated by the following expression:

$$I_{\text{short}}^{\text{cell}} = I_{\text{max}}^{\text{cell}} \frac{1}{\left(1 + \dfrac{R_{\text{short}}}{R_{\Omega,\text{cell}}}\right)} \tag{8.1}$$

where R_{short} is the external resistance, $R_{\Omega,\text{cell}}$ is the internal ohmic resistance and $I_{\text{max}}^{\text{cell}}$ is the theoretical maximum short-circuit current of the cell (at an external resistance of 0 Ω).

The cell's internal resistance comes from the metallic contact resistances in the cell, the ionic resistance of the electrolyte, and the electronic resistance of the anode and cathode. As the ionic resistance is an order of magnitude higher than the electronic resistance, it governs the initial short-circuit current, especially at low temperature. Therefore, in this regime (I), the use of nanosized electroactive materials (Chapters 2 and 3) has, in principle, less impact on the (initial) short-circuit current. Even though the maximum initial current may not be dependent on the nanostructure of the active materials, the duration of regime I can be longer. Nanosized electroactive particles have a much higher surface area; consequently, the cell has higher (double-layer) capacitance. If the ohmic resistance does not change significantly, which is expected, then the capacitive discharge time may increase.

At the end of regime I (Fig. 8.2), the current starts to drop quickly because the double-layer discharges and mass transport becomes the rate-determining step increasingly. Later, it stabilizes (around 10 C in the presented example in Fig. 8.2, which can be close to or within the normal operating current of the cell). Four main types of mass transport losses can be the rate limiting step in this region [77] (Chapter 6): (1) Li-ion diffusion in the electroactive particles, (2) Li-ion diffusion in the electrolyte and through the separator, (3) electric conductivity of the positive or negative active material, and (4) Li-ion conductivity in the electrolyte especially when the forced migration causes near-zero Li-ion concentration in the pores, which leaves a part of the electrode inactive. Nanostructured electrodes discussed in Chapters 1–3 can improve the transport properties of an active material, consequently increasing the current density in regime II. However, more research is needed to understand how the enhanced transport properties obtained for nanostructured electrodes (Chapter 3) can be exploited in commercial batteries [78]. Cell temperature reaches the highest level [73] in regime II, resulting in boiling of the electrolyte, gas formation, and potential rupture of the cell leading to venting and electrolyte leakage. The SOC simultaneously decreases, which reduces the risk of a more severe event, for example, thermal runaway. Potentially better heat conductor nanomaterials may reduce the risks of self-heating or thermal runaway in this region.

After Li continues to be consumed during the forced discharge in regime II, the current drops quickly in regime III leading to the total discharge of the battery. It is expected that this regime occurs earlier for nanostructured electrodes because in the previous regimes higher current discharges the battery faster.

3.4 Overcharge

When a charger or electric motor inverter fails, current may not be interrupted and/or higher potential may be applied unintentionally to a RESS system or to specific cells. At least two different scenarios are worth distinguishing: (1) normal charging is not interrupted and SOC increases above the operating range, and (2) the RESS experiences a jump in potential above its maximum operating value, for example, after a failed communication between a car's BMS and the fast DC charger.

During overcharge the half-cell potential of the cathode increases above the stability of the electrolyte and the cathode material, while the anode's potential approaches the Li plating potential (i.e., zero vs. Li/Li^+). When the anode cannot host more Li, metallic Li deposition occurs. As the cathode loses its structural stability, then exothermic heating occurs (depending on the chemistry) even at slight overcharge, for example, in the case of NCA [79]. As a consequence of the structural stability loss of the active material the internal resistance of the cell increases, hence the Joule effect heats up the battery at a higher pace. At the cathode side the electrolyte starts oxidizing, which further increases the temperature. As overcharge continues, both temperature and SOC increase steadily, which at a certain point can result in self-heating, thermal runaway [67], venting, and then rupture of the cell.

Nanostructuring alone does not change the standard potential of charge transfer reactions (e.g., lithiation) and the heat of chemical decompositions, consequently the safety cut-off voltage and temperature in theory may be the same. However, nanstructures can modify surface energy and hence the reaction kinetics, both of the route and/or their rates, which may surpass (or enhance) side reactions. Furthermore nanomaterials have a considerably higher surface area and the increased volume-to-surface ratio enhances the overall rate of side reactions, that is, the consumption of cyclable lithium during the SEI formation, which in turn may lead to increased heat generation and self-heating at lower temperatures. Overall effects of nanostructured materials on overcharge behavior cannot be predicted at present because of the lack of comprehensive experimental results.

Some works indicate that the heat released on overcharge is proportional to the charging current [80]. Low-charging rates (<C/5) do not result in thermal runaway compared to higher rates (>0.5 C) [68,81]. Tobishima et al. showed that cells overcharged at low rates did not show any venting whereas those cells overcharged at 2 C rate did [82]. Therefore charging current as a source of Joule heat together with heat dissipation rate are important parameters for occurrence and severity of a safety event, and they have an impact on the outcome of the tests. Nanostructured electrodes may have lower resistance, which may prevent them from overheating at higher currents.

Although serious events occur with significant overcharge (e.g., charge up to 2 times the rated capacity), repeated charge/discharge cycling at moderate 110% SOC overcharge can also lead to internal short and failure of the cell after 10 cycles [83]. At severe overcharge, Li plating turns to massive Li dendrite formation, which may penetrate through the separator and create ISC triggering an immediate catastrophic event [84]. Plated lithium also interacts exothermically with the electrolyte [32].

Li plating also occurs during normal charging, for example, at low temperature [85,86] and/or because of nonuniform reaction rate distribution [87,88], but it does not usually lead to dangerous situations because Li dendrites can dissolve during discharge

[89,90]. Guo et al. [89] concluded that a more uniformly deposited graphite layer is less susceptible for Li plating. The surface roughness and porosity depend on the particle size distribution, and it seems that an electrode with broader distribution of size generates a more flat layer, which in turn results in more uniform current distribution. The material is also an influential factor. It was found that hard carbon is more susceptible to Li plating than graphite [91]. A severe Li plating may lead to penetration of the separator, and separators with higher tortuosity (e.g., nanomaterials or ceramic coating) may decrease the probability of an ISC. A geometry with high tortuosity can be reached by the application of nanostructured separator materials.

4 RISK MITIGATION

Risk mitigation is part of any risk assessment exercise. It follows the identification and evaluation of risks that may arise during the battery life and its main objective is to reduce risks (probability of occurrence of harm and severity of that harm) to a tolerable level or eliminate hazards (source of harm) that might injure people, or damage the environment or property [92].

The specific battery application determines the risk mitigation strategy. For this reason, cell design (pouch cell, prismatic, or cylindrical) and optimization of the battery assembly play a relevant role in reducing risks and enhancing safety. For instance, safety issues may easily arise when the cells are working beyond operating conditions set by the manufacturer. In a battery pack the required energy and power may be obtained by oversizing the battery, whereby a single cell can operate in a safer, less demanding profile.

There is no straightforward relationship between degradation of the electrochemical performance (e.g., capacity fade and increase in cell impedance) and decreasing level of safety. However, degradation processes may enhance the risk for critical failure, which will result in a dangerous hazard [93,94]. For this reason risk mitigation must consider also how to control the expected battery degradation to deliver a safe product until its end of life.

4.1 Materials selection

Safety is a serious issue in Li-ion battery technology and consequently, many approaches are under study with the aim to mitigate risks. Unfortunately—as a rule of thumb—all these approaches are expected to decrease the specific energy. Thus, the practical value of an approach depends on whether an acceptable compromise between energy/power and safety can be achieved. Materials selection and use of nanotechnology are two of these approaches. For instance, the adoption of nanomaterials increases the material porosity leading to a lower density and thus lower volumetric energy density, which is generally good in terms of safety. However, the synthesis of nanoparticles and nanoporous materials requires more effort than the synthesis of bulk materials due to potentially costly processes [95] and the introduction of new occupational hazards in the production/recycling phase [96] (Section 2.3).

4.1.1 *Electrodes*

Active electrode materials for Li-ion batteries should fulfill the requirements of reversible capacity, good ionic and electrical conductivity, long cycle life, high rate of lithium diffusion into active material of both electrodes, and safety, ideally combined with low cost and low-environmental impact. Further a series of risk mitigation strategies focused on electrode material selection will be presented.

4.1.1.1 Cathode materials

The chemical composition of the cathode material and its structure defined in terms of porosity, thickness, etc. determine not only the energy density and kinetics of the electrochemical reactions, but also thermal/mechanical behavior of a cell, which may affect its safety performance. The safety level of a cathode is usually assessed by measuring its material's thermal stability by calorimetric methods [97]. From these measurements, and by looking at commercialized cathode materials, LFP appears as the safest cathode, while LCO is considered less safe. Between the two, there are other active cathode materials, such as LMO, NMC, and NCA. However, within each type of cathode material, the precise elemental composition affects the material safety performance. Noh et al. [98] showed, for instance, the effect of Ni content on the electrochemical properties, and structural and thermal stability of different NMC cathode materials: $Li[Ni_xCo_yMn_z]O_2$ ($x = 1/3, 0.5, 0.6, 0.7, 0.8$, and 0.85). An increase in the Ni content results in an increase in specific discharge capacity, but the corresponding capacity retention and safety characteristics, related to thermal stability, gradually decreased (Fig. 8.3).

The major source of safety concerns is considered to be the electrode/electrolyte interface [93]. Several critical reactions may occur at the active material surface in contact with the liquid electrolyte. At the lithium oxides cathode/electrolyte interface the release of oxygen and HF corrosion resulting from decomposition of electrolyte in presence of trace amount of moisture, are reactions directly contributing to cathode thermal instability. Also undesired phase transitions at the cathode surface area produce defects and structural instability that may be a safety concern [99]. A consistent way to mitigate these hazardous situations is the use of coating on the active materials that prevents the direct contact with the electrolyte solution. Carbon coating is a widely employed technique as carbon limits the surface oxidation of active material and provides high resistance to HF corrosion and only shows an electrochemical activity toward the electrolyte at very low potentials [55]. Other coatings are also investigated for the same purpose, such as metal oxides, metal hydroxide, metal fluoride, and metal phosphates [100]. Another way to mitigate the risks related to the cathode/electrolyte interface is the use of nanomaterials. In general, the presence of nanoparticles enhances the probability of

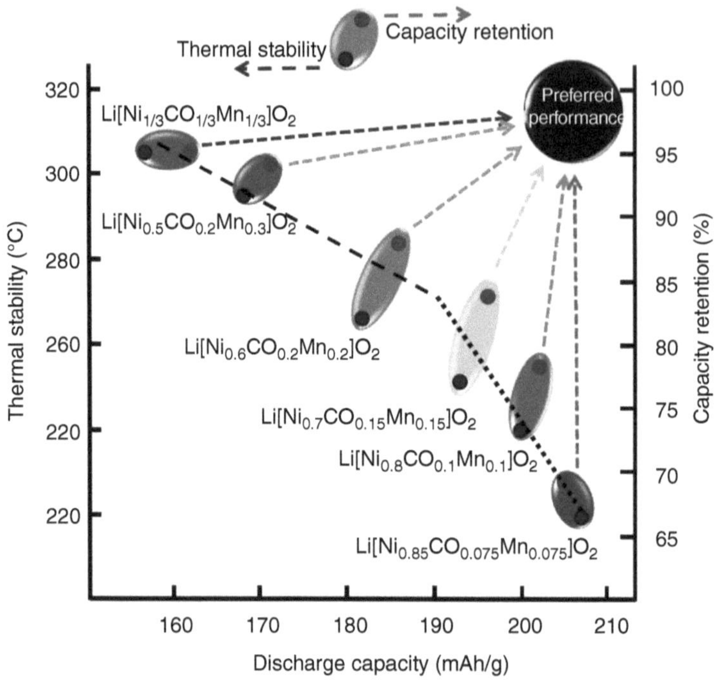

Figure 8.3 ***Thermal stability and capacity retention of $Li[Ni_xCo_yMn_z]O_2$ electrodes in a wide range of Ni concentrations ($0.33 \leq x \leq 0.85$).*** *(Reprinted from H.-J. Noh, S. Youn, C. S. Yoon, Y.-K. Sun, Comparison of the structural and electrochemical properties of layered $Li[Ni_xCo_yMn_z]O_2$ (x = 1/3, 0.5, 0.6, 0.7, 0.8, and 0.85) cathode material for Li-ion batteries, J. Power Sources, 233 (2013) 121–130 [98], Copyright ©2013, with permission from Elsevier).*

undesired side reactions that may result in low-life cycle and poor thermal stability, as slow-bulk transport paths are removed by improving kinetics and extending the contact area between nanometer-sized particles and the electrolyte [101].

However, nanotechnology itself offers the possibility to design the nanoparticles in a way to mitigate these undesired effects. For instance, Yang et al. [102] proposed to stabilize nanospherical $Li[Ni_{0.5}Co_{0.2}Mn_{0.3}]O_2$ core–shell particles through encapsulation with the more thermally/mechanically stable Li_2MnO_3 outer shell. Also Sun et al. [103] improved the thermal stability of the nickel-rich oxide core–shell with a full-gradient nanostructure where the nickel concentration decreases linearly, while the manganese concentration increases linearly from the center to the outer layer of each particle (Fig. 8.4).

4.1.1.2 Anode material

Lithium metal has one of the highest theoretical capacities among anode materials (3860 Ah/kg), and is mostly used in the promising Li–S technology, but its strong reactivity in conventional electrolytes may lead to dendrite formation on the lithium metal

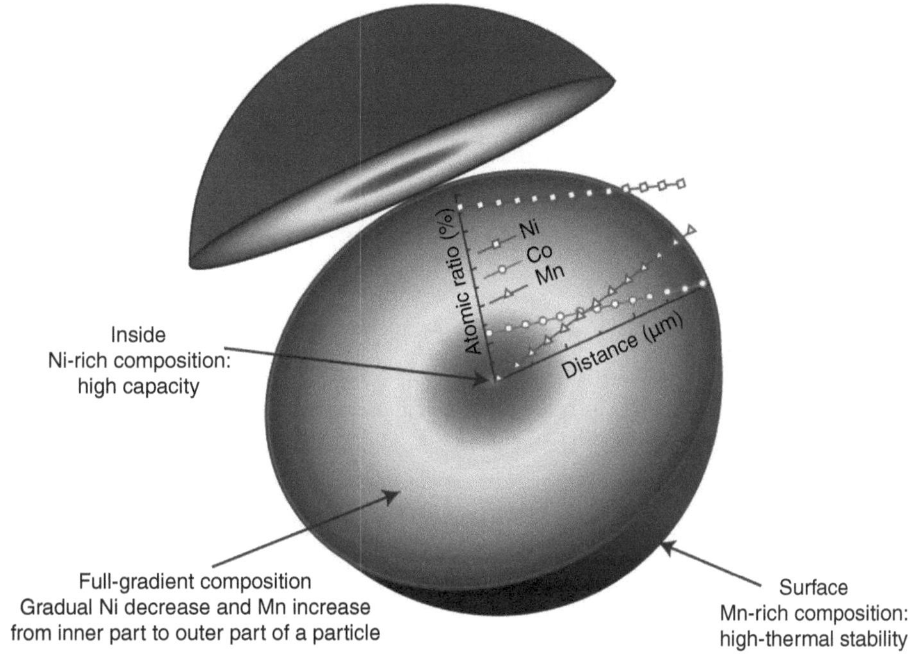

Figure 8.4 ***Schematic diagram of the core–shell lithium transition metal oxide nanoparticle with the nickel concentration decreasing from the center toward the outer layer and the concentration of manganese increasing accordingly.*** *(Reprinted from Nat. Mater. 11, 942–947, Copyright ©2012, with permission from Macmillan Publishers Ltd.).*

that can cause short circuit between anode and cathode. Several strategies are employed to protect the lithium metal surface from the electrolyte, controlling and stabilizing the formation of the SEI. These include nanothin film coatings, using specific additives in the electrolyte (see the following paragraph on electrolyte) or use of ionic liquids.

Graphite is the most-used anode material [104,105] owing to its decent performance, cycle life, and low cost. Unfortunately, its relative low-reversible capacity of 372 Ah/kg (see Chapter 2 on anode materials) limits the cell energy density substantially. Besides that, the carbon-based anodes suffer also from dominant aging mechanisms with associated safety issues, such as SEI instability, high reactivity with electrolyte at relatively low temperature, and metallic lithium plating at low temperature [93] (Section 3). Similar to the cathode-active materials, also for the anode, the major source of safety concerns is the electrode/electrolyte interface. For graphite-based anodes in Li-ion batteries, the carbon coating stabilizes the anode surface, mitigating the risks related to fire and explosion hazards, while improving the anode electrochemical performance at the same time [55].

Several researchers have investigated alternative anode materials for high performance and high capacity, while maintaining a high degree of safety. An implicitly safer alternative to graphite is LTO. Especially the spinel $Li_4Ti_5O_{12}$ allows the design of cells with minor safety concerns and low cost. The high-operating potential of 1.55 V versus Li/Li^+ leads to a reduced achievable energy density, but also guarantees safety conditions by limiting the formation of a potentially unstable SEI and the development of dendrites. However, its low-theoretical capacity (in the range 175–330 Ah/kg) and low-electronic conductivity limit its energy and power density. To overcome these limitations, downsizing the LTO to the nanoscale shows promising results [55].

4.1.2 Binders

In Li-ion battery fabrication, a polymer binder is needed to bind particles of active materials together to ensure sufficient electrical conductivity, mechanical strength, and integrity of the electrodes. The binder plays an important role for battery durability and safety, as it is affecting the overall thermal/mechanical stability of the electrode.

Usually, PVDF is used, which needs to be dissolved in *N*-methylpyrrolidine (NMP) for slurry preparation. The use of NMP leads to concerns related to occupational and environmental hazards and to high cost. Further, PVDF has some drawbacks affecting performance and safety of Li-ion battery, especially due to its poor thermal/mechanical stability resulting in a higher reactivity of the electrolyte at the anode surface. Binders processed in aqueous media, such as carboxymethylcellulose (CMC), styrene butadiene rubber (SBR), or fluorine acrylic hybrid latex are alternative candidates, but unfortunately present new problems from a performance and manufacturing point of view [106].

One partial solution is proposed by Barsykov [107] with the combination of the initial vinylidene fluoride with a copolymer, such as hexafluoropropylene (HFP) that improves adhesion strength and flexibility. Another solution is proposed by Qian et al. [108] where the PVDF is substituted with the promising polyimide (PI) that shows a more stable thermal and mechanical behavior, thus positively affecting safety.

4.1.3 Separators

Although separators are electrochemically inactive components in Li-ion cells, they play a very important role in determining cell safety affecting its thermal stability. Separators commonly employed in Li-ion batteries are polymeric polyolefin microporous films generally made of PE and PP or laminates of them (e.g., PP/PE/PP).

If the cell temperature rises, separators have the inherently protective property to shut down the ionic transport between the electrodes (Section 3.1). However this "shutdown" property presents several drawbacks. Once shutdown occurs, it is possible that due to chemical and thermal inertia the temperature can continue to rise. Under such conditions the separator would melt and short the electrodes anyway [109]. Also, the shutdown separator design for ensuring safety should be tailored to a specific battery size and application. Separators that are acceptable for laptop batteries may not be the correct choice in a plug-in hybrid electric vehicle (PHEV) battery especially because of potential different thermal profile and different cell-to-cell interactions, where not all the separators will shut down at the same time, which can throw the system dangerously out of balance [69,110].

Ceramic separators exhibit excellent thermal stability, blocking the dendrite growth and so enhancing safety, while keeping electrochemical performance and high-ionic conductivity. In the work of Jung et al. [111] ceramic separators based on lithium lanthanum zirconium oxide ($Li_7La_3Zr_2O_{12}$ or LLZO) were proposed in the form of flexible thin films as an alternative to PP/PE/PP.

4.1.4 Electrolytes

Electrolytes (see Chapter 1 on electrolytes) have been shown to be a major source of safety concern in Li-ion cells. The hazards associated with conventional $LiPF_6$ and carbonate-based electrolytes are well documented and include high-volume gas decomposition products at elevated temperature, large-combustion enthalpy, and flammability of solvent vapor (Section 2.3). Besides these safety-related limitations, conventional $LiPF_6$ and carbonate-based electrolytes are limited to an electrochemical window of stability of about 4.5 V [109].

Electrolyte additives are proposed to reduce gas generation and mitigate flammability (flame retardants), while others may be used to enhance the formation of a protective film at the positive and negative electrodes surface. Other additives, so called "redox shuttles," are employed to tackle the overcharge issue (Section 3). Those additives shunt the excess charge injected into the cell during overcharge, acting as a controlled internal short [109]. However, the use of additives is generally limiting the cell performance, affecting the conductivity of electrolyte and the electrochemical window [13].

An approach is the application of a completely new class of liquid electrolytes for Li-ion battery alternative to $LiPF_6$. Some are based on thermally stable lithium salts, such as $LiPF_3(C_2F_5)_3$ (LiFAP), $LiN(SO_2CF_3)_2$ (LiTFSI), and $LiBC_4O_8$ (LiBOB) [109], while others are using room-temperature ionic liquids based on the lithium bis(fluorosulfonyl) imide (LiFSI) and KFSI binary system or on 1-ethyl-3-methylimidazolium (EMI) cations [112]. These latter electrolytes are inherently nonflammable, nonvolatile, and have a low heat of reaction with active materials. They also have favorable electrochemical stability windows, but unfortunately their development is still challenging [109].

Solid electrolytes can provide advantages over liquid electrolytes in terms of safety, reliability, and simplicity of design. Solid electrolytes—compared with liquid electrolytes—show an inherently lower flammability and a higher thermal stability, limiting the reactivity at the electrode/electrolyte interface, one of the most critical areas of safety concerns. In addition, chemical hazards related to electrolyte leakage in case of battery rupture (Section 2.3) are avoided. Both solid ceramic electrolytes (e.g., $Li_{1+x}Al_xGe_2 - x[PO_4]_3$ [LAGP]; $[La,Li]TiO_3$ [LLTO]) and solid or gel organic polymer electrolytes [e.g., poly(trimethylene carbonate) (PTMC); poly(ε-caprolactone) (PCL) homopolymer, PVDF] have been developed as alternatives to the liquid electrolytes currently employed in Li-ion batteries [113], but still their ionic conductivities are generally lower than those of liquid electrolytes. Mindemark et al. [114] improved the ionic conductivity of a solid polymer electrolyte while Wang et al. [115] worked on lithium solid electrolytes, showing their potential to become a suitable alternative to carbonate-based electrolytes. There is one example of batteries utilizing solid electrolyte for automotive applications—the so-called "lithium–metal–polymer (LMP)" battery developed by Bolloré and used since 2005 in Bolloré BlueCar [116].

4.2 Protective devices

Fuses and circuit breakers are designed to protect the RESS system from high current, voltage, and overheating. In principle, circuit breakers and fuses disconnect the RESS in a short time if high current is flowing, consequently limiting Joule heating. The characteristics of a protective device are designed as a function of the battery's short-circuit current. Fisker-Karma cars caught in Hurricane Sandy show a case where short-circuit occurred [117], probably because the environmental circumstances (cold temperature and salty water) prevented the fuses from working properly, finally leading to the burning of several cars.

Commonly positive temperature coefficient (PTC) devices are used, which limit high current. PTC devices are conductive polymers, whose resistance increases 2–3 times above a temperature threshold, consequently limiting current and avoiding overheating [109]. Smith et al. modeled the thermal behavior of PTC devices in different pack configurations. It was shown that PTC devices can work very effectively at cell level; however at module level, when the module terminal voltage is high, they may fail, which may result in severe events, for example, explosion [118]. The same behavior was found for current interruption devices (CID), which break the circuit inside the cell when the internal temperature or internal pressure increases [35]. The mechanism works at cell level, but may fail at module level, as the protection devices of individual cells within a module may be activated at different times: The device that is first activated disconnects the circuit and, consequently, the full-module voltage builds up on its terminal. If the protection device is not designed to withstand high voltage, an arc can be created, which destroys the conductive polymer of a PTC. In turn, the circuit closes again and the current flows through the faulty cell [35,118].

4.3 System-level approaches

The main system that protects the RESS at system level is the BMS, which has several functions:

- It monitors the voltage of each cell and the temperature typically at multiple locations in the battery and when one of the cell reaches the unsafe operational region, it terminates its operation.

- It provides active or passive balancing between the cells, which ensures maximum charge capacity, without overcharging the cells.
- It monitors the RESS current and ensures optimal battery temperature for that current by regulating the heating/cooling system.

Innovative solutions also appeared that are capable of providing early warning of unwanted events. Cui and coworkers [119] developed an ISC protection device, which comprises a very fine-conducting mesh built into the separator. When a dendrite penetrates through half of the separator, a voltage drop is measured between the electrode and the mesh, which can trigger safety mechanisms.

Cell-level safety mechanisms are usually irreversible. Recently Chen et al. [120] developed a special conducting polymer that gets activated at a high temperature and shuts down the current, similar to a PTC, but when the abuse condition ends it is capable to restore the normal operation, similar to an internal circuit breaker.

At the RESS level the thermal management together with the pack design are also extremely important for mitigation of risks related to both excessively high- and low-operating temperatures. For instance, when the battery is operating at too high a temperature, the risks for overheating, combustion, and even explosion may be increased, and when operating at a temperature well below ambient, cold charging may lead to lithium plating (Section 3).

Battery thermal energy management may be performed by passive cooling with air or with liquid cooling systems (water, glycol, oil, acetone, or refrigerants) [121].

5 SAFETY TESTS

Safety tests, also known as abuse tests, are intended to evaluate the performance of any given rechargeable energy storage system (RESS, e.g., batteries, SCs) when exposed to an off-normal abusive scenario, simulating inadequate use or failure [45]. They are defined by organizations, such as the International Electrotechnical Commission (IEC), the International Organization for Standardization (ISO), the European Committee for Standardization (CEN), the European Committee for Electrotechnical Standardization (CENELEC), or the United Nations Economic Commission for Europe (UNECE) to provide assurance that a system can be used in a commercial application.

The definition of an abuse scenario relates both to the likelihood of its occurrence and to the potential severity of its effects. As an example of a frequently occurring scenario, vibration testing is routinely performed as RESS systems are likely to experience vibration loads in a wide range of applications, such as automotive, portable, and generally during RESS transport. Conditions leading to a catastrophic failure (e.g., rupture and thermal runaway) are infrequent, but they are still introduced in abuse testing due to their potentially damaging nature.

The RESS behavior in an unsafe scenario is linked to various parameters. The selection of RESS components, chemistry, and cell construction, to mention some parameters, ultimately define whether a test results in mild benign damage or in irreversible catastrophic failure (e.g., fire, explosion) of the device under test (DUT). Moreover, test parameters, such as SOC or environmental temperature, also have an impact on the outcome of the tests [68]. Also, the level—cell, module, or pack—at which the test is conducted greatly influences the test result. In this sense, immersion of a small coin cell will not result in any reaction as opposed to immersion of a large-automotive battery pack, which will result in, for example, vigorous bubbling due to the electrolysis of water. Similarly, the cell format plays a role—for example cylindrical, pouch, and prismatic cells may result in different performances because the relative rates of heat generation and heat dissipation are fundamental for thermal runaway development [122]. Therefore, the selection of a test procedure needs to be both adequate and proportionate to the DUT size and its field of application.

Safety tests can be classified into three main types of tests: thermal, mechanical, and electrical, according to the type of the abuse. In parallel, monitoring of the chemical hazards can also be performed (e.g., flammability and toxicity). In some cases, safety tests are applicable to a specific chemistry, for example, hydrogen emissions test for open-type traction batteries [123]. At present there are no specific additional tests for nanobased RESSs to evaluate their safety.

Fig. 8.5 presents an overview of the most commonly performed safety tests. A brief description of each type of test is presented in this section (for a detailed description, please refer to Ref. [124]).

In general, standards and regulations set pass/fail requirements for each test. For example, the standard for secondary cells in portable applications IEC 62133:2012 [125] sets no fire/no explosion as acceptance criteria for tests under reasonable foreseeable misuse (except for forced ISC), and no fire/no explosion/no leakage for vibration and thermal shock, which are considered intended use tests. Specific to automotive applications, the response of a technology to an abusive condition can be classified in the following levels (extracted from EUCAR hazard levels [126]): from level 0 (no effect; system maintains its functionality) to level 7 (explosion; mechanical disintegration of the system). Battery and car manufacturers often utilize this classification to evaluate the response of the RESS to each abusive condition. A level of 3 or lower usually represents an acceptable level of performance.

5.1 Thermal tests

Thermal testing exposes the RESS to extreme temperature environments. Under these conditions, volume changes and thermal expansion of active materials, electrolyte, and casing occur. Additionally, failure of the sealing materials, which may lead to leakage, rupture, or even explosion, can occur if the temperature is sufficiently high. Some of the most common tests are summarized.

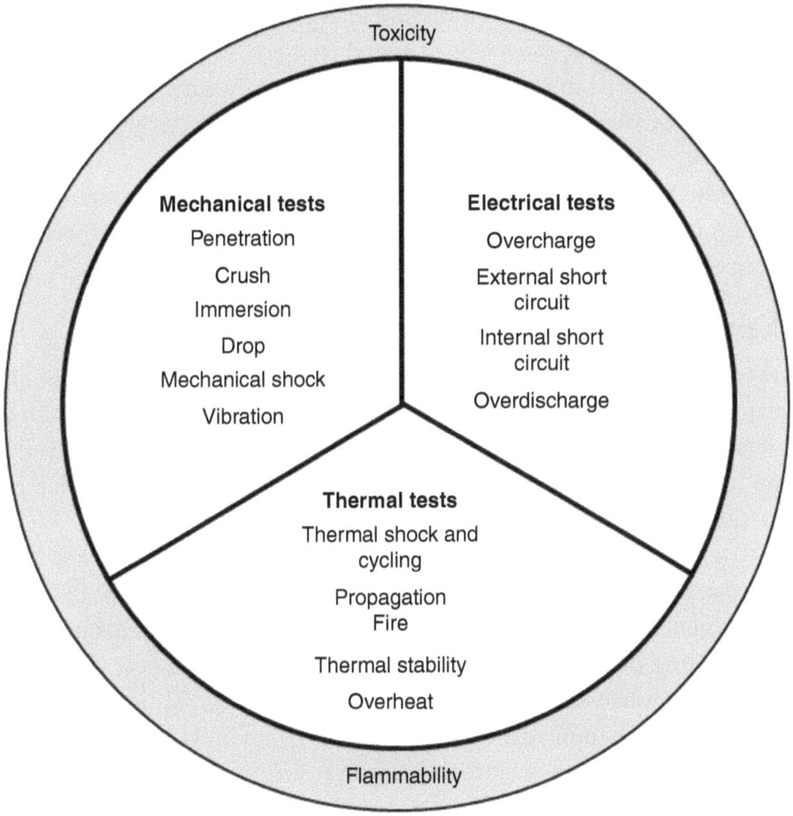

Figure 8.5 ***Overview of commonly performed safety tests.***

Thermal shock and cycling test. In this test the DUT is exposed to sudden temperature changes, which cause expansion and contraction of the RESS components that the system design needs to account for. Extreme temperatures from −40°C to 85°C are required in various standards [48,127]. The RESS is exposed to temperatures within these limits for several thermal cycles (e.g., 5 times [46] or 30 times [48]).

Thermal propagation test. This test is applicable to large-RESS systems (e.g., modules, packs in electrified vehicles, or smart grid applications) and deserves some consideration. In this test, failure of a single cell is induced (e.g., by applying localized heat or penetration), and its propagation to neighboring cells is studied. The DUT is expected not to develop into fire or explosion. The propagation among cylindrical cells has been found to be less probable than in other cell configurations, such as pouch cells [128]. Other parameters of great importance are the cell configuration (parallel vs. series) and heat transfer characteristics of components.

Fire test. In this case, the RESS is exposed to flames for a considerable time leading to high-cell temperature and pressure, which may compromise the integrity of the enclosure, expulsion of electrolyte, and intense energy release [9]. There are several variations to this test, for example, *radiant heat test* and *projectile fire test*. Depending on the test type the RESS is expected not to explode during a certain time (e.g., 70 s exposure to direct flame [123]) or the exploding parts must not penetrate through a mesh that surrounds the battery.

Additional tests that can be relevant are *long-term storage* at elevated temperatures, *low-pressure* (or altitude simulation), *thermal stability* (identification of the thermal runaway initiation temperature), *overheat*, and *extreme cold temperature tests.*

5.2 Mechanical tests

Mechanical tests evaluate the robustness of a RESS under mechanical loads. The potential deformation or degradation of the RESS by external mechanical damage can occur at various levels, from electrode piercing, breakdown of connections or welding points, to enclosure cracking or rupture. A summary of the most common mechanical tests is discussed further.

Penetration test. This test is also called a nail test and is potentially one of the most damaging to the RESS. In this test, a sharp steel rod punctures the RESS at a certain constant speed, generally 8 cm/s [46]. As a consequence, rapid discharge of the system and localized heating occurs, potentially leading to thermal runaway. This test was initially designed to simulate ISCs, as these are very difficult to predict. Despite extensive quality controls during manufacturing between 1 in 10 million [129] and 1 in 40 million [130] cells fail during normal operation. However, various works have questioned this test's appropriateness, as it does not fully represent a spontaneous ISC, as described by several authors [71,131].

Crush test. In this test, the RESS is compressed by utilizing a plate [132] or a crushing bar [127], causing the shorting between cell electrodes. Violent reactions due to short circuits and electrolyte leakage may occur. The force to be applied varies according to the application. For example, 13 kN is required for portable applications [125], whereas 100 kN is required in automotive applications [132].

Immersion test. Complete submersion of the RESS in salted water is carried out in this case. This test aims at simulating a scenario where a RESS is submerged (e.g., underbody of an EV that is flooded or mobile phone battery dropped in water). Due to the conductivity of salt water, a low-impedance short circuit is created. In this case, contrary to the penetration test, the discharge occurs at a significantly lower rate. This test is usually accompanied by vigorous bubbling, often for several hours due to gas generation in the process of water electrolysis [133].

Drop test. In this test the RESS is dropped from a certain height (e.g., 1 m [48]) onto, for example, a concrete floor. This scenario can be encountered, for example, when a RESS falls from a bench top onto the ground, or due to poor handling during installation or transportation. RESS intended to be removed from the vehicle for charging or replacement (e.g., battery swapping principle) are required to carry out this test in triplicate according to UL 2580:2013 [48].

Mechanical shock and vibration tests. These tests are devoted to testing the resilience of the RESS toward bumps and sudden accelerations (e.g., vehicle crash) [125]. These tests are mainly relevant for automotive applications and RESS transportation. Cracking and breaking of the casing, sealing, welded/soldered joints, separator tearing, etc. can occur during these tests. Other mechanical tests, less frequently performed, expose the RESS to shotgun (*bullet impact test*) or dust (*dust test*).

5.3 Electrical tests

Electrical abuse testing of energy storage systems requires exposing the DUT to severe charging/discharging conditions (e.g., exceeding voltage and/or current limits) or to a direct contact of positive and negative electrodes.

Overcharge test. This test is one of the most damaging tests as the DUT is charged beyond the recommended voltage limits. Constant current is applied to the DUT up to a charge limit, for example, 200% SOC [46,123,127,134], 110% SOC [48], 130% [135], or until the DUT (automatically) interrupts or limits the charging. The charging rate plays a significant role in the outcome of the test (Section 3.4), therefore specific test currents for EV, hybrid electric vehicle (HEV), and PHEV applications are required, for example, 5 I_t for HEV [136] and 1 I_t for BEV [127,137]. Other standards recommend the same charging rate regardless of the application, such as 1/3 I_t in UN/ECE-R100.02:2013 [123]. (The current I_t represents the discharge current in amperes during 1 h discharge and C is the measured capacity of a battery pack (or a cell): I_t (A) = C_n (Ah)/1 (h); n is the time base (hours) for which the rated capacity is declared.)

External short circuit test. In this test, both terminals of the DUT are in contact via a low-resistance element and consequently the DUT is discharged. Regulations and standards usually require the application of fixed external resistances (e.g., 5 [123,127,134], 20 [48,132], or 100 mΩ [135]); however, the initial short-circuit current also depends on the internal resistance, which is influenced by the size of the battery [75,76] and its chemistry and electrode structure (e.g., nanostructured). Therefore the assessment of safety performance of novel nanostructured batteries needs new or adjusted testing protocols. In SAE J2464:2009 [46] or FreedomCAR:2005 [126] it is specified, that the external resistance needs to be much smaller than the DUT DC impedance [46] or 1/10 of the minimum resistance of the cell/module for systems with less than 0.9 mΩ/V system voltage ± 0.1 mΩ internal resistance [138]. However these instructions must be applied carefully because in practice it is very hard to apply lower than 1 mΩ external resistance including connections, cables, and contactors.

ISC test. ISCs have been cited as the cause for incidents and recalls of Li-ion batteries [139]. It is an extremely energetic process leading to a rapid heat increase. IEC 62133:2012 [125] provides instructions for ISC testing of cylindrical and prismatic cells.

Disassembly of the cell, insertion of an L-shaped nickel particle, and reassembly of the cell are required. Afterward a certain pressure is applied to induce the ISC. There are experimental methods trying to mimic an ISC, such as *blunt indentation test* and *separator integrity test*.

Overdischarge test. This test can lead to damage of the electrodes and current collectors (e.g., copper dissolution and plating), lithium plating upon recharge, which can ultimately lead to thermal runaway.

Other types of electrical tests are *high-rate cycling* and *cell reversal tests*.

5.4 Chemical hazards monitoring

As a consequence of abuse testing, decomposition and combustion of materials may occur. If the integrity of the energy storage system is compromised hazardous materials, such as gases (e.g., CO, CO_2, H_2, CH_4, and HF) or liquid electrolyte may be released. The hazardous nature of these releases, both in terms of toxicity and flammability, as described in Section 2.3, greatly depends on the composition of DUT components. In this context, multiple standards [46,48] require the assessment of hazardous releases and flammability simultaneously with the mechanical, electrical, or thermal tests previously described.

During and after abuse testing there is a possibility that individuals in the surroundings may be exposed to vapors and mist of electrolyte or even small-particle solids when no appropriate measures are taken. Therefore, testing areas should be designed to avoid human exposure to toxic hazards.

Hazardous substances monitoring. Analysis of gas, smoke, and particulates released during safety testing is required in various standards [46,48,126,134], and for this, analytical techniques or gas sensors are recommended. Some of the possible techniques include gas chromatography/mass spectrometry and/or Fourier transform infrared spectroscopy (FTIR), length-of-stain dosimetry, and use of sorbent or detector tubes. In automotive applications, the requirement set by SAE J2464:2009 [46] is that the volume of hazardous substances released needs to be allowable for human exposure [e.g., below Emergency Response Planning Guidelines (ERPG) level 2 [140]]. (ERPG level 2 is the maximum airborne concentration level below which most individuals could be exposed for up to 1 h without experiencing or developing serious or irreversible health effects or symptoms that could impair an individual's ability to take protective action.) SAE J6469 Part 1:2009 [141] indicates that the maximum allowable quantity accumulated during testing (anywhere in the driver, passenger, and load compartments) shall not exceed the values defined in the latest version of applicable national/international standards or regulations.

Flammability monitoring. Gases being released during abuse testing (e.g., methane, hydrogen, and carbon monoxide) generally have a flammable nature. To evaluate this property, it is possible to install one or several spark sources close to the RESS during safety testing [141]. Alternatively, gas monitors can be used, as mentioned in UL 2580:2013 [48].

5.5 Hazards considerations about safety testing

A parallel aspect that deserves some attention is that related to emissions and waste management when performing abuse testing. As previously mentioned, during safety testing it is expected that the stability of the RESS and its integrity is compromised. Under certain conditions, venting, electrolyte leakage, fire, or even explosion may occur, generating considerable amounts of toxic materials and waste in the gaseous, liquid, or solid forms.

When performing abuse testing, a temperature rise is an indication that the RESS may develop into thermal runaway. Therefore it is a common practice to place temperature sensors and/or infrared temperature probes in the proximity of the RESS. If the system vents or explodes, the testing area needs to be ventilated for long periods of time, allowing the DUT to cool down to room temperature before trained personnel can extract the damaged system and clean the area [9]. In case of fire, an extinguishing agent appropriate to the RESS chemistry (e.g., water is not an appropriate agent in case of lithium metal battery fire) needs to be used. An alternative approach is to let the battery burn out. Then, the generated emissions and waste need proper management (e.g., gas scrubber and waste water collection) following regulations on hazardous waste to avoid contamination of workers or of the environment. A final step after abuse testing relates to the transport, disposal, and recycling of damaged RESS. Each of the steps has its own safety risks, relevant both to nano- and bulk materials. However, in the case of nanomaterials there are still many gaps in our knowledge so following the precautionary principle has been recommended by, for example, the Health Council of the Netherlands [142] or the German Advisory Council on the Environment (SRU) [143], both based on the description set out at the Rio Summit [144]. In any case, proper training and guidelines are needed in the interest of public health. The relevance of nanowaste management is highlighted by the fact that various standardization bodies are actively working in the field, for example, Technical Committee CEN/TC 352 works on the Technical Specification: Guide to the management of waste and the disposal of nanomaterials for the execution of EV standardization mandate M/461 "Standardization activities regarding nanotechnologies and nanomaterials" [145].

6 CONCLUSIONS AND OUTLOOK

RESS technology is crucial and enables many applications in daily life ranging from portable and consumer electronics (e.g., smart phones, tablets, power tools, robotic vacuum cleaners, and electric toys) to e-mobility (e.g., hybrid, plug-in, and pure EVs) and stationary energy storage (e.g., utility and grid storage solutions). During and following deployment of RESS in these and other applications in the recent decades, a significant number of safety events—often battery fires—have been reported [8,10]. As it is important to address these safety concerns, a thorough discussion of hazards related to RESS—mechanical, electrical, and chemical—is necessary.

In this chapter, the main hazards were presented, but it should be noted that not all hazards are equally relevant for all different types of RESS. A selection of failure scenarios—overheating, mechanical deformation, external short circuit, and overcharge—was presented together with the main approaches for risk mitigation. We acknowledge that a significant effort of battery community, as well as of the OEM's in research and innovation aimed at the development of efficient risk mitigation measures, including adoption of the "defense-in-depth" approach with several levels of protection being implemented simultaneously, has yielded consumer products with appropriate safety characteristics. Testing of RESS performance according to the existing standards and regulations under normal operating, as well as under foreseen abuse conditions and development of new relevant legislation aims at ensuring that the required safety level in any given application is met. Additionally, measures, such as protection of RESS software against modification or provision of information or training should be considered to avoid safety events related to unforeseen abuse (e.g., tuning of EVs by amateurs).

Further research and innovation is needed to further improve safety of Li-ion RESS, for example, by developing intrinsically safe materials, less reactive and less toxic active electrode materials, and nonflammable, nontoxic electrolytes, and binders, while maintaining RESS performance and durability. Also, further insight needs to be gained into the effects of nanostructured Li-ion battery materials on the safety of nanomaterial-based RESS toward humans in case of accidental/occupational exposure to these materials. It should also be noted that safety is typically evaluated on a new RESS. As aged RESS are not readily available for all applications and especially not necessarily when a new product is released, the impact of degradation or aging on safety is not always thoroughly examined experimentally. In the studies that have been performed in the literature no common trend was found, as degraded, as well as improved safety performance of aged cells has been observed [94,146]. Consequently, and also because longer operation times can be expected in the future (due to reduced degradation, but also in second life applications [147]), further research on the relationship between safety performance and state of health of a RESS is required.

Overall it has to be concluded that the described hazards are relevant and that the related risks need to be addressed. Large efforts by industry and research community have addressed the development of appropriate mitigation strategies. Risks can be mitigated by the introduction of safer materials or by the adoption of specific protective devices to reach an acceptable level of safety. There is not one univocal most effective mitigation strategy and the right choice should always be chosen based on the application where the battery will be employed and there is still potential to further improve safety. Taking into account fostering clear communication with end users and acknowledging efforts in the development of risk mitigation strategies, it is not expected that safety concerns will hamper or delay uptake of nanomaterial-based RESS technologies.

GLOSSARY

ARC	Accelerated rate calorimeter
BMS	Battery-management system
CEN	European Committee for Standardization
CENELEC	European Committee for Electrotechnical Standardization
CID	Current interruption devices
CLP	Classification, Labelling, and Packaging
CMC	Carboxymethylcellulose
DC	Direct current
DEC	Diethyl carbonate
1,3-DL	1,3-Dioxolane
DME	1,2-Dimethoxyethane
DUT	Device under test
EPA	US Environmental Protection Agency
ERPG	Emergency Response Planning Guidelines
EU	European Union
EU-OSHA	European Agency for Safety and Health at Work
EUCAR	European Council for Automotive Research and Development
EV	Electric vehicle
FTIR	Fourier transform infrared spectroscopy
HEV	Hybrid electric vehicle
HFP	Hexafluoropropylene
IEC	International Electrotechnical Commission
ISC	Internal short circuit
ISO	International Organization for Standardization
KFSI	Potassium bis(fluorosulfonyl)imide
LAGP	Lithium aluminums germanium phosphate
LCO	Lithium cobalt oxide
LFP	Lithium iron phosphate
LiBOB	Lithium bis(oxalato)borate
LiFAP	Lithium fluoroalkyl phosphate
LiFSI	Lithium bis(fluorosulfonyl)imide
LiTFSI	Lithium bis-trifluoromethanesulfonimide
LLTO	Lithium lanthanum titanium oxide
LLZO	Lithium lanthanum zirconium oxide
LMO	Lithium manganese oxide
LTO	Lithium titanate
2-Me-THF	2-Methyl-tetrahydrofuran
NCA	Lithium nickel cobalt aluminum oxide
NiMH	Nickel metal hydride
NIOSH	The National Institute for Occupational Safety and Health

NMC	Lithium nickel manganese cobalt oxide
NMP	*N*-Methylpyrrolidine
OECD	Organisation for Economic Co-operation and Development
OEM	Original equipment manufacturer
PE	Polyethylene
PHEV	Plug-in hybrid electric vehicle
PI	Polyimide
PP	Polypropylene
PTC	Positive temperature coefficient
PTMC	Polytrimethylene carbonate
PVDF	Polyvinilidene fluoride
REACH	Registration, Evaluation, Authorisation and Restriction of Chemicals
RESS	Rechargeable energy storage system
SAE	Society of Automobile Engineers
SBR	Styrene butadiene rubber
SC	Super capacitors
SEI	Solid–electrolyte interface
SOC	State of charge
SRU	German Advisory Council on the Environment (Sachverständigenrat für Umweltfragen)
UL	Underwriters Laboratory
UNECE	United Nations Economic Commission for Europe

REFERENCES

[1] K.T. Lee, J. Cho, Roles of nanosize in lithium reactive nanomaterials for lithium ion batteries, Nano Today 6 (1) (2011) 28–41.
[2] C. Shrikant, Nanomaterials for electrical energy storage devices, in: B. Bhushan (Ed.), Encyclopedia of Nanotechnology, Springer Science+Business Media B.V., 2012, Netherlands.
[3] R. Zhao, S. Zhang, J. Liu, J. Gu, A review of thermal performance improving methods of lithium ion battery: electrode modification and thermal management system, J. Power Sources 299 (2015) 557–577.
[4] S. Bianco, Chiodoni, A. Gerbaldi, C. Quaglio, M. Nanostructures for energy, in: B. Bhushan (Ed.), Encyclopedia of Nanotechnology, Springer Science+Business Media B.V., 2012, Netherlands.
[5] C. Jiang, E. Hosono, H. Zhou, Nanomaterials for lithium ion batteries, Nano Today 1 (4) (2006) 28–33.
[6] S.-T. Myung, K. Amine, Y.-K. Sun, Nanostructured cathode materials for rechargeable lithium batteries, J. Power Sources 283 (2015) 219–236.
[7] H.J.G. Haynes, Fire loss in the United States during 2015, presented to National Fire Protection Association (NFPA), 2015.
[8] B. Vlasic, Car Fire a Test for High-Flying Tesla, The New York Times, NY, (2013).
[9] S.C. Levy, P. Bro, Battery Hazards and Accident Prevention, Plenum Press, New York, NY; London, 1994.
[10] National Transportation Safety Board, Aircraft Incident Report, Auxiliary Power unit battery fire, Japan Airlines Boeing 787-8, JA829J, Boston, MA, January 7, 2013.
[11] M. Ahrens, US vehicle fire trends and patterns, presented to National Fire Protection Association (NFPA), 2010.
[12] D. Aaker, Teslas catch fire less often than gas-powered cars, Harvard Business Review, 2013.
[13] E.P. Roth, C.J. Orendorff, How electrolytes influence battery safety, Electrochem. Soc. Interf. 21 (2) (2012) 45–49.
[14] T.H. Dubaniewicz Jr., J.P. DuCarme, Further study of the intrinsic safety of internally shorted lithium and lithium-ion cells within methane-air, J. Loss Prevent. Proc. 32 (2014) 165–173.

[15] C. Mikolajczak, M. Kahn, K. White, R.T. Long, Lithium-Ion Batteries Hazard And Use Assessment. Exponent Failure Analysis Associates, Inc. Fire Protection Research Foundation, 2011, Contract No.: 1100034.000 A0F0 0711 CM01.
[16] G. Xiong, A. Kundu, T.S. Fisher, Thermal Effects in Supercapacitors, Springer International Publishing, Switzerland, 2015.
[17] R. Groiß, A. Jossen. Sicherheitsaspekte beim Testen von Lithium-Ionen Batterien. Entwicklerforum Batterien und Ladekonzepte, Munich, Germany, 2010.
[18] IEC 62619-1, Secondary Cells and Batteries Containing Alkaline or Other Nonacid Electrolytes—Safety Requirements for Large Format Secondary Lithium Cells and Batteries for use in Industrial Applications CCDV, International Technical Commission. Available from: http://www.iec.ch/dyn/www/f?p=103:23:0::::FSP_ORG_ID:1410
[19] P. Ribière, S. Grugeon, M. Morcrette, S. Boyanov, S. Laruelle, G. Marlair, Investigation of the fire-induced hazards of Li-ion battery cells by fire calorimetry, Energy Environ. Sci. 5 (1) (2012) 5271–5280.
[20] Safety of lithium-ion batteries, Recharge, Brussels, Belgium, 2013. Available from: http://www.rechargebatteries.org/wp-content/uploads/2013/07/Li-ion-safety-July-9-2013-Recharge-.pdf
[21] J. Mähliß, Lithium-Ionen-Batterien—Gefahrenpotential und Umgang, 2013. Available from: http://docplayer.org/5093968-Lithium-ionen-batterien-gefahrenpotential-und-umgang.html.
[22] R. Korthauer (Ed.), Handbuch Lithium-Ionen-Batterien, Springer, Berlin, 2013.
[23] F. Beguin, E. Frackowiak (Eds.), Supercapacitors: Materials, Systems and Applications, Wiley VCH, 2013.
[24] G.G. Eshetu, S. Grugeon, S. Laurella, S. Boyanov, A. Lecocq, J.P. Bertranda, G. Marlair, In-depth safety-focused analysis of solvents used in electrolytes for large scale lithium-ion batteries, Phys. Chem. Chem. Phys., 15 (2013) 9145–9155
[25] S.J. Harris, A. Timmons, W.J. Pitz, A combustion chemistry analysis of carbonate solvents used in Li-ion batteries, J. Power Sources 193 (2) (2009) 855–858.
[26] N.P. Lebedeva, L. Boon-Brett, Considerations on the chemical toxicity of contemporary Li-ion battery electrolytes and their components, J. Electrochem. Soc. 163 (6) (2016) A821–A830.
[27] J. Jiang, J.R. Dahn, ARC studies of the thermal stability of three different cathode materials: $LiCoO_2$; $Li[Ni_{0.1}Co_{0.8}Mn_{0.1}]O_2$; and $LiFePO_4$, in $LiPF_6$ and LiBoB EC/DEC electrolytes, Electrochem. Comm. 6 (1) (2004) 39–43.
[28] M. Takahashi, S.-i. Tobishima, K. Takei, Y. Sakurai, Reaction behavior of $LiFePO_4$ as a cathode material for rechargeable lithium batteries, Solid State Ionics 148 (3–4) (2002) 283–289.
[29] P. Kurzweil, M. Chwistek, Capacitance determination and abusive aging studies of supercapacitors based on acetonitrile and ionic liquids, 16th International Seminar on Double-Layer Capacitors, Deerfield Beach, FL, December 4–6, 2006.
[30] R. Mörk-Mörkenstein, Elektro-und Hybridfahrzeuge, Überlegungen zu Sicherheitsfragen beim Feuerwehreinsatz, ÖBFV Führungsseminar, 2012.
[31] ZEVI, Safe handling of lithium batteries, Guide for Creating Product-Specific Battery Information Sheets, ZEVI, Frankfurt, 2012.
[32] C. Mikolajczak, J.S. Harmon, T. Hayes, M. Megerle, K. White, Q. Horn, et al. (Eds.), Li-ion battery cell failure analysis: the significance of surviving features on copper current collectors in cells that have experienced thermal runaway, in: 25th International Battery Seminar and Exhibit: Primary and Secondary Batteries—Small Fuel Cells—Other Technologies, Fort Lauderdale, FL, 2008.
[33] E. Sahraei, J. Campbell, T. Wierzbicki, Modeling and short circuit detection of 18650 Li-ion cells under mechanical abuse conditions, J. Power Sources 220 (2012) 360–372.
[34] IEC 60479-1, Effects of current on human beings and livestock, Part 1, General Aspects, 2005.
[35] F. Davies, P. Patel, G. Verela, E. Darcy, Series String and Parallel Blank External Short and Overcharge Test Report, NASA Engineering and Safety Center RP-08-75 (Appendix J), 2010, pp. 179–217.
[36] S.J. Dillon, K. Sun, Microstructural design considerations for Li-ion battery systems, Curr. Opin. Solid St. M. 16 (2012) 153–162.
[37] MSDS for lithium cobalt oxide product of SigmaAldrich. Available from: http://www.sigmaaldrich.com/catalog/product/aldrich/442704?lang=en&region=NL
[38] MSDS for Lithium nickel oxide product of SigmaAldrich. Available from: http://www.sigmaaldrich.com/catalog/product/aldrich/757365?lang=en&region=NL&cm_sp=Insite-_-prodRecCold_xorders-_-prodRecCold2-2

[39] MSDS for lithium manganese oxide product of SigmaAldrich. Available from: http://www.sigmaaldrich.com/catalog/product/aldrich/725129?lang=en&region=NL
[40] MSDS for lithium iron phosphate product of SigmaAldrich. Available from: http://www.sigmaaldrich.com/catalog/product/aldrich/759546?lang=en&region=NL
[41] MSDS for mesoporous carbon products of SigmaAldrich. Available from: http://www.sigmaaldrich.com/catalog/product/aldrich/699640?lang=en&region=NL and http://www.sigmaaldrich.com/catalog/product/aldrich/699632?lang=en&region=NL
[42] MSDS for lithium titanate product of SigmaAldrich. Available from: http://www.sigmaaldrich.com/catalog/product/aldrich/702277?lang=en&region=NL
[43] MSDS for silicon dioxide product of SigmaAldrich. Available from: http://www.sigmaaldrich.com/catalog/product/aldrich/204358?lang=en&region=NL
[44] C. Mikolajczak, M. Kahn, K. White, R.T. Long, Lithium-Ion Batteries Hazard and Use Assessment, Springer, New York, (2011).
[45] D. Lisbona, T. Snee, A review of hazards associated with primary lithium and lithium-ion batteries, Process Saf. Environ. 89 (6) (2011) 434–442.
[46] SAE J2464, Electric and hybrid electric vehicle rechargeable energy storage system (RESS), safety and abuse testing, 2009.
[47] SAE J2929, Safety standards for electric and hybrid vehicle propulsion battery systems utilizing lithium-based rechargeable cells, 2013.
[48] UL 2580, Batteries for use in electric vehicles, 2013.
[49] J.A. Jeevarajan, C.S. Winchester, Battery safety qualifications for human ratings, Electrochem. Soc. Interf. 21 (2) (2012) 51–55.
[50] Shinji Koike, Masahiro Shikano, Hikari Sakaebe, H. Kobayashi, Study of generative gas species from lithium-ion battery component under abuse conditions, PRiME, The Electrochemical Society, Honolulu, 2012.
[51] O. Masamichi, K. Shinichi, S. Yuuichi, Y. Miwa, O. Yumiko, U. Makoto, et al. Identification of the source of evolved gas in Li-ion batteries using ^{13}C-labeled solvents, J. Elecrochem. Soc. 155 (11) (2008) A794–A797.
[52] W. Kong, H. Li, X. Huang, L. Chen, Gas evolution behaviors for several cathode materials in lithium-ion batteries, J. Power Sources 142 (1–2) (2005) 285–291.
[53] A. Hammami, N. Raymond, M. Armand, Runaway risk of forming toxic compounds, Nature 424 (6949) (2003) 635–636.
[54] F. Beguin (Ed.), Supercapacitors: materials, systems, and applications, in: New Materials for Sustainable Energy and Development, John Wiley & Sons, New York, NY, 2013.
[55] S. Goriparti, E. Miele, F. De Angelis, E. Di Fabrizio, R. Proietti Zaccaria, C. Capiglia, Review on recent progress of nanostructured anode materials for Li-ion batteries, J. Power Sources 257 (2014) 421–443.
[56] Y.-K. Sun, Z. Chen, H.-J. Noh, D.-J. Lee, H.-G. Jung, Y. Ren, S. Wang, C.S. Yoon, S.-T. Myung, K. Amine, Nanostructured high-energy cathode materials for advanced lithium batteries, Nat. Mater. 11 (2012) 942–947.
[57] A. Pietroiusti, A. Magrini, L. Campagnolo, Mechanisms of nanomaterials toxicity, in: J. Njuguna, K. Pielichowski, H. Zhu (Eds.), Health and Environmental Safety of Nanomaterials: Polymer Nanocomposites and Other Materials Containing Nanoparticles, Woodhead Publishing, Cambridge, 2014, pp. 28–43.
[58] T.S. Peixe, E. de Souza Nascimento, K. Larson Schofield, A.S.A. Arcuri, R.P. Bulcao, Nanotoxicology and exposure in the occupational setting, Occup. Dis. Environ. Med. 3 (2015) 35–48.
[59] Managing nanomaterials in the workplace. Available from: https://osha.europa.eu/en/themes/nanomaterials
[60] Understanding the hazards of nanomaterials. Available from: http://www.hse.gov.uk/nanotechnology/understanding-hazards-nanomaterials.htm
[61] Nanomaterials. Available from: http://echa.europa.eu/regulations/nanomaterials
[62] Nanotechnology. Available from: http://www.cdc.gov/niosh/topics/nanotech/
[63] Research on nanomaterials. Available from: http://www.epa.gov/chemical-research/research-nanomaterials

[64] Publications in the Series on the Safety of Manufactured Nanomaterials. Available from: http://www.oecd.org/env/ehs/nanosafety/publications-series-on-safety-of-manufactured-nanomaterials.htm
[65] A.P.M. Wood, U. Gundert-Remy, C. de Rooij, S.M. Tissot, The AETL methodology as a potential solution to current challenges associated with the development and use of acute exposure levels in Seveso II applications, J. Hazard. Mater. 8 (2006) A133.
[66] R.M. Heinälä, M. Heraty Wood, M. Ruijten, P.M.J. Bos, A. Zitting, S. Bull, D. Russell, E. Nielsen, G. Cassel, P. Leffler, S. Tissot, J.-M. Vincent, T. Santonen, Survey on methodologies in the risk assessment of chemical exposures in emergency response situations in Europe, J. Hazard. Mater. 244-245 (2013) 545.
[67] E.P. Roth, C.C. Crafts, D.H. Doughty, Thermal Abuse Studies on Lithium Ion Rechargeable Batteries, Sandia National Laboratories, Albuquerque, (2000).
[68] R. Spotnitz, J. Franklin, Abuse behavior of high-power, lithium-ion cells, J. Power Sources 113 (1) (2003) 81–100.
[69] E.P. Roth, D.H. Doughty, D.L. Pile, Effects of separator breakdown on abuse response of 18650 Li-ion cells, J. Power Sources 174 (2) (2007) 579–583.
[70] J.L. Murray, Aluminium-copper system, Int. Met. Rev. 30 (5) (1985) 211–233.
[71] H. Maleki, J.N. Howard, Internal short circuit in Li-ion cells, J. Power Sources 191 (2) (2009) 568–574.
[72] A. Kriston, V. Ruiz, T. Kosmidou, J. Ungeheuer, H. Döring, B. Fritsch, et al., Evaluation of external short circuit performance of Li-ion batteries for the Global Technical Regulation of Electric Vehicle Safety, in: 67th Annual Meeting of the International Society of Electrochemistry, The Hague, The Netherlands, 2016, p. 160597.
[73] S. Okazaki, S. Higuchi, N. Kubota, S. Takahashi, Measurement of short circuit current for low internal resistance batteries, J. Appl. Electrochem. 16 (4) (1986) 513–516.
[74] I.J. Ong, J. Newman, Double-layer capacitance in a dual lithium-ion insertion cell, J. Electrochem. Soc. 146 (12) (1999) 4360–4365.
[75] F.V. Conte, P. Gollob, H. Lacher, Safety in the battery design: the short circuit, World Elec. Veh. J. 3 (1.) (2009).
[76] S. Okazaki, S. Higuchi, N. Kubota, S. Takahashi, Predicted and observed initial short circuit current for lead-acid batteries, J. Appl. Electrochem. 16 (5) (1986) 631–635.
[77] A. Franco, A. M. Doublet, Liesse, W. Bessler, G. (Eds.), Physical Multiscale Modeling and Numerical Simulation of Electrochemical Devices for Energy Conversion and Storage, Springer-Verlag, London, 2016.
[78] A. Kriston, A. Pfrang, B.N. Popov, L. Boon-Brett, Development of a full layer pore-scale model for the simulation of electro-active material used in power sources, J. Electrochem. Soc. 161 (8) (2014) E3235–E3247.
[79] C.-K. Lin, Y. Ren, K. Amine, Y. Qin, Z. Chen, In situ high-energy X-ray diffraction to study overcharge abuse of 18650-size lithium-ion battery, J. Power Sources 230 (2013) 32–37.
[80] Yoshiyasu Saito, A. Kiyonami Takano, Negishi, Thermal behaviors of lithium-ion cells during overcharge, J. Power Sources 97 (2001) 693–696.
[81] R.A. Leising, M.J. Palazzo, E.S. Takeuchi, K.J. Takeuchi, Abuse testing of lithium-ion batteries: characterization of the overcharge reaction of $LiCoO_2$/graphite cells, J. Electrochem. Soc. 148 (8) (2001) A838–A844.
[82] S. Tobishima, Y. Sakurai, J. Yamaki, Safety characteristics of rechargeable lithium metal cells, J. Power Sources 68 (2) (1997) 455–458.
[83] F. Xu, H. He, Y. Liu, C. Dun, Y. Ren, Q. Liu, et al. Failure investigation of $LiFePO_4$ cells under overcharge conditions, J. Electrochem. Soc. 159 (5) (2012) A678–A687.
[84] Interim Factual Report, Boeing 787-8, JA829J, Japan Airlines. National Transportation Safety Board, Office of Aviation Safety, Washington, DC, 2013.
[85] S. Theil, M. Fleischhammer, P. Axmann, M. Wohlfahrt-Mehrens, Experimental investigations on the electrochemical and thermal behaviour of $LiCoPO_4$-based cathode, J. Power Sources 222 (2013) 72–78.
[86] M. Fleischhammer, T. Waldmann, G. Bisle, B.I. Hogg, M. Wohlfahrt-Mehrens, Interaction of cyclic ageing at high-rate and low temperatures and safety in lithium-ion batteries, J. Power Sources 274 (2015) 432–439.

[87] M.D.Thomas, F. Fuller, John Newman, Simulation and optimization of the dual lithium-ion insertion cell, J. Electrochem. Soc. 144 (1994) 1–10.
[88] S.J. Harris, A. Timmons, D.R. Baker, C. Monroe, Direct in situ measurements of Li transport in Li-ion battery negative electrodes, Chem. Phys. Lett. 485 (4–6) (2010) 265–274.
[89] Z. Guo, J. Zhu, J. Feng, S. Du, Direct in situ observation and explanation of lithium dendrite of commercial graphite electrodes, RSC Adv. 5 (85) (2015) 69514–69521.
[90] B.L. Mehdi, J. Qian, E. Nasybulin, C. Park, D.A. Welch, R. Faller, et al. Observation and quantification of nanoscale processes in lithium batteries by operando electrochemical (S)TEM, Nano Lett. 15 (3) (2015) 2168–2173.
[91] J. Arai, Y. Okada, T. Sugiyama, M. Izuka, K. Gotoh, K. Takeda, In situ solid state 7Li NMR observations of lithium metal deposition during overcharge in lithium-ion batteries, J. Electrochem. Soc. 162 (6) (2015) A952–A958.
[92] SO/IEC Guide 51:2014, Safety Aspects—Guidelines for Their Inclusion in Standards, 2014, p. 15.
[93] J. Vetter, P. Novák, M.R. Wagner, C. Veit, K.C. Möller, J.O. Besenhard, et al. Ageing mechanisms in lithium-ion batteries, J. Power Sources 147 (1-2) (2005) 269–281.
[94] Lithium-ion batteries: Applying fault tree analysis methodology; Indentation induced ISC test; Advancing Lithium-ion battery standards; Aging effects on Lithium-ion batteries, UL Underwriters Laboratory, 2014.
[95] S. Kraas, A. Vijn, M. Falk, B. Ufer, B. Luerßen, J. Janek, et al. Nanostructured and nanoporous $LiFePO_4$ and $LiNi_{0.5}Mn_{1.5}O_4$-δ as cathode materials for lithium-ion batteries, Prog. Solid State Chem. 42 (4) (2014) 218–241.
[96] NIOSH, Occupational exposure to carbon nanotubes and nanofibers, National Institute for Occupational Safety and Health, 2013. Contract No.: DHHS (NIOSH) Publication No. 2013-145.
[97] C.-Y. Jhu, Y.-W. Wang, C.-Y. Wen, C.-M. Shu, Thermal runaway potential of $LiCoO_2$ and $Li(Ni_{1/3}Co_{1/3}Mn_{1/3})O_2$ batteries determined with adiabatic calorimetry methodology, Appl. Energy 100 (2012) 127–131.
[98] H.-J. Noh, S. Youn, C.S. Yoon, Y.-K. Sun, Comparison of the structural and electrochemical properties of layered $Li[Ni_xCo_yMn_z]O_2$ (x = 1/3, 0.5, 0.6, 0.7, 0.8 and 0.85) cathode material for lithium-ion batteries, J. Power Sources 233 (2013) 121–130.
[99] Q. Wang, P. Ping, X. Zhao, G. Chu, J. Sun, C. Chen, Thermal runaway caused fire and explosion of lithium ion battery, J. Power Sources 208 (2012) 210–224.
[100] S.-H. Lee, C.S. Yoon, K. Amine, Y.-K. Sun, Improvement of long-term cycling performance of $Li[Ni_{0.8}Co_{0.15}Al_{0.05}]O_2$ by AlF_3 coating, J. Power Sources 234 (2013) 201–207.
[101] H. Kim, S.M. Oh, B. Scrosati, Y.K. Sun, High-performance electrode materials for lithium-ion batteries for electric vehicles, in: B. Scrosati, J. Garche, W. Tillmetz (Eds.), Advances in Battery Technologies for Electric Vehicles, Woodhead Publishing, Cambridge, 2015, pp. 191–241.
[102] X. Yang, X. Wang, L. Hu, G. Zou, S. Su, Y. Bai, et al. Layered $Li[Ni_{0.5}Co_{0.2}Mn_{0.3}]O_2$–$Li_2MnO_3$ core–shell structured cathode material with excellent stability, J. Power Sources 242 (2013) 589–596.
[103] Y.-K. Sun, Z. Chen, H.-J. Noh, D.-J. Lee, H.-G. Jung, Y. Ren, et al. Nanostructured high-energy cathode materials for advanced lithium batteries, Nat. Mater. 11 (11) (2012) 942–947.
[104] R. Marom, S.F. Amalraj, N. Leifer, D. Jacob, D. Aurbach, A review of advanced and practical lithium battery materials, J. Mater. Chem. 21 (27) (2011) 9938–9954.
[105] B. Scrosati, J. Garche, Lithium batteries: status, prospects, and future, J. Power Sources 195 (9) (2010) 2419–2430.
[106] D.L. Wood Iii, J. Li, C. Daniel, Prospects for reducing the processing cost of lithium ion batteries, J. Power Sources 275 (2015) 234–242.
[107] V.K.V. Barsykov, The influence of polymer binders on the performance of cathodes for lithium-ion batteries, Sci. J. RTU 21 (2010) 67–71.
[108] G. Qian, L. Wang, Y. Shang, X. He, S. Tang, M. Liu, et al. Polyimide binder: a facile way to improve safety of lithium-ion batteries, Electrochim. Acta 187 (2016) 113–118.
[109] P.G. Balakrishnan, R. Ramesh, T. Prem Kumar, Safety mechanisms in lithium-ion batteries, J. Power Sources 155 (2) (2006) 401–414.
[110] C.J. Orendorff, The role of separators in lithium-ion cell safety, Electrochem. Soc. Interface 21 (2) (2012) 61–65.

[111] Y.-C. Jung, S.-K. Kim, M.-S. Kim, J.-H. Lee, M.-S. Han, D.-H. Kim, et al. Ceramic separators based on Li^+-conducting inorganic electrolyte for high-performance lithium-ion batteries with enhanced safety, J. Power Sources 293 (2015) 675–683.
[112] C. Liu, X. Ma, F. Xu, L. Zheng, H. Zhang, W. Feng, et al. Ionic liquid electrolyte of lithium bis(fluorosulfonyl)imide/*N*-methyl-*N*-propylpiperidinium bis(fluorosulfonyl)imide for Li/natural graphite cells: effect of concentration of lithium salt on the physicochemical and electrochemical properties, Electrochim. Acta 149 (2014) 370–385.
[113] J.W. Fergus, Ceramic and polymeric solid electrolytes for lithium-ion batteries, J. Power Sources 195 (15) (2010) 4554–4569.
[114] J. Mindemark, B. Sun, E. Törmä, D. Brandell, High-performance solid polymer electrolytes for lithium batteries operational at ambient temperature, J. Power Sources 298 (2015) 166–170.
[115] Y. Wang, W.D. Richards, S.P. Ong, L.J. Miara, J.C. Kim, Y. Mo, et al. Design principles for solid-state lithium superionic conductors, Nat. Mater. 14 (10) (2015) 1026–1031.
[116] D. Marginèdes (Ed.), Bolloré Blue Solutions, in: Advanced Automotive Battery Conference (AABC), Strasbourg, France, 2013.
[117] S. O'Malley, D. Zuby, M. Moore, M. Paine, D. Paine (Eds.), Crashworthiness testing of electric and hybrid vehicles, in: The 24th International Technical Conference on the Enhanced Safety of Vehicles (ESV), Gothenburg, Sweden, 2015.
[118] K. Smith, G.-H. Kim, E. Darcy, A. Pesaran, Thermal/electrical modeling for abuse-tolerant design of lithium-ion modules, Int. J. Energy Res. 34 (2) (2010) 204–215.
[119] H. Wu, D. Zhuo, D. Kong, Y. Cui, Improving battery safety by early detection of internal shorting with a bifunctional separator, Nat. Commun. 5 (2014) (Article 5).
[120] Z. Chen, P.-C. Hsu, J. Lopez, Y. Li, J.W.F. To, N. Liu, et al. Fast and reversible thermoresponsive polymer switching materials for safer batteries, Nat. Energy 1 (2016) 15009.
[121] Z. Rao, S. Wang, A review of power battery thermal energy management, Renew. Sustain. Energy Rev. 15 (9) (2011) 4554–4571.
[122] S.-I. Tobishima, J.-I. Yamaki, A consideration of lithium cell safety, J. Power Sources 81-82 (1999) 882–886.
[123] UN/ECE Regulation N°10002, Uniform provisions concerning the approval of vehicles with regard to specific requirements for the electric power train, 2013.
[124] V. Ruiz, A. Pfrang, A. Kriston, N. Omar, P.V.D. Bossche, L. Boon-Brett, A review of international abuse testing standards and regulations for lithium-ion batteries in electric and hybrid electric vehicles, submitted to Renew. Sustain. Energy Rev.
[125] IEC 62133, Secondary cells and batteries containing alkaline or other nonacid electrolytes—safety requirements for portable sealed secondary cells, and for batteries made from them, for use in portable applications, 2012.
[126] D.H. Doughty, C.C. Crafts, SAND 2005–3123, FreeDomCAR Electrical Energy Storage Systems Abuse Test Manual for Electric and Hybrid Electric Vehicle Applications, 2005.
[127] IEC 62660-2, Rechargeable Cells Standards Publication Secondary Lithium-Ion Cells for the Propulsion of Electric Road Vehicles, Part 2: Reliability and Abuse Testing, 2011.
[128] J. Lamb, C.J. Orendorff, L.A.M. Steele, S.W. Spangler, Failure propagation in multicell lithium-ion batteries, J. Power Sources 283 (2015) 517–523.
[129] M. Jacoby, Burning batteries, Chem. Eng. News 85 (51) (2007) 26–28.
[130] T. Reddy, Linden's Handbook of Batteries, fourth ed., McGraw-Hill, Inc., New York, (2011).
[131] C.J. Orendorff, E.P. Roth, G. Nagasubramanian, Experimental triggers for internal short circuits in lithium-ion cells, J. Power Sources 196 (15) (2011) 6554–6558.
[132] ISO 12405-2, Electrically Propelled Road Vehicles—Test Specification for lithium-Ion Traction Battery Packs and Systems, Part 2: High-Energy Applications, 2012.
[133] T. Chatwin (Ed.), Abuse testing of sodium nickel chloride batteries, in: Telecommunications Energy Conference "Smart Power and Efficiency" (INTELEC), Proceedings of 35th International, Hamburg, Germany, 2013.
[134] T. Unkelhaeuser, D. Smallwood, SAND99-0497-USABC, United States Advanced Battery Consortium Electrochemical Storage System Abuse Test Procedure Manual, 1999.

[135] ISO 12405-1, Electrically Propelled Road Vehicles—Test Specification for lithium-Ion Traction Battery Packs and Systems, Part 1: High-Power Applications, 2011.
[136] IEC 61434, Secondary Cells and Batteries Containing Alkaline or Other Nonacid Electrolytes—Guide to Designation of Current in Alkaline Secondary Cell and Battery Standards, 1996.
[137] IEC/(CD) 62660-3, Rechargeable Cells Standards Publication Secondary Lithium-Ion Cells for the Propulsion of Electric Road Vehicles, Part 3: Safety Requirements of Cells and Modules, 2013.
[138] AIS-048, Battery Operated Vehicles—Safety Requirements of Traction Batteries, 2009.
[139] B. Barnett, S. Sriramulu (Eds.), A Perspective on Li-Ion Safety and Opportunities for Portable and Electric Vehicle Applications, Presentation at the 27th International Battery Seminar and Exhibit, Fort Lauderdale, FL, 2010.
[140] ERPG level 2. Available from: https://www.aiha.org/get-involved/AIHAGuidelineFoundation/EmergencyResponsePlanningGuidelines/Documents/2014%20ERPG%20Introduction.pdf
[141] ISO 6469-1, Electrically Propelled Road Vehicles—Safety Specifications, Part 1: On-Board Rechargeable Energy Storage System (RESS), 2009.
[142] Health Significance of Nanotechnologies, Health Council of the Netherlands, The Hague, April 27, 2006. Report No.: Publication No. 2006/06E.
[143] Precautionary strategies for managing nanomaterials, German Advisory Council on the Environment (SRU), 2011.
[144] The Rio Declaration on Environment and Development, The United Nations Conference on Environment and Development, 1992.
[145] American and European Standards Organisations agree to strengthen Transatlantic Cooperation on Standards for Electric Vehicles. The European Committee for Standardization (CEN) and the European Committee for Electrotechnical Standardization (CENELEC), 2012. Available from: http://www.cencenelec.eu/news/press_releases/pages/PR-2012-18.aspx
[146] E.P. Roth, D.H. Doughty, Thermal abuse performance of high-power 18650 Li-ion cells, J. Power Sources 128 (2) (2004) 308–318.
[147] J.S. Neubauer, E. Wood, A. Pesaran, A second life for electric vehicle batteries: answering questions on battery degradation and value, SAE Int. J. Mater. Manuf. 8 (2) (2015).

CHAPTER NINE

Application of the Energy Storage Systems

P.N. Borza
Transilvania University of Brasov, Brasov, Romania

Contents

1 INTRODUCTION: ENERGY STORAGE SYSTEMS AND THEIR APPLICATION

The advent of the transistor (1940) marked the beginning of the electronics era, which was dominated by the modern communication technologies based on semiconductor devices. A similar situation unfolds nowadays as this is a new era of modern technology for energy storage systems, focused on improving overall performance. The main areas benefiting from the addition of energy storage systems are the fields of electric power production and distribution of electricity, especially based on renewable energy sources; the electric or hybrid transport sectors; telecommunication basic stations; home automation systems; medical field devices; and the domain of sensors and autonomous systems. With the world's growing population of 7.4 billion (as of 2015), and with the increasing demand for electronic devices worldwide, energy consumption is growing at a steady rate [1]. During the past 20 years we have seen an accelerated evolution of renewable technologies represented by the development of solar, wind, thermal, and hydropower generation. The insertion of these new power sources into the electrical power networks generated many integration and operation problems because of the dependencies of these sources of natural conditions and their variations in time. These disturbances must be solved by the solutions that permanently balance supply and demand in power generation. As mentioned

Emerging Nanotechnologies in Rechargeable Energy Storage Systems

in Ref. [2] "the key issues of the energy transition in the power sector are the interaction between strongly increasing amounts of variable renewable electricity generation and the mix of flexible resources to integrate them into the power system." As mentioned by Katie Fehrenbacher [3], "A next-generation smart grid without energy storage is like a computer without a hard drive: severely limited." Starting from parallel analyses of informational and energetic systems, a clear analogy between these two systems can be identified. A similar role in informational and energetic networks is played by the cache memory and the storage facilities, respectively. Inserting storage facilities into the power networks will facilitate the power flow control in a similar way that cache memory performs in the case of informational systems. Based on this, smart grids can be designed, including energy storage facilities that will permit improved control of an electric grid parameters. Therefore, like the "cache memories" that are inserted close to a computer CPU and peripherals elements, the storage facilities will be installed close to generators and also close to the location of consumers, smoothing the power and energy variation of power lines.

2 CHARACTERIZATION OF STORAGE CELLS AND DEVICES, PARAMETERS, AND FEATURES

Electric storage systems (ESS) represent a high degree of complexity and variety. The majority of storage technologies, their advantages and disadvantages, and also their typical applications are discussed in Ref. [4]. This is the effect of many factors and phenomena, all characterizing the resulting storage system performance. The first problem is the hierarchical organization of the ESS; starting from the cells, these are grouped to form a module, and the modules are grouped to form a storage system.

At the cell level, a complex structure is encountered formed on multiple interfaces oversimplified in the following lines:

- porous activated carbon zones form the armature of a capacitor having a huge area. On this are accumulated ions of the same sign. Because the activated carbon surfaces are very thin the distance that separating the charge carriers having opposite sign is extremely small. This is correlated with very large surface and will determine an extremely high electric capacitance. As result of the close interaction forces that appear between atoms of electrode and ions of electrolyte, the ions are very close arranged on porous activated carbon surface
- the active zones formed by porous material are compacted using a "binding material" that ensures the mechanical integrity of the electrodes, improving in the same time the electrical conductivity and
- the electrical connection between the active areas and the cell electrodes is implemented by graphene, graphite, carbon nanotubes, or other structured carbon materials or constituents.

In general, these interfaces form a complex mixed and structured matter in various states of aggregation—solid, liquid, or gas and their combinations: solid–liquid, solid–gas, or liquid–gas, which are called electrode–electrolyte interfaces. The shape of the stationary phase and the electrode matter crystallization affect the properties of the interface. These interfaces are living entities because during cell functioning, or as a result of time passing, their properties and features are affected. This phenomenon is known as aging of the ESS. Charge carrier mobility depends on the electrode–electrolyte interface as an expression of integrative interaction of the two phases that are adjacent to this level. The degree of consistency of the interfaces and the minimum distance ensured between the opposite charges carrier determine their electrical properties. The variety of chemical compounds and substances that can interact at this level is particularly high. The main part of electrodes can be implemented using activated carbon, or MnO_2 based materials that includes a large variety of porous areas [5–7].

The International Union of Pure and Applied Chemistry (IUPAC) felt the need to define the methodology for determining the material's porosity by using Brunauer, Emmett, and Teller (BET) method [8]. They also have suggested a first classification of pore sizes. The classification of pores as a function if their average diameter is given as: (1) nanopore materials, in which the average diameter of the pores is less than 0.2 nm, (2) micropores when diameter is below 2 nm, (3) mesoporous when diameter of porous is between 2 and 50 nm, and (4) macropores when diameter is greater than 50 nm [8]. The theory of Langmuir, Brunauer, Emmett, and Teller explores the method that underpins the determination of the surface of a material that adsorbs a polarizable gas [9].

$$\frac{p}{v \cdot (p_0 - p)} = \frac{c-1}{c \cdot v_m} \cdot \left(\frac{p}{p_0} \right) + \frac{1}{c \cdot v_m} \tag{9.1}$$

where v is the total volume of the gas absorbed, v_m is the volume of gas adsorbed when entire adsorbent surface is covered with a complete unimolecular layer of adsorbed gas, p_0 is the saturation pressure of the gas, and p/p_0 is the relative pressure.

Based on the BET method (theory about adsorption of gas molecules), the specific surface area (*SSA* = the total area of a material per unit of mass) of the active porous zones is determined. For determination of *SSA* in many cases, the gas used is nitrogen. Thus, the BET formula adapted for measurements of W, weight of gas adsorbed, is given by:

$$\frac{1}{W.\left(\frac{p_0}{p} - 1 \right)} = \frac{1}{W_m \cdot c} + \frac{c-1}{W_m \cdot c} \cdot \frac{p}{p_0} \tag{9.2}$$

where, p/p_0 is relative pressure, W_m is weight of adsorbate as monolayer and c = BET constant. In the particular case of nitrogen with a cross section, A_{cs} is adsorbate cross-sectional area, the total surface S_t is given by:

$$S_t = \frac{W_m \cdot N \cdot A_{cs}}{M} \tag{9.3}$$

and

$$SSA = \frac{S_t}{W} \tag{9.4}$$

where W is the sample weight. In the case of activated carbon, the SSA can vary as a function of porosity, precursors and materials used.

For example, in the case of activated carbon-based interfaces, the electric double-layer capacitor (EDLC) is very thin, and therefore the distance that will separate the opposite sign charge is very short. Consequently, the intensity of the electrical field can reach very high values.

For this reason, the fundamental constraints applied to voltage and temperature in the case of EDLC are essential to preserve the integrity of the electrode–electrolyte interface. The energized charged particles or molecules (ions) can sometimes break the nanostructure of the interface, causing degradation of the cell electrical parameters, accelerating the aging process, and even fully damaging the storage cell.

The supercapacitors or ultracapacitors are based on nonfaradic phenomena having high-power density and comparatively higher life cycles than batteries. There are a few main advantages of supercapacitors over other storage systems: (1) it has virtually unlimited life cycle (millions of times); (2) it has very low impedance; and (3) it supports a very high rate of charging and discharging, the limitation being the accepted device temperature domain. A classical overall image (Fig. 9.1) is generated by Ragone's plot.

In the case of batteries, the faradic phenomena that occurs as a result of the cyclical charging–discharging processes will have similar effects like in the case of supercapacitors, affecting the shape and size of the solid part of the cell interface. Therefore, the variation of molecular dimension of reactants at the interface level will affect the functional capacity of the battery cell over time, causing aging of the battery [11].

The "thickness of electrodes and separators" influence the performance of the ESS. The phenomena that occur at the interface level are essential for the parameters and characteristics of the storage cells, power and energy densities, cell aging, time constants of their storage capacity, and the storage temperature range. The ESS implementation technologies must ensure a high reproducibility of the geometry of the interfaces and a constant degree of uniformity for the stationary phase layers. Another important factor related to performance is the adequate choice for the interface reactants. The perfect adaptation occurs in the shape of the local near field configuration between the porosity corresponding to the shape of the stationary phase and the ions of electrolyte (mobile phase) that will be hosted by the interface.

In general, energy storage cells have a finite capacity and their electrical parameters fit only partially with the requirements of most applications. Batteries and supercapacitors present an integrative character. In the case of batteries, theoretical maximum capacity may be determined as the integral of all reactant molecular link energy, at the interface

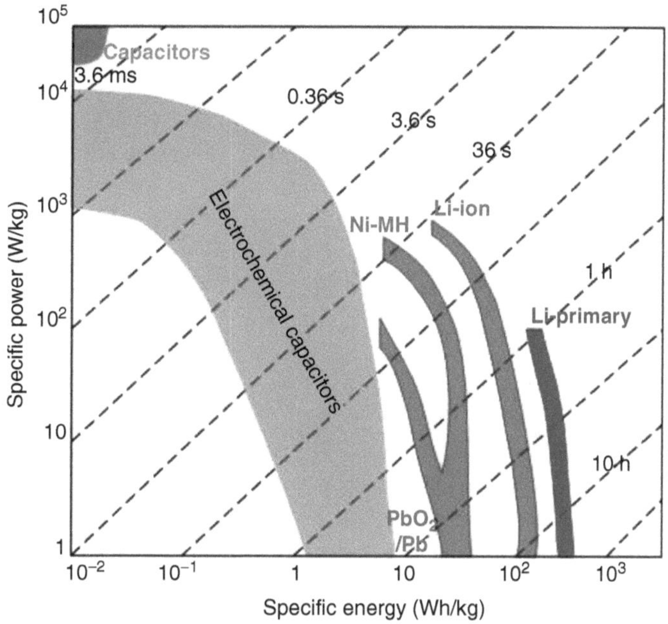

Figure 9.1 ***Specific power as a function of specific energy for various electrical energy storage devices logarithmic scales.*** Characteristic times correspond to lines with unity slope see [13].

electrode–electrolyte, able to react in a finite time. The chemical reaction speed at microscopic level is illustrated at macroscopic level by the steady-state internal series resistance. In the case of supercapacitors, their capacity depends on the maximum area that can be used to host adsorption phenomena at the level of electrode–electrolyte interface.

It is not feasible to build customized cells for each type of application based on its requirements. The simplest strategy consists of grouping standard cells and building up the necessary capacity as a modular system adapted by link topology to create an appropriate storage facility in accordance with the application requirements. Therefore, these elements can be "grouped" as an assembly of cells connected in series, parallel, or series–parallel to implement the storage modules. The size of most elementary cells determines the "granularity of the achieved storage system." At a storage system level, the wires that link the system elements and the way they are linked denotes the topology of devices and storage systems. These properties play an important role in meeting the energy efficiency targets and in achieving the technical performance and system requirements (ecofootprint, shape, gauge, etc.).

The uniform distribution of matter at the interface level, and how to "package" the cells as a storage module and the storage modules as storage system are important design aspects. The uniformity of the thermal and electromagnetic fields obtained during the operation of the module and system in conjunction with the application are also key

Table 9.1 Supercapacitors and batteries: features and extreme values

Cell type	**Supercapacitor [12]**	**Battery [13,14]**
Energy density	*Moderate* 10–20 Wh/kg	*High* 80–300 Wh/kg Dependent on technology Peak value Li-O_2 3,500 Wh/kg [13]
Power density	*High to very high* 1–20 kW/kg	*Moderate to high* 0.25–1.3 kW/kg Dependent on technology [14]
Domain of temperature	*Extended* −50°C to +70°C	*Medium* 0°C to 45–50°C Dependent on technology
Cyclability	*Very high* 500,000–1,000,000 cycles	*Moderate* 200–5,000 full cycles (DoD 95%)
Voltage	1–2.7 V Inorganic electrolyte 1–1.2 V Organic electrolyte 2.7–3 V Asymmetric electrodes 4.5 V	1.2–4.5 V Lead acid 1.2 V Li-ion 3.7 V $LiFePO_4$ 3 V
Self-discharge	*High*	*Low*
Energy efficiency	90–98%	85–92%

factors. These factors are among the most important as they determine the reliability, availability, and lifespan of the whole storage system. All matters mentioned earlier are part of the structural design of the storage system and specifications determined by the deployed application.

The features and properties of supercapacitor and batteries are compared in Table 9.1. Very often cells are reported with better features.

In Fig. 9.2 (supercapacitor box) a practical implementation solution for a locomotive starting system is illustrated. In Fig. 9.2, A represents the supercapacitors used for starting the diesel engine, B is the electronic switching device with its cooler, and C is the connection rail between supercapacitors and electrical starting machines. The electrical

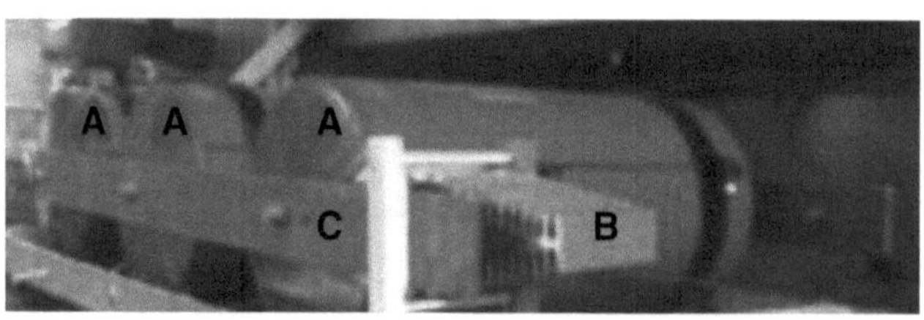

Figure 9.2 ***Integration of supercapacitors storage with power electronics switching device on a starting system of diesel hydraulic locomotive.*** *A*, Supercapacitors 12 F/110 V stacked with inorganic electrolyte, *B*, thyristor 1900 A/1200 V, and *C*, electric connection wires: cooper massive with 100-cm^2 section area).

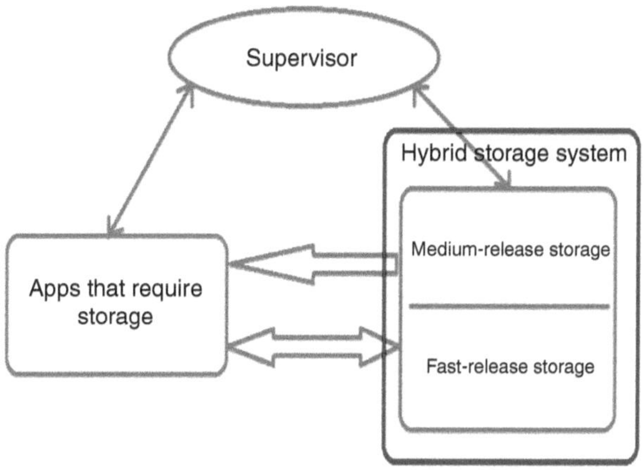

Figure 9.3 ***Hierarchy of a system endowed with a hybrid electric storage system.***

resistance between supercapacitors and acting motors must be minimized to preserve the high level of energy efficiency. As the supercapacitor equivalent series resistance (ESR) is very low, the distance between system components should be minimized. Also the material used for the connections must have a very low resistivity.

In the case of power electronics, it is possible to control the power flow function, fulfilling some objectives that should be implemented for the entire system. So, as illustrated in Fig. 9.3, two levels of control can ensure the overall system control integration. The embedded power electronics will manage the preservation of the hybrid energy storage system (HESS), protecting against overvoltage, overtemperature, overcurrent, variations in voltage, and charge balancing between cells.

The supervisor level will implement complex functionalities, such as scheduling, intelligent traffic control, adaptive driving systems, intelligent management of energy, compensation of electric circuits power factor, and so forth.

The combination of batteries and supercapacitors has the potential to dramatically increase the performance of any hybrid system, especially insulated by combining the energy density of the battery with the power density of the supercapacitor. At the same time, a hybrid system increases the complexity of the system and presumes a more attentive design for the system control.

The spectrum of the ESS is very large, with advantages and disadvantages, and therefore a single system type does not exist that is able to fulfill all the requirements of all applications.

From a chemist point of view, the power and the energy densities of storage cells are important parameters.

The engineers are focused, in principal, on features and parameters required by the applications, such as overall energetic efficiency and the functional and ergonomic features. At least as important is the development of a functional energy storage system according to the specific requirements of the application. In this regard, the complexity of the requirements, the variety of "use cases that must be satisfied" by designed application.

The variety of situations encountered while implementing and running the applications make the design process highly complex. More than that, it is imperative that the design process takes into consideration even rarely occurring situations. Storage cells and devices are very sensitive to the exceedance of the nominal values or occurrence of other exceptions, which can damage the cell completely or can significantly reduce their reliability, availability, and performance. Unfortunately, occurrence of peaks of functioning parameters can fully damage the storage cells and the worst cases are not easy to anticipate. The variety of cell interfaces is described by the variation of the equivalent resistance and capacitance with direct consequences to the "time constant" of the cell, module, or system.

Two viewpoints—of chemists and engineers—require an effort to accommodate and adapt to each other, and this synthesis can only be achieved by adopting a common vision.

3 OVERVIEW OF STORAGE CELLS, MODULES, AND SYSTEMS

A holistic picture of energy exchange circuits including storage systems is presented in Fig. 9.4. The structure of the storage system can be relatively complex and many variants of storage systems can be implemented as a combination of conversion subsystems and reservoirs or buffers of the converted energetic agent, which could be compressed air, water, hydrogen, or many other materials.

The generator can directly or indirectly power the network to be connected to consumer to deliver electrical energy. The balance between produced and consumed energy must be preserved to keep the accepted variations relatively small. If the electric power generator produces more power than needed, the supplementary power will be stored by the network storage facilities. The network storage facility shown in Fig. 9.4 (right)

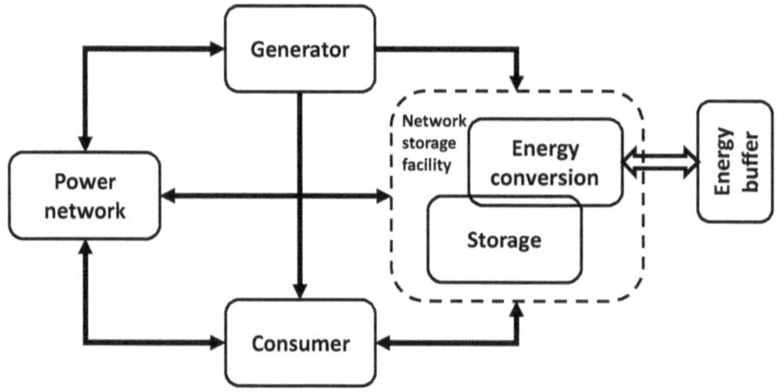

Figure 9.4 ***Ontology of energy generation, transport, consumption, conversion, and storage.***

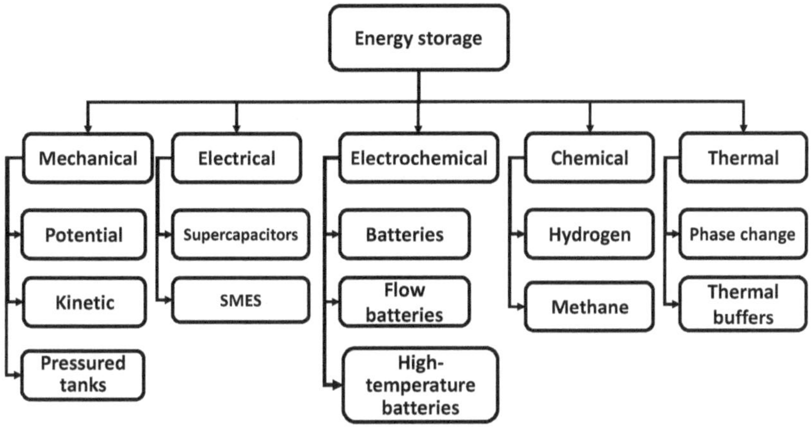

Figure 9.5 ***Taxonomy of storage systems*** (SMES, Superconducting magnetic energy storage).

can have two different forms, a simple electric energy storage system (e.g., battery system or batteries combined with supercapacitors) or a similar role can be played by the energy converters, such as the electrolyzer in conjunction with oxygen and hydrogen tanks that form the energetic buffer and the appropriate fuel cell, which transforms back to electricity when energy is introduced into the system. Therefore, the described system, formed as a combination of three different elements, implements a similar function: the storage of energy for the network or for the consumer.

The networks storage facility ensures a supervised control of a stored quantity of energy as a function of the power network status and the demands of consumers. Such facilities are included in smart grid solutions and also perform as secondary and tertiary reserves; compensation, active or passive, of the power factor; and preservation of voltage frequency supplying.

The taxonomy of storage elements is presented in Fig. 9.5.

Only few cases are briefly explained below, as representative examples of the principles of operation for different types of storage systems.

3.1 Mechanical storage

Transforming electricity into mechanical energy and vice versa is among the most common conversion processes. The stored electrical energy in batteries or supercapacitors is transformed into mechanical form by using an electromechanical converter like an induction machine or a complex system, which integrates the electrical machine with the power converter. Before each case is analyzed, the corresponding physics transformation equations should be written. The expression of kinetic energy is given by:

$$W_k = \frac{1}{2} \cdot mv^2 \tag{9.5}$$

and the expression of potential energy [Eq. (9.6)] is given by:

$$W_p = m \cdot g \cdot \Delta h \tag{9.6}$$

where m is the mass, g is gravitational acceleration, and Δh is the difference between the elevation of a lake and its base for water accumulation in transversal dams placed on river valleys, and v is the speed of the considered body.

Another possibility to store mechanic energy is to use the pressure of gas or fluid beings in reservoirs.

In the case of liquids, the compressibility (volume change due to change of pressure) is inversely proportional to its volume modulus elasticity, which is defined as follows:

$$E_v = -v \cdot \frac{dp}{dv} = -\left(\frac{v}{dv}\right) \cdot dp \tag{9.7}$$

where v is the volume, and dp is the pressure variation of liquid at the same temperature. E_v is named "bulk elasticity modulus" and is similar to the definition given for solids [13], for elasticity expression in the case of solids.

Considering the variation with temperature of bulk elasticity coefficient of water, Eq. (9.8) can be written using finite difference as:

$$\frac{\Delta v}{v} \approx -\frac{\Delta p}{E_v}, \text{ respectively: } \frac{v_2 \sim v_1}{v_1} \approx -\frac{p_2 \sim p_1}{E_v} \tag{9.8}$$

In the case of compressed air reservoirs, the potential energy depends of the type of process (a state transition of a thermodynamic system) that takes place in a pressurized tank [15].

$$dW = p \cdot A \cdot dx \tag{9.9}$$

where dW is the work performed by a piston into a cylinder, p is the pressure, A is the transversal array of cylinder, and dx is the displacement of the piston [15]. Integrating Eq. (9.9) results in:

$$W = \int_{V_1}^{V_2} p \cdot dV \tag{9.10}$$

In polytrophic processes the analytical equation that describes the p–V law is given by:

$$p \cdot V^n = \text{const} \tag{9.11}$$

or in the case of mass change, Eq. (9.11) is replaced by:

$$p \cdot V = n \cdot R \cdot T \tag{9.12}$$

where p is pressure of the gas, T is the temperature in Kelvin, V is the volume of tank (cylinder), and R is the gas ideal constant (8.314 J/K/mol).

3.2 Electrical storage

In the case of ESS based on supercapacitors, the energy accumulated by these is shown by Eq. (9.12):

$$W_C = \frac{1}{2} \cdot CV^2 \tag{9.13}$$

(where C is the supercapacitor capacity and V is the voltage). The charge is stored in nanostructured active material and the basic phenomena (EDLC) were discovered by Helmholtz in 1853. He also has provided the first model of the distribution of ions in the vicinity of the surface of a metal electrode in 1874 [5].

When the processes that govern the accumulation and releasing of charge do not involve chemical reactions then the phenomena are named nonfaradic. This is the case for supercapacitors.

In the case of superconductive magnetic systems, the energy of the system:

$$W_L = \frac{1}{2} \cdot L \cdot I^2 \tag{9.14}$$

where L is coil inductance and I is the current through coil, W_L represents the magnetic energy what can be stored in an inductance maintained under the superconducting temperature point. In this case the coil resistance can be neglected.

3.3 Electrochemical storage

In the case of batteries and flow batteries, the main phenomena consist of chemical reactions that take place between the electrodes and the electrolyte in a reversible reaction that depends on the type of reactants. The energy produced is given by reactants Gibbs free energy (thermodynamic potential at constant volume) and their mass involved in reversible reactions [16]. The Gibbs free energy [17] (isothermal and isobaric free energy) is determined by Eq. (9.15):

$$G = H - T \cdot S \tag{9.15}$$

where H is enthalpy (measurement of energy in thermodynamic system), T is absolute temperature in Kelvin, and S is entropy (measure of energy divided by temperature). Differentiating Eq. (9.15) results in:

$$dG = -S \cdot dT + V \cdot dp - \delta W' \tag{9.16}$$

where $\delta W'$ is other form of work except expansion work expressed by $p \cdot dV$.

In the case of closed systems, the simplified expression of Gibbs free energy is:

$$dG = -S \cdot dT + V \cdot dp \tag{9.17}$$

The Gibbs free energy in the case of processes preserving T and p constant G variation can illustrate the equilibrium of thermodynamic processes. So, if $\Delta G < 0$, the process is spontaneous; when $\Delta G = 0$, the process is in equilibrium; and when $\Delta G > 0$, the process is spontaneous but in the opposite direction.

To characterize the features and performance of storage cells, the standard test methodologies for interface characterization are voltammetry, galvanostatic, and electrical impedance spectroscopy methods [18]. The capability to supply energy by the batteries is given by their capacity depending on type of interface electrode–electrolyte and size of main elements and battery state of charge. "State of charge (SoC) is the percentage of the maximum possible charge that is present inside a rechargeable battery"[18]. The method used to determine SoC is based on the measurement of open circuit voltage (OCV) that was proposed by Christianson [19].

Knowing two points of SoC, an intermediate value will be calculated by linear interpolation, respectively, by coulomb counting of charge transferred. The voltage of the battery terminals is:

$$\text{OCV} = V_{\text{bat}} + I \cdot R \tag{9.18}$$

where V_{bat} is the voltage at battery terminals, I represents the current measured in external circuit (positive in the case of discharge and negative in the case of battery charge), and R is the ESR.

In the case of flow batteries, the capacity of the generators is related to the volume or mass of reactant reservoir and the effective surface of ion exchange membrane that ensures the release of stored energy.

Another challenging aspect is related to finding out new combinations that can form efficient interfaces that enable storage of electrical charges. It is important to develop new technologies that represent new ways for the implementation of these interfaces to meet a higher level of energetic and functional efficiency.

3.4 Hybrid concepts

A first proposal for a hybrid electric storage system was done in Ref. [20] as a European patent. It proposed a configuration of HESS formed by three kinds of storage devices, having spread in time constants: fuel cells (FC), batteries, and supercapacitors. By building up a dynamic changing link topology between storage devices, a control system dedicated to HESS can implement adequate control laws based on the function of the application in which the HESS is integrated. The large variety of supply demands observed in almost all applications does not allow an optimal solution that is implemented by a single type of storage device, but by combining two or more different electric storage elements, it is possible to obtain more appropriate solutions to satisfy the application requirements.

For chemical storage, such as electrolysis or methane generation/energy storage, the energy stored is proportional to the reservoir volume or mass.

In the case of thermal storage, two technologies are relevant: the phase change systems based on latent melting or vaporization constants and the calorific storage based on temperature variation of a material used as thermal buffer. In the first case, the storage is produced at the same temperature and in the second case, the mass of the material used for energy storage is proportional to the storage capacitance.

An overview of these kinds of energy storage technologies is discussed in Refs. [2,4,21]. These are analyzed in conjunction with the applications that require storage facilities to improve the overall energy efficiency.

4 APPLICATIONS THAT USE STORAGE FACILITIES

ESS integration with applications involves the reciprocal adaptation of energy and informational flows. This ensures the applications high-energy efficiency, which is mainly based on minimization of the power flow excursion.

In Refs. [2] and [22], a large palette of applications is described along with types of storage systems used for increasing the power quality and ancillary services in the grid, and for mobile applications. The block diagram of an ESS is illustrated in Fig. 9.6. The system storage interface implements the dual link of energy and information between the network supervisor and the storage subsystem. On the basis of information exchange between both systems, their functioning is reciprocally adapted. The storage subsystem controls the charging and discharging processes locally with respect to storage element constraints.

Based on the current status of ESS, the integrated system (power network, vehicle, communication, or any other system that receives the benefit of storage facilities) will modify the scheduler strategy used to supervise the energy transfer. In the case of stationary applications, such as power networks, the main role of energy storage system is to provide ancillary services, such as frequency regulation, spinning reserve, peak reduction,

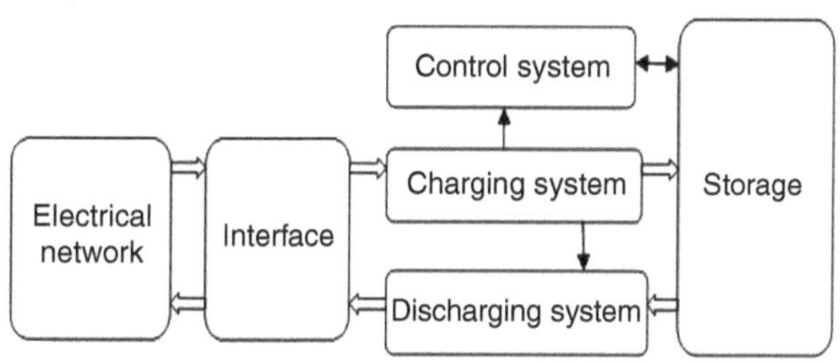

Figure 9.6 ***Components of an energy storage system.***

and load leveling, enhancement of reliability and power quality, and enabling greater penetration of renewable energy resources. In the case of urban transportation systems, the application of ESSs will avoid the disturbances produced by moving EVs and also in the majority of cases it will increase the system efficiency. In the case of communication systems, the role of ESS is to smoothen the power flow variations during communication sessions.

The interfacing of the storage systems with power networks involves the control, the power and the information flows. The power bidirectional interface (see Fig. 9.6 interface) ensures the dispatch of power flow between the storage system components as a function of their capacity to store or to supply energy, respectively. How this energy is transferred is also very important because the storage system is very sensitive to the variations of current, voltage, and temperature. In this sense, the charging and discharging processes are controlled with respect to maximum instantaneous values for voltage, current, temperature, and state of charge of ESS. At the same time, ESS is supervised by a power network controller that will communicate with the ESS controller to inform it about the storage needs and also to receive its current status. So, the coherency of the whole system is preserved.

A first classification of applications that benefit from storage systems can be made depending on the characteristics of the system that incorporates storage devices: stationary and mobile systems. Function of their characteristics, storage systems are adapted. Applications that involve a permanent connection with the power grid, or involve a temporary connection, or are disconnected full time from the power grid. These details present major consequences for the sizing design of the storage system and also for its configuration. A possible classification of storage systems that is useful for application design is as follows:

1. *Permanently connected to the grid system* (energy storage system technology for sustainable/smart power networks). In this case, an energy storage system stocks the electricity from a local power generator and provides it for peak demand. For example, in the case of PV power plant facilities or wind power plants, the storage systems ensure the smoothing of power flow provided and it can compensate short- or medium-term variations due to natural factors, such as sun radiation or wind speed variation. Using hybrid storage systems, we can address all kinds of variations in the system with maximum energy efficiencies and reliability and ensure the quality of power in the network.
2. *Insulated application.* Insulated systems present an autonomy, which is ensured by harvesting energy from a system environment.
3. *Temporarily connected to the power grid.* All the hybrid (plug-in) or electrical vehicles can be considered as temporarily connected to the grid systems. Periodically, these systems are charged from the power grid. In this case, if the infrastructure allows the bidirectional transfer of the energy to the power network, the usage of existing

storage facilities on electrical vehicles would develop the ancillary services for low-voltage or even for the local distribution grid.

It is well known that the electricity produced should be consumed at the same moment. This process is called "balancing" of the network or "demand response control," and implements the dynamic equilibrium control between supply and demand of electrical energy to ensure its stability. The power propagation along power lines, the topologies, the type of consumers (inductive, pure resistive, or capacitive), and other factors influence the stability of the power networks and are between the parameters determining the equilibrium of power flows. This process presents a very short-time window for control of the network parameter variation. The international regulations and standards that must comply for supplying electrical energy are IEC 60038 [23] and EN 51060 [24]. These standards define parameters of the voltage supplied by the public power distribution systems and the electricity quality parameters. (e.g., voltage tolerance $\pm 10\%$ of nominal value). The function of the nominal voltage and the types of networks are classified as follows: low voltage is 230 V/400 V [23] in a monophase/three-phase system, domain 1 kV and less than 33 kV is defined as the medium-voltage system, and the high-voltage system presents voltage in the range between 33 and 230 kV. The extra high-voltage system is considered if voltage value exceeds 230 kV. The frequency can vary from 47 to 52 Hz for a time period that is less than 0.5% of a week, but the normally allowed variation in 10-min intervals should be between 49.5 and 50.5 Hz for more than 99.5% of the week time. The regulations in this sense can differ from one country to another and remain to be standardized. The storage system facilities inserted into the grid near the main consumers will offer the "temporal window," which will facilitate the power management avoiding large fluctuations of the voltage, current, and power factors.

Nowadays, the dissemination of renewable energy sources (RES) is a reality. RES integrated inside power networks generate significant variations in providing and consuming power with important consequences for stability and power quality. By inserting electric storage facilities at each level of the power grid, and especially at the local level new possibilities of power network control will open.

For the classical consumers adding and using the new facilities based on renewable energy sources in order to meet their need for energy will transform them in prosumers. That means that these new actors into the power grid (prosumers) they will become able to transfer in both senses the power flow assuring so better ancillary services for themself and the grid.

By inserting the energy storage facilities, it becomes possible to implement, "time shift" control strategies that allow smoothing the load or generation curves (see Fig. 9.7 for an illustration of long-term smoothing and Fig. 9.8, showing short-term smoothing).

The continuity of electrical power supply is a fundamental goal for the controlling system of the power network. This is ensured by organizing the control system on three different levels (primary, secondary, and tertiary [2]).

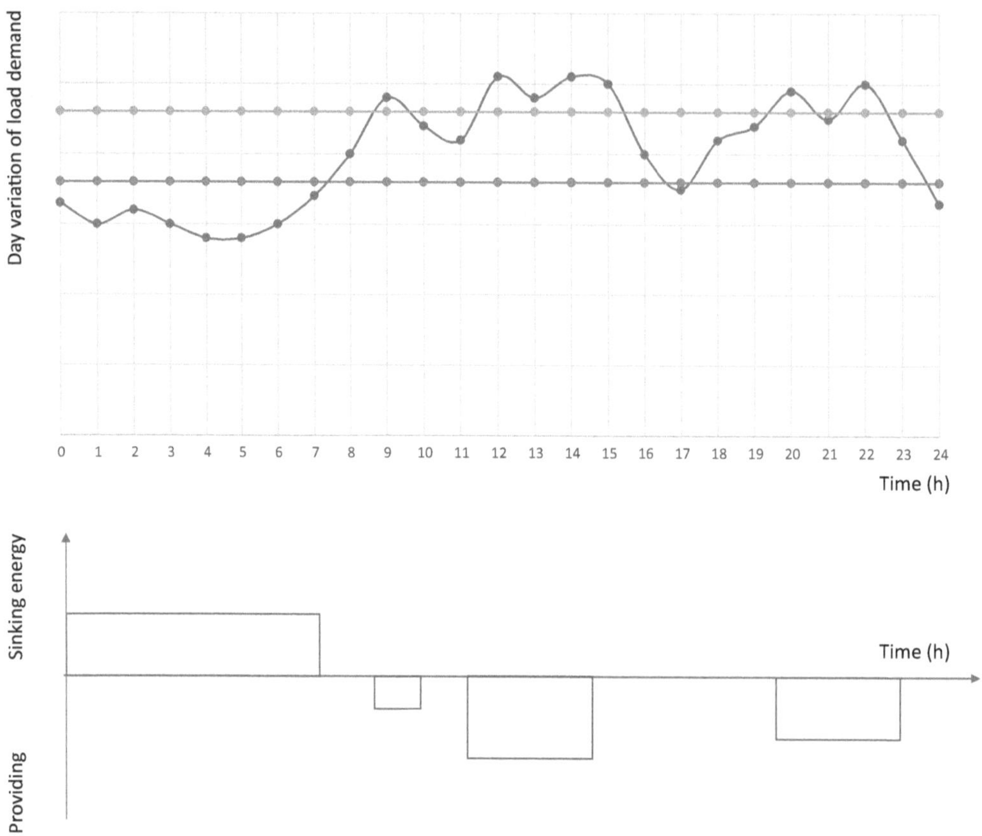

Figure 9.7 ***Daily profile of-demand (daily load leveling).***

Thus, it ensures the stability, reliability, and availability of the power grid. Behind the automatic control, the electrical generators or the storage facilities are integrated with the power converters or spare generators. For the alternative current (AC) power networks the energy is mainly produced by synchronous machines that are driven by turbines or motors. When the ratio of demand to supply is unbalanced, the voltage frequency will reflect this change, as it is the most sensitive signal. Therefore, when a relative decrease of the load occurs, the voltage frequency will increase and when the power supplied does not meet the demand for power, the consequence is a slight decrease in the voltage frequency. The automatic control of power grid balancing is based on voltage frequency monitoring and is called "primary control of the power network." This function is implemented by connecting quickly to the power grid of the electrical generators or to high-capacity storage facilities (Fig. 9.7). The primary control is replaced by the secondary control after a certain amount of time (30 or 60 s). The tertiary control systems "minute reserve" will be triggered after an imbalance is

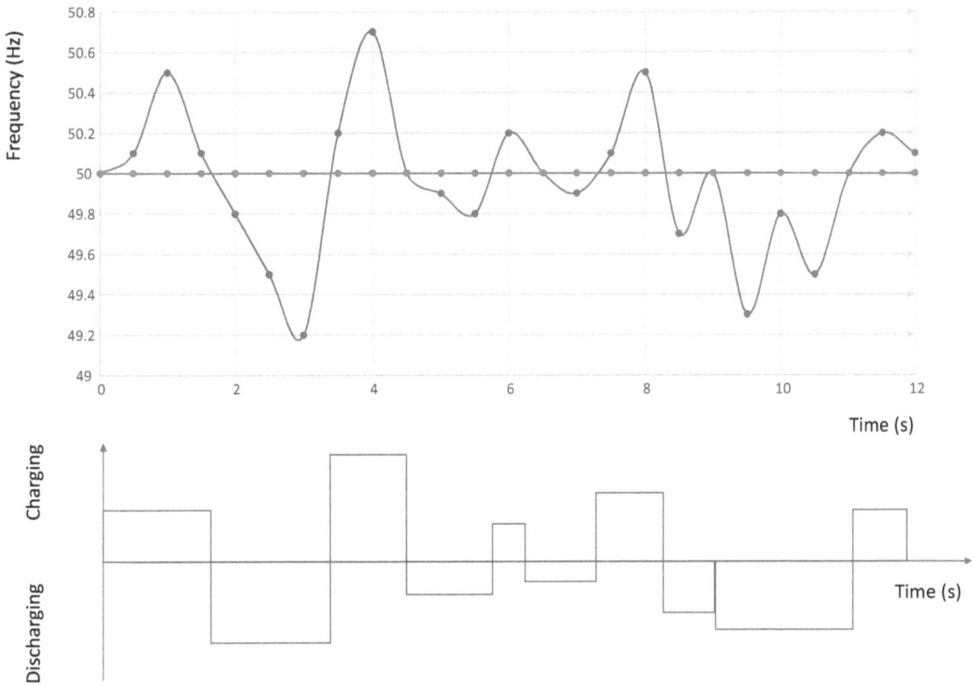

Figure 9.8 ***Fast-voltage frequency variation and its compensation.***

detected for several minutes. During a 24 h period, the ratio between supply and demand will vary significantly (Fig. 9.7). If the variations are related to very brief voltage and power fluctuations (Fig. 9.8), only the active power filters can compensate them. The compensation to ensure "power quality" is also reflected by spectral components of the power delivered to the consumer. For vulnerable consumers, such as data centers, this characteristic should be well understood. This is the result of particular demands and also is the result of the energy-supplying variation of the renewable power supplies (RES), which are correlated with the natural factors (wind, solar, and water). Currently, there is a proliferation of distributed energy generation based mainly on RES installations. These facilities are in consumers' neighborhoods, and therefore the energy produced by such facilities is mainly transferred to local consumers and only the overproduction flows to the power network. Therefore, by looking at the connection point of consumers with the power network, one can discover an important change of view. Former consumers have become a complex element integrated into the power network, able to absorb and also to supply power through their connection line with the power network. These new actors on the power grid are termed *prosumers* (Fig. 9.9). Currently, these actors exchange electrical energy using the low- or medium-voltage power networks. The classical role of the regional network, based on an optimized scheduling implementation, is assumed

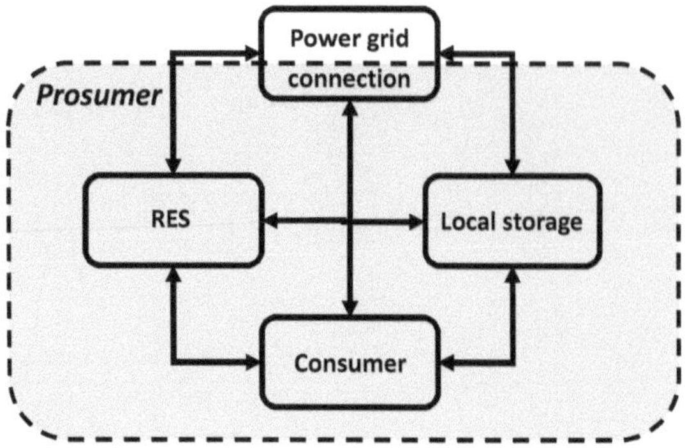

Figure 9.9 ***Ontology of a prosumer.***

at the level of each prosumer, who will control and schedule the consumption and distribution of power generated, consumed or exchanged between the power network and local prosumer grid and also inside the prosumer microgrid. A large variety of topologies and types of networks, AC and DC microgrids, are implemented at the prosumer level. Also a large variety of local storage facilities are implemented. For example, battery-based storage facilities, hybrid storage facilities formed by batteries and supercapacitors, flywheel facilities, hydrogen-based storage facilities, and so on have been developed. The capacity of the local storage, the instantaneous parameters, the energetic interfaces (power converters or other power electronics facilities) all must be adequate to meet a specific prosumer requirements.

The aggregation of these new actors is easier to implement because of a deeper link created between the information processed by prosumer subsystems and the energy that is exchanged between prosumer components. Also, this fusion between energy and information is revealed by each grid subsystem or elements and also is available for the central grid control point (dispatch point). These new networks actually include a majority of "smart" or "intelligent" components that are able to provide very complex functionality not only on scheduler, but also with protection, energy managers, monitoring, and other processing systems that generate higher stability, reliability, availability, and performance for the new "smart grids."

The complex integration of these grids with new cybernetic integrated cities with heating or district heating facilities, lighting, transportation, and community services, forms the final target in the development of "smart cities."

This new trend has modified the control hierarchy inside power networks and the shift from centralized architecture to distributed control architectures. Therefore, the focus of the supply–demand balance was changed from central points to every distributed prosumer to determine the necessary balance between supply and demand. This situation generates another kind of equilibrium, which will reduce the overall power

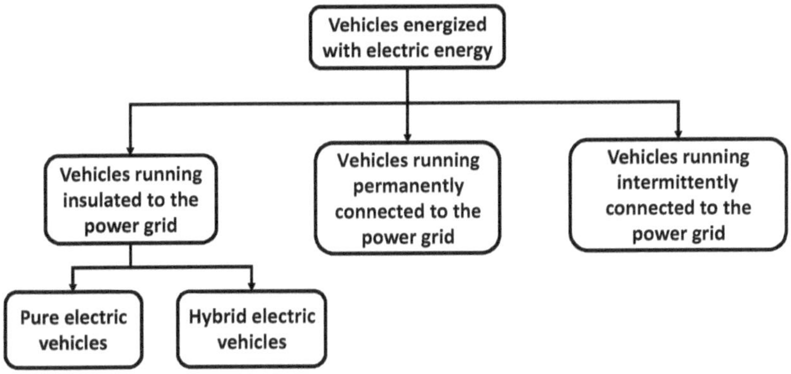

Figure 9.10 ***Taxonomy of vehicles energized with electrical energy.***

excursion inside the networks. Recent implementations prove that insertion of power storage systems into the grid improves significantly the energy efficiency not only of the whole system, but also the economic efficiency of such investments [e.g., Schwerin, SWEMAG power facility (Li-ion batteries) with 5 MW/5 MWh with the average revenue of 3810 €/MW/week [25–27]].

A second main type of applications for electric energy storage facilities is associated with the development of electromobility. The kind of energy used to energize vehicles can be characterized by three types of EVs: those using only their own electricity during transport, those that are permanently connected to the electricity network, and those that connect only intermittently to the power grid. The proposed taxonomy is illustrated in Fig. 9.10. In the case of insulated vehicles in relation with the power network, their energetic capacity directly influences the vehicle range. Storage system hybridization implemented by introducing devices allowing fast-release storage devices improves energy efficiency by regenerative braking and reconversion of the vehicles kinetic energy. Also, implementation of the hybrid system improves the stability and thermal behavior of the overall HESS. In the case of EVs permanently connected to the grid, the fast- or medium-release energy storage systems minimize the power excursion flows on catenary at the acceleration and the deceleration of the vehicle. This improves the power grid stability, vehicle thermal stability, and vehicle energy efficiency.

In the case of hybrid EVs, different levels of "electrical assistance" can be identified. The PHEVs are vehicles that work based on ICE classical engines but connected in series or parallel with the electric motor. The later one is connected with the batteries or even a hybrid storage system.

These types of applications suppose an intermittent connection between the vehicles and the power network during charging. During transport mission, the EVs are isolated systems, which will benefit from having storage facilities and will manage the energy in a permanent exchange with the vehicle reserves (mechanical to electrical and vice versa). In this sense, a main role is played by the regenerative braking system implementation [28]. The energy management of EVs will extend their range between two charging processes and will improve their dynamics.

In the same category as autonomous EVs are light-rail vehicles (LRV) (tramways and metro trains) and trolleys that are permanently connected to the power grid by an overhead contact system (OCS) (catenary). To improve energy efficiency, an LRV is equipped with on-board storage systems, which can significantly improve the overall energy efficiency and also the dynamics of the vehicle [29]. The "locality" principle illustrates the relative adequacy and optimization of power system elements, as a result of their rational placement into the network, which can improve the overall power network efficiency significantly. This aspect is particularly important in the case of fast-release storage systems because these elements are designed to compensate fast variation of power demands and to have an important echo-related energy efficiency.

In Fig. 9.11 two situations are illustrated: first, a case where LRVs do not have storage facilities on board; as a result the losses along the catenary even in the case of regenerative breaking are important and affect the overall efficiency. In the second situation, when the LRV is endowed with on-board storage facilities, the power excursion will be significantly reduced because the power excursion, in the case of regenerative breaking, will

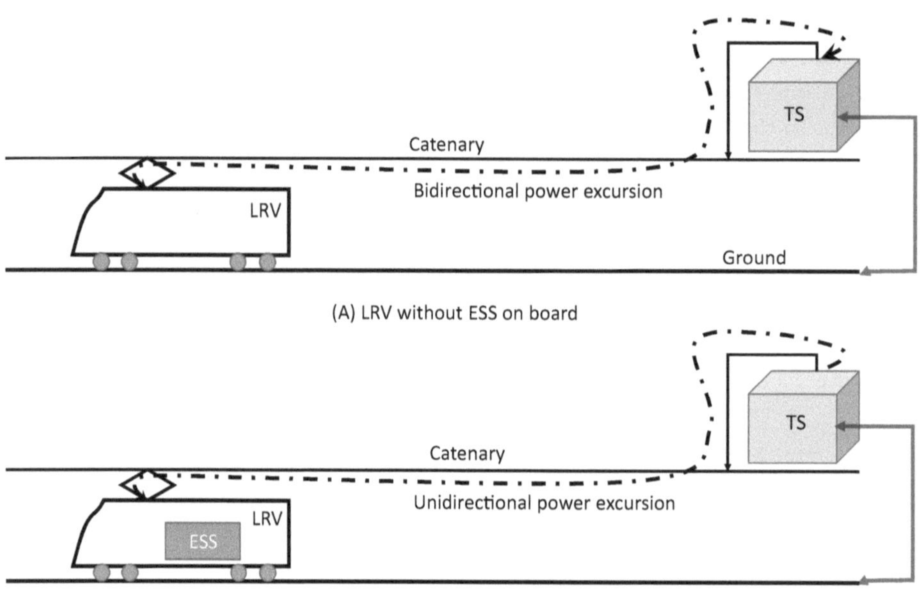

Figure 9.11 ***Light-rail vehicles*** **(LRV)** ***connected at traction network, traction power station*** **(TS),** ***on-board electric storage facility*** **(ESS).** (A) Power excursion LRV TS in the case of classical LRV without on-board ESS, and (B) power excursion LRV TS in the case of configuration with on-board ESS facilities.

have a very short pathway between the traction machine and the local storage system (ESS). In the case of acceleration of LRV, the supplementary amount of energy will be provided by local storage facility (ESS) and so the power fluctuations at traction station level, and also grid disturbances, will be smoothed.

Implementation of the on-board ESS on LRV, reduces LRV own consumption, and reduces catenary losses by 50% as a result of minimization of power transfer along the catenary. The dynamics improvement of LRV (MITRAC Energy Saver system) is around 24% on domain 0–50 km/h. The vehicle dynamics will increase with 36% on the speed domain 50–100 km/h [29,30]. The overall energy saving of MITRAC Energy Saver achieved around 28% energy savings [29].

A third type of applications is related to the mechatronics and acting systems, such as fast robots, cranes, elevators, and other systems that require an important amount of power or energy for a short- or medium-time interval. Nowadays, vacuum cleaner technology benefits from battery development and robotic cord–free vacuum cleaners function by using only their integrated batteries. Also, the cord-free handheld drill needs batteries in order to be energized. In all these kinds of applications, to achieve maximum energy efficiency it is necessary to provide a detailed analysis of the main power flows inside the systems and to understand how one form of energy is transformed into another form of energy, for how long, and how efficient are these transformations.

In the case of EVs or PEHVs, the variation of supply demands presents a significantly large variety as a result of current transfer between ESS and the acting system (DC series, DC parallel, BLDC, and AC asynchronous or synchronous motors), which reaches a very high level. In general, this phenomenon affects the life span of the ESS. The variety of applications generates a variety of demands and constraints for storage systems.

Thus, in the field of electric mobile applications, two aspects are essential: to ensure vehicle dynamics and to maximize the range of the vehicle between successive battery charges using regenerative breaking. For example, in this case when using only batteries, the stress would be relatively high because each time the vehicle changes speed, it generates significant variations in the current that passes through the battery. Also, batteries will overheat when used exclusively, and thus the life span of the batteries will be reduced. Vibration resistance of batteries and supercapacitors devices is different. So, supercapacitors can support a higher vibrational level in comparison with the vibration level tolerated by batteries.

For example, in the case of an EV, the combination of supercapacitors and batteries can avoid overstressing of the batteries and fulfill the application constraints. In this sense, the forecasting methods can help to anticipate demands for ESS.

In the case of HESS, the appropriate sizing combined with the adequate combination of HESS systems will prevent the voltage exceedance of the current and the temperature for the storage system, and will significantly increase the availability and reliability of the application. Electrical energy storage systems can not recover all statements obtained through regenerative braking energy systems because the vehicles have limited storage capacity for reasons of size and cost rationalization. In such situations, the kinetic and potential energy of the vehicle will be lost as heat, which will dissipate.

In the case of tramways or trolleybuses, the hybrid or simple ESS can improve the energy efficiency of the whole system by limiting the power excursion between injection points, the catenary and the vehicle that is in traffic [30]. The improvement of efficiency will depend on vehicle type, traffic conditions, and the driving style of the driver. The constraints that must be considered are domain of temperatures, voltage limits, overcurrent limit, mechanical stress, depth of discharge for battery systems, inclination of the roads, and dynamic performances required for the vehicle.

In Ref. [31], a review of the HESS used for electrical vehicles is presented. The advantages and disadvantages of each device type are presented together with their component models. The main methodologies for the control of the hybrid storage system are also described. In Ref. [32], a recent review of electrical storage technologies and the application potential in power system operation is illustrated. The critical analysis of an actual implementation proves that the hybrid electric storage system can represent a viable, energy efficient, and reliable solution for balancing local power networks and for the assurance of the ancillary services.

For example, in the case of elevators and cranes, the reversal of their direction of motion can generate significant disturbances in the voltage and power consumed by the system, which propagates in the power grid and affects the power quality. The frequency of occurrence of such situations is at least several times per hour. Combating these perturbations can be achieved by implementing "active power filters," which are based on batteries and supercapacitors attached to lifting installations. Details about these systems are given in Ref. [33]. Frequency spectrum that appeared is the result of elevator system functioning and this will be extremely variable. It is dependent on the electric machine, the power electronics, the power line, and all the other consumers connected to the same network. It is very important to emphasize that some "critical load" such as data centers, some facilities existing in industry, like silicon foundries, the heavy industry and some military facilities need very accurate wave form for power supplying. The high power, active filters are used for this purpose.

Another class of applications that can benefit from using ESS is the mechatronic systems and actuators, which need short-term energy pulses for moving their mobile elements. In this category are the "start–stop" systems that control the internal combustion engine of a car. In the case of start–stop systems, technical specifications require a high number of cycles for the energy storage system and an extended operating temperature domain. In addition, it is necessary to ensure a good dynamic stability of the system across a wide range of vibrations.

The European Wind Energy Association (EWEA) estimated that "the world electricity needs could be met by harvesting 10% of the global wind energy resources" [34]. RES are intermittent and unpredictable, which makes them vulnerable to "load following." Also, integrating different types of renewable energy sources in the same power grid might affect the electrical power networks. Thus, integration of storage system into the grid would mitigate the effects of intermittency of the renewable resources and will facilitate higher injections of renewable electricity.

In Refs. [27] and [35], several implementations based on Li-ion batteries storage facilities are presented, such as KEPCO, JeJu Island Korea (4 MW/8 MWh), Duke Energy, Notrees, and Texas (36 MW/13.6 MWh). These ensure the backup power of the substation and the wind stabilization supply. All of these are modular and conceived with integrated cooling systems and battery-management systems. This aspect is very important because in the case of faults, the damaged module can be replaced. Specific configurations of such modules can be implemented. Some of the applications are energy shifting, curtailment, peak cut, peak shift, and frequency and voltage regulation. In the cases of peak cut or peak shift and voltage and frequency regulation from the storage systems, the hybrid solutions are mandatory.

In the case of communication networks, an important role is played by the distributed basis stations ensuring the distribution of mobile communication services between the network and mobile terminals. In many cases, these stations are placed in faraway or isolated zones where the power grid is lacking or is very weak. In such situations, it is mandatory to use "back-up" facilities that are able to prevent the interruption of services. The implementation uses ESS or HESS solutions, sometimes in combination with parallel-connected internal combustion engine–based generators.

Weather-monitoring systems and mobile communication base stations benefit from HESS or ESS, sometimes in combination with ICE-based small generators or harvesting systems, such as small-PV panel or wind-based power supply facilities.

The mobile phone communication systems benefit from the use of improved battery technologies to preserve the energy supply for mobile phone as long as possible, allow short battery charging and high volumetric energy density.

Nowadays, another class of applications that is fast growing is portable systems, especially for medical purposes that ensure the preservation and monitoring of vital functions. In this case not only are aforementioned constraints necessary but also the reliability, availability, and control of temperature must be ensured. For example, in the case of cardiac defibrillators the apparatus will include a battery capable of delivering a certain number of defibrillation cycles, and also a supercapacitor (high-voltage) defibrillation waveform.

The development of nanotechnologies, nanotubes, and graphene technologies created the foundation to implement microsupercapacitors (MSC); the majority of these are pseudocapacitors, respectively functioning on both faradic and nonfaradic phenomena.

A holistic presentation of all of these developments is offered by Andrea and Ferrari in Ref. [13].

The smart and autonomous sensors are intelligent and durable fully insulated systems and are examples of the deployment of nanotechnologies in the domain of distributed data acquisition systems. Using harvesting elements (EHE) to energize the sensors, it is possible to combine sensors with efficient generators, such as photovoltaic, thermoelectric, chemical, vibration-based, or concentrated coherent light (laser) elements on smart sensors. The use of received energy is controlled with maximum parsimony to ensure continuity of sensor functionality and also communication into the network frame. Fig. 9.12 depicts the block diagram of an autonomous smart sensor. MSC implements the energy storage functionality mandatory to pass over a period when the natural sources are unable to offer the necessary energy for sensor operation. The instantaneous parameters of the electric power are adapted using the boost converter that might affect the overall system energy efficiency in a reasonable measure. The local storage element (LSE) (MSC) will facilitate the functioning of the RF transceiver and its controller, responding better to the variable power request in reception and data transmission. Data processing unit will perform several functions: processing and temporary storage of information from sensors, energy management for smart sensor, as well as its communication into the network. New versions of the sensor are compliant with the internet of things (IoT) standards, so cumbersome tasks related to automatic network integration are achieved by using standard processor facilities. Using advanced processors, multicore processors are powered from floating voltage generators. The variable supply demands are followed by fluctuations of power provided by local generators that are dependent on natural factors. As a result, the variation of power exchanged with LSEs presents a higher frequency and a larger power variation. Therefore, the ideal in implementation of the storage elements is to use MSC having a high cyclability (more than millions of cycles)

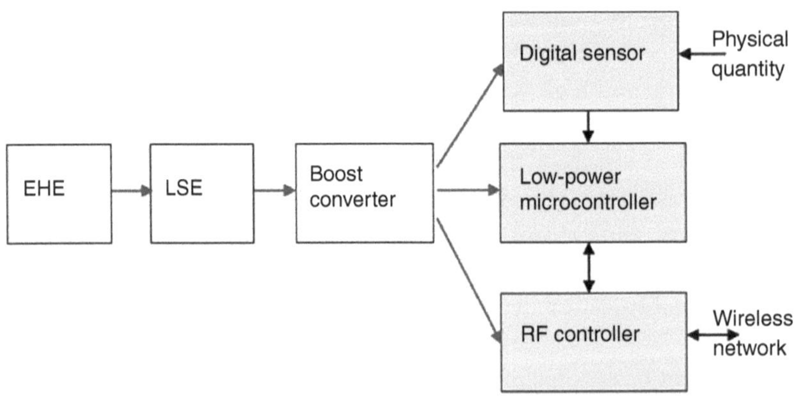

Figure 9.12 ***Block diagram of an intelligent for perpetuity sensor energy-harvesting element*** **(EHE)** ***and local storage element*** **(LSE).**

and also very good elasticity and reliability, small size, and a wide operating temperature range. Using floating voltage for sensors supplying a boost converter can be eliminated from the system and this will increase energy efficiency.

The most common solution to energize the perpetuity sensors is to use the batteries, especially Li-ion batteries that have high-energy densities and acceptable temperature tolerances. Unfortunately, for perpetuity sensors such solutions do not work because these need longer life span. Another important aspect is the specific supply demands of wireless sensors, where the load profile presents important peaks. These peaks accelerate the normal aging of batteries and their degradation. A holistic and detailed analysis of batteries behavior used for wireless endowed with embedded systems is discussed in Ref. [36]. For perpetuity sensors, some self-services processing are important. For example, the system should realize a deep management of energy in which the most important steps are represented by determination of the depth of discharge (DoD) and the state of health (SoH) of batteries. The monitoring of these parameters is resource consuming (computing resources and also energy resources). A detailed evaluation of energy consumed is dependent on structure, organization, and type of hardware implementation. It supposes a permanent functioning of the services and presents a high complexity in system design. Even if the successive stages reached by the system in functioning are well known, the intercorrelation between load and temperature of environment generates large variations in predictions about the remaining quantity of energy. It needs a more complex algorithm for the management of the system. In Ref. [36] a complex algorithm of battery management is described which reaches 95% of life span accuracy. However, it has difficulty in relating to the evaluation of the consumption of system transceivers.

The implementation with LSE of energy supply does not need a complex analysis of status, and the life span and temperature domain are larger in comparison with the batteries. The main disadvantage of MSC is related to the restrained energetic density capacity.

A holistic vision about these kinds of solutions is included in Refs. [37,38]. Thus, the function of the template of the stationary phase is discussed in Ref. [38] and the type of electrolyte for the specific capacitance can vary significantly. In the case of liquid-phase exfoliation (LPE) of graphene, LPE reaches 351 F/g, 110 kW/kg, and 12.5 Wh/kg, and in the case of MnO_2 nanorods electrodeposited onto CNPs, the capacitance reaches 389 F/g [39]. The data in Ref. [40] should be seen in the context of thermal, chemical, or ultrasonic treatment applied on stationary phase and the function of electrolytes. Some data are relevant to define the limits of the energy and maximum power that could be stored and provided. The MSC based on EDLC reaches a significant performance, which makes it ideal to integrate with the energy-harvesting elements.

Sometimes the energy demands of the perpetuity sensor do not satisfy the application requirements; in this case, a hybrid storage solution might increase the performance of the systems. This solution can be implemented using different hybrid storage solutions:

- By combining two or more small batteries and a supercapacitor (as independent entities) in conjunction with an electronic control system, it is possible to maximize the depth of power cycling processes to be done on MSC, which will protect the high-supply demands from the battery.
- Using devices or storage cells presents capabilities for balancing a different ratio of the fast- and medium-time release charges. The storage solutions used for perpetuity sensors could be:
  - hybrid supercapacitors (simultaneously EDLC and pseudocapacitor) and
  - pseudocapacitor.

Both present the different ratio between the faradic and nonfaradic storage that can satisfy the complex energy/power demands of applications.

This solution is not so common till now, for perpetuity sensors, the simplification of functioning for the system can be obtained by relaxing the variation of power supply voltage domain, which should be acceptable to processor and transceivers. So, a simple, energy-efficient system can be built.

An emerging case is represented by "wearable" electronic applications. These are sensors, microactuators, data acquisition, and communication systems or systems with therapeutic purposes with small dimensions that are integrated into a person's clothing. The dimensions, the mechanical properties (such as flexibility and vibration resistance), and the weight requirements are very specific and important. In these cases, the flexibility and implementation of storage elements and electronics made with "plastic materials" and also the details of custom shapes are typical requirements.

5 CONCLUSIONS

The variety of potential applications that can benefit from storage systems is now extremely wide. It is certain, given the specifications of applications, that we cannot discuss a universal solution or ideal, but the progress made in recent years in the field of nanotechnologies gives hope that we can design devices with properties that are more balanced (power density and energy), more efficient, and more available and reliable.

Relative to design methodologies, they are highly diverse and appeal to the "multiphysics" simulation, not only for their electrical behavior, but also for the thermal, mechanical, and chemical aspects which are all important to obtain more accurate and reliable simulations for the functionalities.

Structural design involves a first and essential step, that is, determining the proper dimensions of the storage system facility in strict correlation with the supply demands that affect the system. The structural design establishes the main constraints of application and the control design specifies and implements the functionality and behavioral constraints in detail.

GLOSSARY

AC	Alternative current
BET	Brunauer–Emmett–Teller method
BLDC	Brushless direct current motor
CPU	Central processing unit
DC	Direct current system
DoD	Depth of discharge
EDLC	Electric double-layer capacitor
EHE	Energy-harvesting element
ESR	Equivalent series resistance
ESS	Electric Storage System
EV	Electric Vehicle
EWEA	European Wind Energy Association
FC	Fuel Cell
HESS	Hybrid Electric Storage System
ICE	Internal Combustion Engine
IoT	Internet of Things
IUPAC	International Union of Pure and Applied Chemistry
LRV	Light-rail vehicle
LPE	Liquid-phase exfoliation
LSE	Local storage element
MSC	Microsupercapacitor
OCS	Overhead contact system
OCV	Open circuit voltage
PHEV	Plug-in hybrid electric vehicle
RES	Renewable Energy Sources
SoC	State of charge
SoH	State of health
SSA	Specific surface area
TS	Traction power station

REFERENCES

[1] N. Kularatna, Energy storage devices—a general overview, Energy Storage Devices for Electronic Systems: Rechargeable Batteries, Supercapacitors, Elsevier, Academic Press, London, UK, 2015, ISBN: 978-0-12-407947-2.

[2] G. Fuch, B. Lunz, M. Leuthold, D.U. Sauer, technology overview on electricity storage: overview on the potential and on the deployment perspectives of electricity storage technologies, on behalf of Smart Energy for Europe Platform GmbH (SEFEP), Institute for Power Electronics and Electrical Drives (ISEA), RWTH Aachen University, Germany, 2012.

[3] Facts and figures. Available from: http://energystorage.org/energy-storage/facts-figures

[4] Abbas A., Akhil, Georgianne Huff, Aileen B. Currier, Benjamin C. Kaun, Dan M. Rastler, Stella Bingqing Chen, Andrew L. Cotter, Dale T. Bradshaw, William D. Gauntlett, "DOE/EPRI Electricity Storage Handbook in Collaboration with NRECA", SANDIA REPORT, SAND2015-1002 Supersedes SAND2013-5131 Unlimited Release, February 2015, Sandia National Laboratories Albuquerque, New Mexico 87185 and Livermore, California 94550.

[5] B.E. Conway, Electrochemical Supercapacitor, Kluwer Academic Publishers/Plenum Press, New York, (1999).
[6] B.E. Conway, V. Birss, J. Wojtowice, The role and utilization of pseudocapacitance for energy storage by supercapacitors, J. Power Sources 66 (1–2) (1997) 1–14.
[7] F. Beguin, E. Frackowiak (Eds.), Carbons for Electrochemical Energy Storage and Conversion Systems,, CRC Press/Taylor & Francis Group, 2010.
[8] K.S.W. Sing, Reporting physisorption data for gas/solid systems with special reference to the determination of surface area and porosity, Pure Appl. Chem. 54 (1982) 2201–2218.
[9] S. Brunauer, P.H. Emmett, E. Teller, Adsorption of gases in multimolecular layers, J. Am. Chem. Soc. 60 (2) (1938) 309–319.
[10] Deleted in review.
[11] H.A. Keihne, Battery Technology Handbook, second ed., Marcel Dekker, New York, Basel, (2003).
[12] U.K. Sur, Recent trends in electrochemical science and technology, in: Yu. M. Volfkovich, A.A. Mikhailin, D.A. Bograchev, V.E. Sosenkin, V.S. Bagotsky (Eds.), Studies of Supercapacitor Carbon Electrodes with High Pseudocapacitance, InTech, 2012.
[13] A.C. Ferrari, F. Bonaccorso, V. Fal'ko, K.S. Novoselov, S. Roche, P. Bøggild, S. Borini, F.H.L. Koppens, V. Palermo, N. Pugno, J.A. Garrido, R. Sordan, A. Bianco, L. Ballerini, M. Prato, E. Lidorikis, J. Kivioja, C. Marinelli, T. Ryhänen, A. Morpurgo, J.N. Coleman, V. Nicolosi, L. Colombo, A. Fert, M.G. Garcia-Hernandez, A. Bachtold, G.F. Schneider, F. Guinea, C. Dekker, M. Barbone, Z. Sun, C. Galiotis, A.N. Grigorenko, G. Konstantatos, A. Kis, M. Katsnelson, L. Vandersypen, A. Loiseau, V. Morandi, D. Neumaier, E. Treossi, V. Pellegrini, M. Polini, A. Tredicucci, G.M. Williams, B.H. Hong, J.H. Ahn, J.M. Kim, H. Zirath, B.J. van Wees, H. van der Zant, L. Occhipinti, A. Di Matteo, I.A. Kinloch, T. Seyller, E. Quesnel, X. Feng, K. Teo, N. Rupesinghe, P. Hakonen, S.R.T. Neil, Q. Tannock, T. Löfwanderaq, J. Kinaret, "Science and technology roadmap for graphene, related two-dimensional crystals, and hybrid systems", Nanoscale 7 (2015) 4598–4810, doi: 10.1039/C4NR01600A.
[14] A. Burke, H. Zhao, Applications of Supercapacitors in Electric and Hybrid Vehicles, Research Report–UCD-ITS-RR-15-09, presented at EESCAP2015, Brasov, Romania, 2015.
[15] M. Moran, H.N. Shapiro, B. Munson, D. Dewitt, Introduction to Thermal Systems Engineering, John Wiley & Sons, (2003).
[16] E. Fermi, Thermodynamics, Prentice-Hall, Englewood-Cliffs, NJ, (1937).
[17] J. Wang, Modern Thermodynamics Based on the Extended Carnot Theorem, Springer, Berlin, (2012).
[18] V. Pop, H.D. Bergveld, D. Danilov, P.P.L. Regtien, P.H.L. Notten, Battery Management Systems, Accurate State of Charge Indication for Battery Powered Application, Philips Research, Springer, 2008.
[19] C.C. Christianson, R.F. Bourke, Battery state of charge gauge, US Patent 3,946,299, 1975.
[20] P. N. Borza, Electric power cell, EPO Patent, EP 1796199A1, Application no. 05026704.6, 2005.
[21] J.B. Goodenough, Basic Research Needs for Electrical Energy Storage, Office of Basic Energy Sciences Department of Energy, 2007. Available from: http://www.sc.doe.gov/bes/reports/files/EES_rpt.pdf
[22] J.I. San Martin, I. Zamora, J.J. San Martin, V. Aperribay, P. Eguia, Energy storage technologies for electric applications, in: International Conference on Renewable Energies and Power Quality, Las Palmas, Spain, 2011.
[23] International Standard, Horizontal Standard IEC 60038, Edition 7.0.
[24] A. Klajn, M. Ba̧tkiewicz-Pantuła, Application Note Standard EN 50160 Voltage Characteristics of Electricity Supplied by Public Electricity Networks, ECI Publication No. Cu0147. Available from: www.leonardo-energy.org/node/145851
[25] IEC, Electric Energy Storage, IEC, Switzerland, 2011.
[26] L. Stephan, Mapping the future of energy storage in European energy markets, in: Energy Storage Update Conference, Younicos, November, 2015.
[27] E. Russo, E.ON battery solutions, in: Energy Storage Update Conference, Younicos, November, 2015.
[28] J.W. Dixon, M. Ortúzar, E. Wiechmann, Regenerative Braking for an electric vehicle using ultracapacitors and a buck-boost converter, in: Proceedings of 17th International Electric Vehicle Symposium, Montreal, Canada, 2000, pp. 1–6.
[29] M. Fröhlich, M. Klohr, S. Pagiela, Energy storage system with ultracaps on board of railway vehicles, in: IEEE European Conference on Power Electronics and Applications, Aalborg, Denmark, 2007.

[30] EcoActive Technologies Propulsion & Controls MITRAC Energy Saver Bombardier Transportation, Zurich, Switzerland.

[31] M.A. Hannan, F.A. Azidin, A. Mohamed, Hybrid electric vehicles and their challenges: a review, Renew. Sustainable Energy Rev. 29 (2014) 135–150.

[32] Xing Luo, Jihong Wang, Mark Dooner, Jonathan Clarke, Overview of current development in electrical energy storage technologies and the applications potential in power system operation, Appl. Energy 137 (2015) 511–536.

[33] M. Macia̧z˙ek, D. Grabowski, M. Pasko, M. Lewandowski, Compensation based on active power filters: the cost minimization, Appl. Math. Comput. 267 (2015) 648–654.

[34] H. Chen, T.N. Cong, W. Yang, C. Tan, Y. Li, Y. Ding, Progress in electrical energy storage system: a critical review, Prog. Nat. Sci. 19 (3) (2009) 291–312.

[35] B. Ricci, Samsung SDI energy storage systems—Li-ion solutions, Energy Storage Update Conference, Samsung, November, 2015.

[36] B. Buchli, D. Aschwanden, J. Beutel, Battery state-of-charge approximation for energy harvesting embedded systems, wireless sensor networks, in: Lecture Notes in Computer Science, vol. 7772, Springer, 2013, pp 179–196.

[37] Z.S. Wu, X. Feng, H.M. Cheng, Recent advances in graphene-based planar micro-supercapacitors for on-chip energy storage, Nat. Sci. Rev. 1 (2) (2014) 277–292.

[38] X. Xiao, T. Li, P. Yang, Y. Gao, H. Jin, W. Ni, W. Zhan, X. Zhang, Y. Cao, J. Zhong, L. Gong, W.C. Yen, W. Mai, J. Chen, K. Huo, Y.L. Chueh, Z.L. Wang, J. Zhou, Fiber-based all-solid-state flexible supercapacitors for self-powered systems, ACS Nano 23 (6) (2012) 9200–9206.

[39] G. Yu, L. Hu, M. Vosgueritchian, H. Wang, X. Xie, J.R. McDonough, X. Cui, Y. Cui, Z. Bao, Solution-processed graphene/MnO_2 nanostructured textiles for high-performance electrochemical capacitors, Nano Leters 11 (7) (2011 Jul 13) 2905–2911, doi: 10.1021/nl2013828 Epub 2011 Jun 13.

[40] L. Yuan, X.-H. Lu, X. Xiao, T. Zhai, J. Dai, F. Zhang, B. Hu, X. Wang, L. Gong, J. Chen, C. Hu, Y. Tong, J. Zhou, Z.L. Wang, Flexible solid-state supercapacitors based on carbon nanoparticles/MnO_2 nanorods hybrid structure, ACS Nano 6 (1) (2012) 656–661.

INDEX

M

FSC
www.fsc.org
MIX
Papier | Fördert gute Waldnutzung
FSC® C083411

Kolibri 360 GmbH
Ferdinand-Jühlke-Straße 7
99095 Erfurt, Deutschland
produktsicherheit@kolibri360.de